An Instinct for Truth

An Instinct for Truth

Curiosity and the Moral Character of Science

Robert T. Pennock

The MIT Press
Cambridge, Massachusetts
London, England

This book was set in ITC Stone by Westchester Publishing Services. Printed and bound in the United States of America.

Library of Congress Cataloging-in-Publication Data

Names: Pennock, Robert T., author.
Title: An instinct for truth : curiosity and the moral character of science / Robert T. Pennock.
Description: Cambridge, MA : The MIT Press, [2019] | Includes bibliographical references and index.
Identifiers: LCCN 2018046588 | ISBN 9780262042581 (hardcover : alk. paper)
Subjects: LCSH: Science—Social aspects. | Science—Philosophy. | Science—Moral and ethical aspects.
Classification: LCC Q175.5 .P456 2019 | DDC 174/.95—dc23
LC record available at https://lccn.loc.gov/2018046588

10 9 8 7 6 5 4 3 2 1

For my children, Laura and Grant, who have grown up in tandem with this project, demonstrating every day the wonder that inspires.

Contents

Preface: Facts and Values

Truth vs. Post-Truth

Excellence in science involves not only its methods and practices but also the character of its practitioners. Exemplary scientists have a characteristic way of viewing the world and their work. Everything they do aims to discover truths about nature. This leads them to care about certain things and to value certain traits in other researchers. My contention is that this scientific mindset has a tacit moral structure.

I came to appreciate this in the mid-1990s as I was writing *Tower of Babel*,[1] which rebutted the arguments of so-called Intelligent Design creationism (IDC). In doing the research about the evolution/creationism controversy I was struck by the vehemence of scientists' reaction to IDC. They saw IDC not just as bad science but as counterfeit science. They saw it not just as wrong but as dishonest. I noticed in their reaction to creationism what could only be called a kind of moral indignation. They saw it, no, they felt it, as a violation of a deep sense of what it means to be a scientist—a violation of scientific integrity.

Some may view creationism as a minor issue on the radical fringe of the culture wars, but denying evolution was symptomatic of a larger philosophical issue and a broader politicized rejection of science. It was not just that creationists denied well-confirmed truths of evolution—they attacked the modern scientific worldview. Though identified with the far right, IDC had co-opted a radical philosophical view from the far left that claimed we were now in a postmodern period and that scientific truth was just a story—a sham foisted upon us by cultural elites who suppressed "alternatives" to evolution. Climate science denial is a closely related example.

In the mid-2000s, bills proposed by conservative legislators to allow IDC in schools began to include language that would require criticisms of global warming as well as evolution. As with anti-evolution, ideology in opposition to evidence drove that legislative agenda, as climate science was politicized as a "liberal" view.[2] The National Center for Science Education, the watchdog group that for years focused exclusively on creationism, decided in 2011 that it needed to expand its mission to include climate change denial.[3] The situation has only gotten worse in the intervening years. In his presidential campaign, Donald Trump tweeted that "the concept of global warming was created by and for the Chinese in order to make U.S. manufacturing noncompetitive" and continued to dismiss climate change as "bullshit" and a "hoax."[4] One could give many similar examples where scientific facts are ignored or dismissed in favor of alternatives that one or other ideology prefers.

There are, of course, cases in any age where particular truths are denied for ideological reasons, but it is fair to say that this problem has been growing in recent decades. Scientists are dismayed by the blatant denial of science findings, the blithe dismissal of evidence-based conclusions, and the bald-faced fabrication of "alternative facts." This reached a tipping point recently; the Oxford Dictionaries Word of the Year for 2016 was *post-truth*. The word was most often used in commenting upon politics; truth had lost its value as the gold standard and had become "a worthless currency" in political debate. Citing cases such as the campaign in Britain to exit the European Union and the presidential race in the United States, the Oxford philologists defined post-truth as "relating to or denoting circumstances in which objective facts are less influential in shaping public opinion than appeals to emotion and personal belief."[5] In fact, this definition is too narrow, for the current trouble extends beyond the usual political mendacity. It is not just that emotion and belief predominate over objective facts but rather that the idea of objective facts is itself attacked.

Sadly, it is no surprise that postmodernism led the way to post-truth, as it discarded the possibility of objective truth along with science. Indeed, I had first noticed this in an early talk I gave about ID creationism at a science studies conference when a postmodernist professor in the audience made just such a point in the question and answer period, suggesting that evolution was just the dominant narrative, and that there was no reason to think that it was any truer than a religious narrative. However, if truth is no more than a story told by those in power, then in the limit one has no choice but to go to war to establish the facts. The very notion of *establishing* facts is inverted on such a view; no longer does this term refer to discovering what the facts are by shared, appropriate methods but rather to forcing them to be so on the personal say-so of someone in authority. As we shall see, this kind of thinking subverts the possibility of free and just cooperative action. If we are to hold the center against this or other forms of authoritarian dishonesty we cannot allow the integrity of science to be undermined from the right or the left, from without or within.

However, to defend the integrity of science requires that we recognize it as a practice to which the notion of integrity applies; that is, we must recognize that science has a moral structure. My claim is that science does have such a structure and that scientists feel it even though they may not be able to explain it. Articulating this structure in virtue-theoretic terms is the main goal of this book. Most of philosophy of science deals with the structure of scientific theories and the evidential methods that support them. I have spent much of my career thinking about the nature of science in this way, but this book shifts the focus to the character of science, which includes the character of the scientist. It takes seriously the idea that one should ask not only "What should I do?" but also "What kind of person should I be?" The two questions are interconnected; the book is an examination not of the philosophy of science in the traditional sense but of the philosophy of the scientist. It argues that the character of science has a normative structure that links epistemic and ethical values and that the flourishing of science depends not only on the soundness of its methods and practices, but also on the excellence of its practitioners.

Methodology

For the last couple of decades, I have developed and presented the views expressed here in many articles, dozens of talks, and several courses. The ideas also formed the basis of a curriculum developed and piloted at BEACON, an NSF Science and Technology Center that I am a coprincipal investigator of, to advance responsible conduct of research and to foster a culture of integrity. The BEACON Center has provided a valuable research environment to advance other aspects of this project as well. BEACON investigates evolutionary processes as they operate in real time—evolution in action in both natural and digital model systems. My own scientific research in this area involves the evolution of intelligent behavior—how organisms evolve to become knowers who are capable of individual and cooperative action, including moral action. That work informs the evolutionary perspective that pervades this book, and a few specific elements appear in passing, but a fuller account of that research will be a topic of its own book. Similarly, my discipline-based education research about how to improve student learning in science, especially with regard to evolution, the nature of science, and scientific values, provides an empirical foundation for some thoughts on the philosophy of science education that are presented in chapter 9, but details and elaboration of that have been and will be presented elsewhere.

The most significant empirical component of this project is my study of the views of exemplary scientists about the values that are important for excellence in research. I began doing informal interviews along these lines starting in 2007 and was able to pilot a formal study in 2012, followed by a full national study beginning the following year. In 2014, I spent half of my sabbatical as a visiting scholar at the University of Michigan's Institute for Social Research working to advance my social science research skills with the help of my collaborator Jon Miller. We completed data collection in 2017, and as I write this in the spring of 2018 we are preparing results for journal publication. We presented a preliminary analysis at the American Association for the Advancement of Science annual meeting in 2016.[6]

All these aspects of the project fall under the heading of what has become known as experimental philosophy, which aims to supplement traditional "armchair" analysis with various sorts of empirical data. Because the project is highly interdisciplinary, it is worth saying a bit about its methods. In most ways, the book uses standard methods in the field of history and philosophy of science (HPS), but for readers who are unfamiliar with HPS, let me say a little about that before describing its more unique elements.

History, Philosophy, and Sociology of Science
History of science concerns itself with the development of science over time. It focuses on scientific figures, ideas, institutions, and so on in different periods and places. It asks descriptive questions about the practitioners, practices, and products of science, seen either in terms of their internal development within science (what is known as "internalism") or in relation to how they affect or are affected by external, contextual factors ("externalism"). This book takes an internalist approach to the history because it is

primarily interested in what philosophers call a "rational reconstruction" of scientific concepts, which is the attempt to make them systematically more precise in a rationally justified manner.

Philosophy of science concerns itself with the conceptual foundations and structure of science. It focuses primarily on scientific concepts such as theory, model, hypothesis, evidence, explanation, and the like, as well as scientific methods and practices such as experiment, induction, model-based reasoning, confirmation, and disconfirmation. It asks prescriptive questions involving what we should do to improve these for the benefit of both theory and practice. Philosophers of science begin with concepts as they are used in science and then they work to sort out possible ambiguities in their definitions and applications in order to make them more precise, so they fit rationally within a systematic theoretical framework. In a similar way, I developed my virtue philosophy of science as a rational reconstruction of the scientific mindset, considering the ideas in light of touchstone, historical cases as well as contemporary research. My approach is an extension of Imre Lakatos's recommended method for the evaluation of rival methodologies.[7] My own inclination is toward what Lakatos called the "Euclidean" philosophical side of the street, but I fully endorse the value of "two-way traffic" between philosophy and history of science. My extensions to his method involve additional constraints from sociology of science and from evolutionary biology, which make for a more complicated four-way traffic pattern. I think that he would not have objected to these methodological additions.

Sociology of science looks at the institutional foundations of science and its social structures and practices. It focuses on its professional norms and practices such as the organization of professional societies, relationships among members in a lab, patterns of reward, and so on. Sociologists ask descriptive questions about what goes on in research groups and other scientific settings and how these relate to other issues in the broader social context. They try to answer them empirically, either by quantitative or qualitative methods with data drawn from surveys, experiments, ethnographic studies, interviews, or other observations in field or lab settings. The field was founded by Robert Merton, and this book may be seen as an extension of his pioneering work on what he called the "scientific ethos," though with a philosophical emphasis. However, as will be seen, I reject the radical sociological approach that adopted a strong social constructivist philosophy and went under the name of "sociology of scientific knowledge (SSK)." Under that umbrella term, the so-called "Strong Program," together with dogmatic forms of cultural studies and science studies, took a postmodern view that saw no distinction between true and false theories (except when put in scare quotes) and treated both "symmetrically," that is, solely with social explanations in terms of cultural context, self-interest, and power dynamics. I concur with other philosophers of science who found that approach "internally incoherent and factually mistaken" and seek an approach that recognizes that role of values and virtues in science without discarding truth.[8] As with history of science, I have no objection to externalist studies, but I view these as a pluralist realist.

There are still some remnants of SSK and extreme social constructivist Cultural and Science Studies that are taught, especially in departments of communication, education, English, and others that absorbed them second-hand, but I think that scholars in these fields are already coming to recognize the problems. Within philosophy of science, the problems with the philosophical underpinnings of these views had already been identified by the late 1990s so I will only deal with this in passing as part of a general review of radical postmodernism as it relates to scientific truth, particularly as that stemmed from the work of Thomas Kuhn and Paul Feyerabend. I have followed Lakatos in rejecting the notion of irrational scientific change that he attributed to these thinkers; I suggest that a virtue philosophy of science helps make sense of scientific change, even changes in particular scientific methods, by way of the rationality of instrumental values. With regard to the views of those two figures, I offer a more charitable take on Kuhn than Lakatos did; my approach connects to Kuhn's own later work where he tried to correct misinterpretations of his views and that emphasized the role of values in science.[9] However, I mostly concur with him about Feyerabend. Though he would no doubt have differed with various aspects of my account, I like to think that the parts of this work that confront Feyerabend's views stand in to a small degree for the rebuttal that Lakatos did not have the chance to complete.

Again, my method is Lakatosian in spirit, aiming to use not just history but empirical data more generally to help judge the adequacy of a philosophical account—here my virtue philosophy of science. In focusing on character traits within this normative framework, I have gone beyond Lakatos's recommendations, but I do not claim completeness; there are other empirical questions that I am setting aside. Intelligence, for example, is part of the stereotype of the scientist and can be important for highly technical scientific research, but it is not a character virtue and so falls outside the scope of this project. Similarly, I am not concerned with personality or psychological traits. Scientists are more likely to be introverted rather than extroverted, and they lean more toward conscientiousness and openness to experience. They may or may not be neurotic or agreeable. Such personality traits are interesting and may be significant, but they are not our issue.

My focus here is on character virtues as normative ideals. In this rational normative reconstruction the task is to begin with ordinary concepts and try to systematize them in a theoretically justified way. To do philosophy of science well requires close familiarity with the science; similarly, for this philosophy of scientists, one needs to know scientists. Philosophers typically operate from their own armchairs or from one or a few case studies, but my aim has been to expand the base to provide a broader test of my philosophical theory and to give a more nuanced explication than is possible when we rely just on our own intuitions. My national study is grist for that philosophical mill, so let me provide a few more details about those data.

The Scientific Virtues Survey

The information comes from a five-year empirical study of over 1,100 scientists. About 600 of these are from a sample population drawn at random from lists of scientists recognized by their peers as exemplary, such as Nobel Laureates, members of the National Academy of Sciences, or Fellows of major professional societies in the natural sciences. To assess possible changes of view over career stages, the remainder were early-career scientists. We did not have the resources to mount a global survey but captured a fair number of international scientists employed at American institutions in both groups. Most of the data came from hundreds of hours of phone interviews. (Scientists who did not have the time or inclination to be interviewed could fill out a paper or online written survey instead, which accounted for about a fifth of the total.) The interviews averaged forty-five minutes in length, with many running to well over an hour for scientists who went into greater detail when answering the open-ended questions.

My collaborator Jon Miller and I had previously developed the survey instrument in a pilot study. We were interested not only in what values scientists took to be important for scientific practice but also how they were acquired and exemplified and how they may have changed over time. We collected both quantitative and qualitative data, including stories that illustrated specific applications of the virtues and gave flesh to what might otherwise have seemed to be rather abstract concepts. We now have a treasure trove of information about the values that underlie scientific practice, and it will take several years to publish its various findings. I had originally intended to include more specific anecdotes drawn from our interviews in these chapters but laying out the theoretical account made the book long enough as is, so these will have to wait for another venue. Although it does not report specific results, this book is informed and shaped by the study. Needless to say, it was heartening to find that philosophical expectations were substantiated by the data.

More remains to be done, of course, both philosophically and empirically. I would like, for example, to do a follow-up study of the virtues of another vocation to serve as a comparison case. This is needed in order to support the claim that the virtues discussed in relation to science are appropriately distinctive of that vocation and not simply virtues that *any* vocation would affirm in the same way. On my theoretical account, different vocations should be characterized by differently weighted sets of values that reflect their distinctive goals. Preliminary data from focus groups we conducted with regard to the virtues of engineering, with members of the National Academy of Engineering, and to those of the health professions, with members of the National Institute of Medicine and other exemplary practitioners, do support this view. I hope to be able to conduct full-scale studies of these and other vocations not only for purposes of comparison to science, but for their own interest.

We might also wonder how close the descriptive practice is from the prescriptive ideal, but that is a separate and secondary question. The goal here is to understand the structure of the ideals, which is a necessary prerequisite for that kind of study—one cannot measure how far we are from excellence or investigate how we might better approach it if we have not first thought through what it should look like. Most studies along those line have been on cases of obvious scientific misconduct, especially falsification or fabrication of data, that involve violation of the basic value of honesty. With our more complete account of scientific values, we are now in a better position to study other kinds of misconduct and questionable research practices that have been overlooked or neglected.

Perhaps even more important is what having a better understanding of the ideal puts us in a position to do on the positive side, namely, improving science education and perhaps scientific practices themselves. This is another sense in which this is a normative project; it is aspirational. One may think of it as an examination of excellence in scientific practice. To seek excellence in science is to cultivate the kind of character—the scientific mindset—that leads to a vocational form of what Aristotle spoke of as practical wisdom. Developing the habits and judgment that make for this kind of excellence helps us navigate the difficult choices that inevitably arise in any real-world practice, making it more likely to achieve our ultimate aim. Again, the primary goal of this book is to elucidate this normative structure, highlighting its moral character. That is why it is framed in terms of virtue theory, which is a philosophical theory of excellence and what we ought to do to pursue it.

Theory

Virtue ethics has ancient roots, and I will refer often to Aristotle, who we may trace the most important pioneering work to, but this is not an exercise in ancient Greek scholarship. I owe my understanding of Aristotle to James Lennox and Jennifer Whiting, and I do not claim to offer any special insight into his specific views. Rather, my approach is Aristotelian in spirit, though with a Darwinian twist; I will usually introduce virtue-theoretic concepts in Aristotelian terms but then reframe these within an evolutionary framework that does not assume the kind of typological essences or fixity of natural kinds that Aristotle did. For instance, I will retain a teleological element in my account but one that is explicated in evolutionary terms. Similarly, I will occasionally draw inspiration from (and, again, put a Darwinian spin on) some elements from David Hume, whose notion of virtue I came to appreciate as a student of Annette Baier, but I offer no analysis of his views here.

Neither will I attempt to review contemporary virtue theory, either in ethics or epistemology, as I could not hope to do justice to the general theoretical issues that concern the field in the space available, especially in relation to other philosophical views. For instance, I will not engage here in the debate over what is known as the "situationist challenge" to virtue ethics, which holds that situational factors play a larger role in moral agency then do character dispositions. I think that an evolutionary account may help resolve this kind of dispute, but this is not the place to make that argument. With regard to that issue and others, I here simply acknowledge my debt to contemporary virtue ethicists who are addressing them and will mostly bracket those aspects of the current debate.[10]

Beyond the practical issue of keeping this book to a somewhat reasonable length, another reason for setting these larger issues aside is that my project is, at least initially, narrower in scope. Virtue ethics, both historically and currently, mostly deals with what it means to be a moral person. What is different here is the emphasis on what I call the *vocational virtues*, that is, the virtues that are partially constitutive of particular disciplinary practices such as science, engineering, medicine, teaching, and so on. One could also call these "professional" virtues, but that puts the weight on formalized institutional norms, which I hold are derivative of the more basic values of a vocation. Thus, in the present case, I want to explicate first what it means to feel science as a calling. Although we will need at the end to integrate science and other vocations into the larger picture of human flourishing, it is a mistake to think that understanding excellence of vocations is simply a matter of applying general human virtues. Understanding excellence in science, for example, requires more than just mapping general human virtues onto scientific practice. We will make better progress, I think, by first considering the more constrained notion of scientific flourishing.

Moreover, focusing initially on this narrower topic allows us to examine issues that are of special interest and import within vocational contexts that might otherwise be overlooked. To mention just one example, to be taken up in chapter 3, there is an interesting tension between the emotional joy of scientific research and its rigorous heartlessness that has implications for the issue of disinterestedness in science. Being disinterested is not the same as being uninterested. Scientists are passionately disinterested—they want to know what really is so about the world. Examining this in virtue-theoretic terms involves the relationship between curiosity and objectivity and provides a framework within which one may develop judgment about what constitutes excellence of practice. This allows us to think about core issues in philosophy of science from a different perspective. As there is now a robust field of virtue ethics, and a newly developing one of virtue episte-mology,[11] one may think of this as an initial foray into developing a virtue philosophy of science. This book deals with what we might call a philosophy of the scientist, as it focuses on the conceptual structures that constitute the scientific mindset.

This takes a different tack with regard to values than has been typical in philosophy of science. After hearing the virtues that emerge, they may seem obvious in hindsight, but when I ask other philosophers beforehand what they would expect, they start with things like simplicity, fruitfulness, or other values related to theory choice. For various reasons, theoretical, his-torical, and practical, philosophers of science have not thought as much in terms of the virtues of the scientist compared to those of scientific theory. I hope this book will help stimulate others to go further into the fascinating complexities of this topic than I have been able to in my initial explorations that I summarize here.

While concepts such as curiosity, objectivity, skepticism, and so on are all familiar to science, we should acknowledge that scientists and others will find it disconcerting at first to speak of these as virtues. One needs to recognize that "virtue," as it is used in philosophy, is a technical term. As we will see, it refers to character traits; more specifically, a virtue is a mental disposition to act in ways that are characteristic of excellence. Virtues confer normatively admirable abilities or powers, and the concept thus helpfully connects the moral and the causal. As I will analyze the notion, character traits have a biological basis, but they must be cultivated and exercised to become habitual, so it can be useful to see them in terms of mental skills. Once developed, they are enduring mental characteristics that constitute a reliable attitude or state of mind. As virtue ethicists Rosalind Hursthouse and Glen Pettigrove put it, "To possess a virtue is to be a certain sort of person with a certain complex mindset."[12] It is for this reason that I speak of scien-tific values and virtues as the "scientific mindset." However, I will not shy away from using virtue and other ethical terms, for one of my goals is to get scientists to recognize the values that structure scientific methods and to give them a vocabulary that will help them speak of their practice in moral terms.

I have aimed to frame my account in terms that would engage not just other philosophers but also contemporary scientists. My national study pro-vided the empirical grounding for that broader goal. Some scientists initially

expressed hesitation about being interviewed about scientific values and virtues, saying that they did not know anything about ethics. However, once
interviewers began asking questions in these terms, scientists quickly felt at
home and realized that they had quite a lot to say after all. I have mostly
retained the terms that scientists themselves used, even while beginning
the task of making them more precise. I do this under the assumption that
it is not the terms that are critical but the dispositions that they name. I thus
do not care to argue about words and will accordingly pay little attention up
front about fine distinctions that may become important later. This is not
an exercise in Ordinary Language Philosophy, a movement that attempted
not to solve but to "dissolve" philosophical problems by paying attention
to nontechnical, everyday use of language. Indeed, as I discuss starting in
chapter 6, my view is that the fundamental relationships that are important
for making empirical discoveries are prior to language.

Terminological Points

Still, communicating through language is important, so it is worth making
a few additional points about the terminology of virtue and excellence to
address some obvious questions. Why these terms? Why not insist upon precise definitions immediately? Why just this number of virtues, rather than
more or fewer?

First of all, we must accept that usage and grammar do not always easily
mesh. Some virtue terms are regularly used as nouns or adjectives (e.g., curiosity-
curious; humility-humble). Others may have a noun form but are more commonly expressed in some other grammatical form (e.g., attentiveness-being
observant). In some cases, as Aristotle pointed out, there are virtues that do
not have common names. On the other hand, there are also constellations
of closely related virtue terms that may be thought of as synonymous in
some circumstances or at least "of a kind," even though they have different
senses or connotations in different contexts (e.g., patience, perseverance,
persistence, tenacity, assiduity).

Such examples of terminological irregularity provide an excellent argument for introducing technical vocabulary. There are advantages to this in
any field, and I do a little of that, but in this book, I mostly try to start with
the terms that scientists use, as evidenced in our survey. For example, scientists themselves spoke plainly of honesty as the most important virtue.
I will sometimes refer instead to veracity because the word is more directly

rooted in truth (from the Latin *verus*) and because the concept encompasses not only the quality of being honest, but also the qualities of accuracy and conformity to facts, which are of special significance in science. However, while the latter is more precise for the scientific context and is thus arguably better as a technical term, in most settings the former is preferable for the sake of initial clarity. It is simpler to begin with the more common term and then explain ways in which it takes on a distinctive character in science. Given contemporary usage, one is more likely to be understood if one argues that scientists should be honest than that they should be veracious. Similarly, I mostly stick with the simple term truth even though I will be advancing an analog rather than a binary understanding of it, which might make verisimilitude a better term, though that has similar problems that I want to bracket here. For the most part, I try to resist the philosophical temptation to introduce new terms when drawing distinctions. However, in one case in the survey we coined a phrase—humility to evidence—because it captured the scientific form of intellectual humility; scientists did not find humility in general to be very important but endorsed this special concept.

What about other possible virtues that are omitted or are given short shrift? It may not be surprising that charity is not especially important in science, but one could argue, with a nod to John Dewey, that open-mindedness should be featured separately. In fact, scientists did not use this particular term on its own as often as one might have expected, but I would argue that it may be thought of as covered under a combination of skepticism and humility to evidence. Creativity did not make the top ten in our study, and, in chapter 8, I offer a possible way to understand why. It can have an important helping role, but is not a central virtue in science in the way that it is in literature and art. In principle, any virtue could be of value in some circumstances, but the initial task here is to make sense of those that have special weight in science because of its distinctive goals. One reason for this approach is to set up a framework that may help connect science to other vocations, including literature, art, and the humanities generally.

C. P. Snow and the Culture of Science

In 2009, Jon Miller and I organized a conference—Science and Liberal Education: C. P. Snow and the Two Cultures @ 50—for the fiftieth anniversary of Snow's famous Rede lecture at Oxford, where Snow, who was both a scientist and a novelist, highlighted a serious communications gap between what he called "the Two Cultures" of scientists and literary intellectuals. How can we hope to address wicked social problems such as systemic economic inequality, he said, without a unified effort? Lyman Briggs College, the residential program where I teach at Michigan State University, was founded a few years after Snow's lecture with the explicit goal of bringing together the sciences and the humanities to help form a "third culture." Snow's own analysis focused on the circumstances of his day, and there are some clear oversimplifications in his view.[13] Nevertheless, the general issue he identified remains a serious problem, as speakers at our conference highlighted from a variety of perspectives—art, biology, chemistry, engineering, history, journalism, philosophy, physics, and social science—while also describing their personal attempts to bridge the gap. I used the opportunity to discuss how a virtue philosophy of science might help link the cultures. Without committing to Snow's particular way of advancing the claim, I see the scientific virtues project as a systematic attempt to explicate and support his contention that "there is a moral component right in the grain of science itself."[14] It is my hope that elucidating the values that structure the culture of science will put us in a better position to connect excellence in scientific practice with the excellences of other vocations.

Again, I aim to do this as a rational reconstruction, not as radical deconstruction of the kind that too often inadvertently undermined the value of cultural and science studies. For the record, many scholars in science studies rejected or came to reject the relativist postmodern social constructivism that denies science and speaks of "truth" and "reality" only in scare quotes; even many of the leading figures associated with that extreme, dogmatic view have explicitly backed away from it or declared that they were misunderstood and never held such a view in the first place. My goal in this book is to articulate a progressive, forward-looking account and not to reignite old battles, so in the places where these come up, as cannot be avoided, I will speak generically of "dogmatic social constructivism" to refer to the extreme science-denying and truth-denying conceptual positions to avoid pointing fingers. Scholars who have recognized the problems with and unintended consequences of that view should be welcomed back to the reality-based community, as we will need the expertise of the humanities as well as that of the sciences to solve the sorts of wicked social problems that Snow identified. Though I mostly focus on scientists and scientific examples, I also bring in figures and examples from literature and the arts to show the value of integration with the humanities. If we cannot reach in fellowship "across the aisle" to close the gap between these two cultures and others, we will have little chance of closing the climate gap or other dangerous social divisions whose solution will require a unified effort.

There are always alternative ways to present ideas, but the chapters are organized to highlight intersecting conceptual points about how curiosity and other virtues provide a moral structure to science and provide a new perspective on how to connect science to other vocations for the general advancement of human flourishing. To help the reader get a sense of the organization of the book, let me now preview each chapter, highlighting themes and philosophical goals of each.

Chapter Outline

Charles Darwin wrote that he felt within himself "an instinct for truth, or knowledge or discovery" that was "of the same nature as the instinct of virtue."[15] The first chapter unpacks Darwin's radical idea, arguing that organisms have an instinctual curiosity, which has both epistemic and ethical aspects. It introduces the notion of vocational virtues as practiced dispositions and argues that science has a deep moral structure that arises in relation to its central, guiding purpose of discovering empirical truths about the natural world. Understanding science in virtue-theoretic terms may help us find ways to communicate across the gaps that seem to divide it from the humanities and other practices.

Chapter 2 introduces the philosophical notion that cultures may be usefully analyzed in terms of their ideals—their aspirational values—and argues that veracity (honesty) and curiosity are the core values of scientific practice. It lays out some of the basic concepts of virtue theory and ways that a virtue approach can help analyze scientific norms. I modify some standard concepts to better account for how scientific values help us understand and assess its methods and practices. Not all scientists are always honest, but through an example of scientific fraud we see why dishonesty is inimical to scientific curiosity. Another example shows what scientific integrity looks like in practice.

In the third chapter, I examine skepticism and objectivity in relation to the notion of trust in science, showing how they are not just individual traits but are connected to the structure of the scientific community. This discussion allows us to review and extend the work of Robert Merton. Science's institutional ethos, as analyzed by Merton, is a critical but incomplete part of the picture. The kinds of social controls that he was interested in are justified by the underlying epistemic and ethical values of scientific practice, making it appropriate to think of science as systematic curiosity.

Chapter 4 focuses on the virtue of attentiveness through a discussion of the role of friendship and emotion in science. The stereotype of science as cold and emotionless is much mistaken. Science is, properly understood, a passionate activity, with scientists as attentive lovers of nature. As with the other virtues, an evolutionary perspective helps explain the importance of attention and links the emotional and the cognitive aspects of a virtue-theoretic approach to ethics. Darwin himself exemplified what is properly thought of as an "emotion of the intellect."

The fifth chapter begins with a discussion of the notion of a vocation and discusses science as a discipline in various senses of the term, before moving to a discussion of the importance of habits in a virtue-theoretic analysis. It argues that scientists may be thought of as instruments that should become reliable by developing disciplined mental and physical habits relevant to scientific methodology, especially its requirement of reproducibility. Darwin explained the importance of the habits he developed, particularly while on his voyage of exploration on *HMS Beagle*. The virtues of meticulousness, perseverance, and patience are introduced by way of this discussion of what it means to be a disciple of scientific truth.

In chapter 6, I turn to the virtues of intellectual humility and courage, using these to connect the virtue-theoretic account we have been developing to more traditional topics in philosophy of science having to do with scientific methodology. As a warning to the reader, I should note that the middle section of this chapter is rather dense. It is my attempt to quickly review some important history having to do with the logic (or purported lack thereof) of scientific reasoning. To do this, I move quickly through the work of important figures whose names—Karl Popper, Carl Hempel, Wesley Salmon, Thomas Kuhn—may or may not be familiar, highlighting lessons learned over several decades of philosophical debate. The payoff is that it allows us to make sense of an insight from another philosopher of science—Imre Lakatos. I argue that science requires a particular kind of intellectual humility and intellectual courage that embraces uncertainty and failure as essential components of scientific method. It also briefly shows the connection to my evolutionary contention that knowing *how* is more basic than knowing *that*, and that we ought to ground scientific reasoning in demonstration rather than in language.

Curiosity is the most idiosyncratic and the most characteristic of the scientific virtues, and all the other scientific virtues take on their distinctive character by virtue of their relationship to it. This sometimes puts science at odds with other cultural practices that do not hold curiosity in such high esteem. Chapter 7 turns to the relationship between curiosity and a contrasting virtue—creativity. I examine this conflict as it arises in premodernism and postmodernism. Dogmatic religion is a case in point for which curiosity has been seen as a sinful rejection of God's power and authority. One finds a similar devaluing of curiosity in the postmodern rejection of science, which rejects claims of objective knowledge in favor of power. Bridging the gap between science and literature, as with ending the war between science and religion, requires reconciling their value differences, and analyzing these conflicts in virtue-theoretic terms may be helpful for dialogue.

Chapter 8 examines the flip side of scientific virtue and considers the characteristic vices that scientists can be prone to. As Aristotle argued, in most cases the vices arise from either a deficiency or exaggeration of what is a virtue in moderation. Even curiosity can be taken too far. The chapter also considers problems of scientific arrogance, indifference, and what might be called gluttony. A proper understanding of the epistemic and ethical values in science should make it clearer when scientists overstep their expertise. It also provides a framework that may help prevent scientific misconduct by fostering a culture of scientific integrity based upon internal values.

Chapter 9 turns to the topic of education. How is the value structure that constitutes scientific culture passed on? How does the scientific character reproduce itself? For science to properly reproduce itself, I argue that science education needs to explicitly nurture the scientific mindset. Too often science classes crush a child's natural curiosity. It is not enough to teach scientific literacy and methodology. If we hope to encourage students to see science as a vocation to pursue, we need to put scientific values at the root of STEM education and think of how to create an environment where those traits can grow and bloom. The best science teachers are those who themselves exemplify curiosity, making them inspiring role models.

Scientific understanding is not the only worthy goal in life; for human beings the good life involves a variety of life projects. If a life of virtue is to be a coherent one, scientific values will have to be integrated with the values of our other pursuits, including our diverse social, religious, and political ones. This is part of the reason why, as Aristotle argued, virtue is connected to politics. The final chapter discusses science as an aspect of our role as citizens and argues that scientific habits of mind are an important basis for democratic action. Through a rebuttal of Paul Feyerabend's view, it shows why the radical postmodern attack on truth, which has recently been taken up by the far right, inadvertently undermines the values of freedom and justice. Scientific virtue must be reintegrated with other human virtues, ultimately aiming at general human flourishing. This implies a duty of social and political responsibility. The chapter concludes by returning to the original point that curiosity and the other scientific virtues have a basis in our common evolutionary history and are part of what it means to be human.

Welcome to the Wunderkammern

To make these points in a time of polarization and to attempt to find a common path that disparate audiences can tread is no simple task. Cultural norms are quintessentially intrinsic structures—they constitute what it means to belong. In so doing they simultaneously distinguish members from nonmembers, insiders from outsiders, home cooking from foreign fare. However, my aim here is not to exclude but to include, for the solution to wicked social problems will require the unified efforts of a broader community. I write, therefore, for both my fellow scientists and humanists. I write to realign their perspectives and aims under a banner of liberty, equality, and reality, and to rally their collective energies for the common defense of truth and truth-seeking against authoritarian attacks from the political extremes.

However, I write not just as an academic for other scholars. I write to welcome a broader audience of citizens with whom we must work if we are to defend the center and the core notions of truth and facts. With that in mind, I have tried to write with the recognition that some readers will take certain things for granted and that others will want a dive guide, especially as we get into deeper philosophical waters. Ideas make a difference, and my hope is that I can interest general readers in these academic debates. I try to do this by presenting this book as a philosophical Wunderkammern—a conceptual cabinet of curiosity. Filled with a collection of useful concepts, telling anecdotes, and nuggets of practical wisdom, it is organized into a system to exemplify my theme and theses about curiosity and the inter-relationship of facts and values. In this way, perhaps even those who do not care about the theory will find reason to pause and wonder, organize and act. I hope readers of all sorts will find it useful for bridging the gaps in our culture and for promoting the flourishing of science and the natural human spirit of inquiry.

Acknowledgments

My primary appreciation goes to my wife, Kristin, who began encouraging me to write this book two decades ago and kept it up until it was done. Many individuals have helped in the subsequent years as it simmered while I worked on other parts of the project. I am indebted to all the scientists I have been fortunate to collaborate with, to those I interviewed during a stimulating and productive summer in Oxford in 2007, and to the thousand more who agreed to be interviewed or surveyed as part of my national study. I'm grateful to my graduate student interviewers and data coders: Karen Meagher, Tony Givhan, Zachery Piso, Brian Robinson, Ike Iyioke, Anna Malavisi, Lori Hale, Brittany Tucker, and especially to Eric Berling and Chet McLeskey, who exemplified the virtues of meticulousness and perseverance over five years of data collection. I owe a special debt to Jon Miller, who has been an exemplary collaborator and practically wise mentor for my apprenticeship in social science research.

The University of Michigan's Institute for Social Research welcomed me as a visiting scholar, as did the American Association for the Advancement of Science (AAAS), where I spent my sabbatical in 2015, when the book moved back to the front burner of my attention. For long-term institutional support, I want to thank Michigan State University and its Lyman Briggs College for an environment that provided the freedom to cross disciplinary boundaries; to enlightened deans and directors, especially George Leroi, Diane Ebert-May, and Elizabeth Simmons, who fostered the goal of creating a third culture; and to my faculty colleagues in HPS and STEM, who continue the work of connecting the cultures, especially Michael O'Rouke whose own interdisciplinary expertise has been invaluable as we develop ways to put these ideas into practice. I send a general shout-out to students in my

Lyman Briggs HPS courses who engaged in lively discussions about chapter drafts. The bacteriobot research described in chapter 1 was done with my graduate student Laura Grabowski and postdoc Wesley Elsberry. I especially want to thank my co-PIs and collaborators of the BEACON Center for the Study of Evolution in Action, who are shining exemplars of the sort of scientific excellence that this book attempts to describe.

Material in this book is based in part upon work supported by grants from the National Science Foundation (NSF) under Cooperative Agreement No. DBI-0939454 for the BEACON Center for the Study of Evolution in Action and from the John Templeton Foundation (JTF) under Cooperative Agreement No. 42023 for the Scientific Virtues national survey. Any opinions, findings, conclusions, or recommendations expressed in this book are my own and do not necessarily reflect the views of NSF or JTF. Judith Andre, Owen Flanagan, Noretta Koertge, Renee Leone, and Percy Pierre provided expert feedback on the entire manuscript. Errors that remain are my own.

1 A Virtuous Instinct

I believe there exists, & I feel within me, an instinct for truth, or knowledge or discovery, of something of the same nature as the instinct of virtue...
—Charles Darwin[1]

Reason Enough

Beginning in Wonder

Science, like all great philosophy, begins in wonder. The goal of science is to make discoveries, but the announcement of a scientific discovery occurs only at the end of an investigative process that started when a scientist observed something strange, interesting, different, or puzzling, and thought: "Hmmm, that's curious." Science is usually presented as a set of answers, but this misrepresents the nature of the practice, which lies in its process more than its product. Science is better characterized as a series of questions pursued by inquisitive minds that just cannot stop wondering what or why.

The secret to science is wondering in a special way. It is well understood that not all answers are scientific—in their search for scientific answers, religious and nonreligious scientists alike recognize that "It was a miracle" is a nonstarter. Similarly, neither are all questions scientific: "Should capital punishment be abolished?" "Are you my soul mate?" "Yes, but is it art?" Such questions call for modes of thinking very different from the methodologies of science. Nor is distinguishing the scientific questions just a matter of picking the right interrogatives. It is not as easy as saying, for example, that science deals with the *whats* and religion with the *whys*. "What is the meaning of life?" is not a well-posed scientific question even though it begins with "What." Science cartoonist Sydney Harris nicely satirized the idea that one could scientifically calculate such an answer by depicting a couple of self-satisfied researchers looking at an intricate equation for the meaning of life they had formulated on their blackboard; the punchline is that the result is a zero.

On the other hand, there can be a perfectly reasonable scientific answer to a question like "Why are we here?" (which, at first glance, looks very similar), provided that its different senses are disambiguated. Science cannot address this question if the *why* is understood to refer to ultimate metaphysical purpose. Leave that to the philosophers or theologians. It can, however, very well address the more mundane *why*-question that refers to the causal processes that brought us here. Evolutionary science provides satisfying scientific answers to that latter sense of the question. Some might think that one merely needs to sort out the philosophical questions from the scientific ones, but that is not a simple matter. Most of what we think of now as scientific questions began as philosophical questions. Indeed, in a basic sense, scientific questions are a subset of philosophical questions.

Some may find it curious to speak of science as a philosophy. Are not the airy notions of head-in-the-clouds philosophers the very opposite of the hard facts produced by feet-on-the-ground scientists? This may be a common view, but it arises mostly from misconceptions about both philosophy and science. There are actually very good reasons that an advanced degree in science is called a PhD—doctorate of philosophy.

The word "science" comes from the Latin term *scientia*, which means knowledge or understanding, construed in a broader philosophical sense than the modern usage. The word "scientist" did not come into use until the mid-nineteenth century, when it was coined by William Whewell, an English philosopher, historian, Episcopal priest, and "man of science." Previously, cultivators of science thought of themselves as natural philosophers. Science itself was referred to as *natural philosophy*.

That term made sense because it highlighted science as a philosophy of how we can know about the natural world through methods that refer to natural phenomena.[2] Natural philosophy stood in explicit contrast to occult philosophy, which tried to explain the world in terms of supernatural powers and agencies. Natural philosophy did not really blossom until the scientific revolution of the seventeenth century, but its roots go back much further. They can be found among the ancient Greeks, such as in Hippocrates's view that epilepsy should be seen not as a "sacred disease" caused by divine possession but as a natural affliction with a natural cause.

This natural methodological approach is part of what distinguishes scientific thinking. It is not an airy metaphysics, but a thoroughly grounded empirical philosophy. Scientific explanations may be "occult" only in the archaic sense of the term, which refers to causes that may be "concealed" and "hidden" from view. Part of the wonder of science is discovering how the secret, intricate, causal structures of the world produce the wonderful effects we observe.

It may have been Socrates who originally spoke of philosophy as beginning in wonder. Plato gives him this line in one of the important dialogues where he discussed epistemology, the nature of knowledge.[3] As is typical in these Socratic dialogues, readers are quickly led to realize that concepts they had taken for granted are not so straightforward after all. What it means to have and acquire knowledge, which is the subject of epistemology, is a particularly complicated question. Plato's own answer—that we are born with knowledge but need to use reason to tease it out in a form of remembering—does not initially seem very plausible to our modern sensibilities. We are more in the epistemological lineage that traces back to Aristotle, who argued that all knowledge has its source in sensation. Science sits most naturally in this empiricist camp. However, a modern-day Socrates would again point out that things are not so simple.

Perhaps we are not born with innate knowledge in the way that Plato imagined, but neither is the mind just a blank slate, as one strict form of empiricism held. To take an obvious case, think of the instincts of animals— innate behaviors that are pretuned to the environments in which they will live. Sometimes these are specific hardwired behaviors, such as fixed action patterns in insects or the compulsion for salmon to return to their home streams to spawn, but other behaviors are programmable, making them flexible in an environment that may be variable. The human animal tends more toward the latter sort of programmable instincts, but the point is the same: we are born ready to learn about the world with a mind that is partially preshaped by evolution.

A Darwinian Instinct for Truth

In *The Descent of Man*, Charles Darwin explored the evolution of mental traits including attention, memory, imagination, and reason. He showed that even the most advanced human mental characteristics could be found in incipient forms in other animals, and he offered explanations of how these originally simple instincts could have been shaped by natural selection.[4] Without instincts that ready them for the world, organisms would not stand as good a chance of surviving. An organism that was unable, for instance, to figure out critical cause-effect relationships in its environment would fare poorly in life's tough competition. Human beings are no different. Though our mental powers are greater than those of other animals, as well as more plastic and adaptable, at a basic level we, too, are compelled to figure out the causal structure of the world.

Given such a view, Darwin's reflections on his own scientific motivations and character in the years during which he formulated his theory of evolution should come as no surprise. In a letter to his friend and scientific colleague John Henslow, he wrote:

> I believe there exists, & I feel within me, an instinct for truth, or knowledge or discovery, of something of the same nature as the instinct of virtue, & that our having such an instinct is reason enough for scientific researches without any practical results ever ensuing from them.[5]

Much more than just a personal sentiment, this is an astonishing observation with radical implications. Darwin noticed something highly significant about himself as a scientist. He recognized himself as an epistemic agent—a knowledge seeker and actor—at a very deep level. It is worth looking in detail at some of the key elements in the statement with an eye to unpacking their import for how to think about the character of science. Let us begin with truth.

Truth   Darwin is completely typical in focusing on truth as a scientific goal, but our scientific Socrates would caution us to try to be clear about what we mean by this. One reasonable answer would be to say that science aims at propositional knowledge, which involves well-defined logical notions of truth. Propositions are expressed as descriptive statements and are either true or false, though we may not know which. Presumably, the job of science would be to find out. If scientific hypotheses are just statements that express propositions, then maybe science is just about determining their truth or falsity, hypothesis by hypothesis. Attractive as this logic-based approach is, making it work was more difficult than expected. In philosophy of science, the Logical Positivists made a valiant attempt to analyze scientific knowledge in terms of propositional logic, but for reasons that do not concern us here, the project ultimately failed.[6]

Something different makes better sense for science. Consider a broader notion of truth whereby we may refer to a more or less true likeness. A picture is the simplest case, and we often speak of science as providing a picture of the world, perhaps because we default to thinking of science in terms of visual observations. There is nothing wrong with speaking loosely in this way, so long as we keep in mind that we may be informed about the world by other senses; there are limitations to the picture metaphor, and science is certainly not limited to visual models. When scientists speak of seeking truth, they refer to their search for the best models of the world. To model something means to reproduce it—to produce a likeness—to a greater or lesser degree, in ways that are relevant to one's interests.

It may be helpful to think of this as the difference between a binary and what may be called an analog notion of truth. Classical logic has just two states—true or false—so if a proposition is not true, it has to be false. Scientists, however, know that their models are never perfect, which on a binary notion of truth would imply that they are all false. One might argue that, strictly speaking, this is an accurate account of our situation, and we just need to learn to live with it, for only the gods could know what is "truly true." But such a view seems perverse and not especially useful for understanding the practice of science.

Let us take a different approach. The truth of classical logic might be thought of as being based upon one extreme of a scale of accuracy or likeness—truth is 100 percent, with falsity being anything less than 100 percent. But this way of parsing things conflates something that is very nearly correct, say 99.9 percent, with something at the opposite end of the scale. Instead, what if we take truth and falsity as being at the upper or lower ends (i.e., the upper or lower regions) of the scale and then think of how a likeness may be truer and truer (or vice versa) as one moves up (or down) the scale. This makes sense of how we often say that something is "kind of true" or "very true" even if it is not "perfectly true." In a similar fashion, one can speak of degree of fidelity: How good a likeness to reality is model M? This, too, fits with common ways of speaking, such as when we ask how accurate a rendering model M is, or how faithful a reproduction M is.

Couching truth in this way and connecting it to notions of modeling and the idea of faithfulness of reproduction is also useful for understanding another key aspect of science's methodology. Specifically, it relates to the idea of reproducibility in the very practical sense of being able to replicate (i.e., reproduce) one's results. Scientists will not trust experiments they cannot replicate. As an experimental result is found to be more robust—meaning that it continues to hold up under various, repeated tests—they reasonably increase their confidence in how true their model is. This is a very practical notion of truth. Unlike the classical notion, it is not necessarily tied to language. For example, models may be true in the same way that someone's aim is true, such as that of a skilled archer. Individuals whose aim is true can shoot or throw something so that it goes where they want it to go, hitting or coming sufficiently close to the bullseye or mark each time. Truer models will, similarly, make more accurate predictions with greater regularity. This shows the connection between truth and replicability.

Much more could be said about how and why this makes sense in science (and philosophers of science are more than happy to do so at the slightest instigation), but we need not pursue that now. The important point is that when scientists say they are seeking for truths about the world, they are not talking about absolute truth, but a humbler sort of empirical truth that helps us make our way about in the world and that necessarily comes in degrees—there are always p-values, confidence intervals, degrees of likelihood, or other indicators of accuracy attached to all scientific findings.

Knowledge   The concept of knowledge (the second element Darwin referred to) for a scientist is of a kind. Philosophically, analyzing knowledge requires that one carefully draw various important distinctions, such as the difference between *knowing-how* and *knowing-that*. As mentioned above, the latter is what epistemologists, philosophers of knowledge, call propositional or descriptive knowledge—that is, knowledge that can be stated in the form of descriptive sentences. This is quite different from the more basic notion of knowing how to do something, which philosophers call procedural knowledge; I may know how to throw a rock, but not be able to articulate that knowledge in any sort of descriptive form. The relationship between knowing-how and knowing-that is important, but for the moment, let us just stick with the latter and ask what it means for someone to know *that* P (which is to say, to know some particular proposition). As before, the classical philosophical notion of this goes back to Plato, who suggested that for something to count as knowledge it had to be not just a true belief, but a justified one as well. A lucky guess does not count as knowledge. Knowledge comes from appropriate evidence and reasoning. Even that may not be all there is to it, for epistemologists have found cases where even true beliefs that are rationally justified may still not count as knowledge if the process by which the knower's belief was formed was unreliable. Scientific knowledge is a type of knowledge that has been justified by scientific testing. Again, not every question is scientific or has a scientific answer, but the methodologies of science have proven themselves to be reliable for answering empirical questions about the natural world.

Here, too, we need to reiterate that scientific knowledge is not absolute. To say that science is a reliable process is not to say that it is perfect or that it inevitably reaches the exact answer after some given amount of effort. Empirical knowledge of any sort is fallible, and science is no exception. In keeping with our analog notion of truth, scientific knowledge also comes in degrees.

Unlike the deductive formal certainty that math provides, science is an inductive enterprise. That means that even if all the premises of one's argument are true, there is still a chance that the conclusion is false. An ornithologist may have observed a hundred white swans, or a thousand, or ten thousand, and so conclude with increasing confidence that all swans are white, but it remains possible that a black swan will someday be discovered. Scientific knowledge thus always comes with some degree of confidence attached.

Scientists speak this way naturally, qualifying their claims not only by reporting statistical levels of significance but indicating appropriate inductive caution with phrases like "the data suggest," "it appears likely," "from what we have seen so far," and so on. Nonscientists often regard such phrases as signs of equivocation, if not confessions of ignorance. In fact, what such indicators of degree of confidence represent is the basic epistemology of science; they are linguistic tags that mark the strength of accumulated evidence, and the ever-present recognition that further evidence may necessitate a change. That is the nature of inductive evidence and part of the reason that scientific discoveries rarely conform to the stereotypical eureka moment.

Discovery   This brings us to the third element in Darwin's quote—discovery. Every scientist dreams of making a new discovery...of being the first to observe a phenomenon, identify a new species, devise a new process, or uncover some secret law of nature. Our overarching goal in science is to understand the workings of the natural world. To return briefly to procedural knowledge, scientists do not just seek to discover true facts; they also seek to discover useful methods. One may discover not only *that* X, but also *how to* Y. In either case, the core of the notion of discovery is contained in the word itself; to discover is to reveal and make apparent something that before was shrouded. It aims to bring light to what before was obscured.

Discovery is rarely so simple as the stereotype suggests, in part because of the points noted above about the nature of scientific justification. On observing something extraordinary, for instance, one may at first just say "I have discovered something interesting" in a sense that no one would quarrel with. However, scientists will rarely be satisfied with that statement; they will consider it in relation to other observations and extant theories, and subject it to relevant tests. For such reasons, scientific discoveries usually happen slowly. Can you fairly say, for instance, that you have discovered something if you do not really understand exactly what you have found? A scientific discovery requires not just knowledge of something in the world, but some account of what makes it so. In its robust scientific sense, discovery requires understanding—not just the *what*, but the reason *why*.

Reason Philosophically, we know that reason demands this higher standard be met before one dubs something a real scientific discovery. To lay the groundwork, let us return our attention to the prescientific, instinctual level that Darwin investigated.

Darwin pointed out that mental traits evolved. Rudimentary forms or precursors of the full-blooded human traits are to be found in other animals. Given this discovery, we should similarly expect that scientific knowledge arose out of simpler forms of knowledge that preceded scientists and even the evolution of human beings. At this basic evolutionary level, discovery need not be articulated in language. From an evolutionary point of view, *knowing-how* precedes *knowing-that*. The first order of business for an organism is to just be able to know how to survive in its environment. How much does it need to know? It only needs to be good enough in practice. How well does it have to reason? Just enough. As mental traits become more complex the concepts of knowledge and reason also become more sophisticated, but in earlier stages what we are dealing with is prelinguistic knowledge and gut-level reasoning prompted by an instinctual drive to seek, explore, and find out. This may be experienced initially as a vague feeling of irritation and dissatisfaction—akin to an itch that must be scratched or a hunger pang that moves one to search for food. Sometimes this is actually felt not only in the furrowing of a brow but also in the gut, as a slight but uneasy tightening of the stomach. Perhaps it is not surprising that an especially puzzling question is spoken of as something that has one scratching one's head or tying one's stomach in knots. Even before we can put words to it and form it into an actual question, we feel a basic confusion that must be resolved or a need that aches to be satisfied.

For organisms with higher-order mental traits, and especially for big-brained, linguistic primates such as ourselves, the ways in which problems are solved and questions answered are more sophisticated. Some problems may always remain too difficult, however, even for intelligent beings like ourselves. Our rationality is bounded, as the pioneering cognitive scientist and economics Nobelist Herbert Simon put it, so we should not expect optimal decisions. However, evolution does not require perfection of organisms; it just requires them to be good enough for the competition of the day. As mental traits improve, so may our cultural evolution, including science. At first, we just need to have enough reason to begin to do science; standards will be raised thereafter. The point is that, at base, these traits that constitute the seeds of science are instinctual.

So, what is that instinct for truth, knowledge, and discovery? Darwin used the word elsewhere, and we recognize it easily—it is *curiosity*.

Curiosity as an Instinct

A More Powerful Drive Than Feeding

Medical researcher and essayist Lewis Thomas described some observations about scientists and their characteristic behaviors in his essay "Natural Science" that are relevant to this point.

> Scientists at work have the look of creatures following genetic instructions; they seem to be under the influence of a deeply placed human instinct. They are, despite their efforts at dignity, rather like young animals engaged in savage play. When they are near to an answer their hair stands on end, they sweat, they are awash in their own adrenaline. To grab the answer, and grab it first, is for them a more powerful drive than feeding or breeding or protecting themselves against the elements.[7]

Thomas did not use the term curiosity in this evocative passage or any-where in his essay, but that surely was what he had in mind in speaking of scientists' drive to find the answers they seek. Thomas recognized that this kind of inquisitive behavior is instinctual but professed not to know how it works. It cannot be planned, he said. It takes the right sort of environment. It requires that something be right in the air for a swarm of scientists to make the "pure honey" of science—a new truth about nature.[8]

I think we can say more than this. Indeed, Thomas himself was already on the right path. He was deliberate in his points of comparison: the instinct of curiosity, he said, can involve competition and can be stronger for scientists at times than the needs to eat, breed, and be sheltered. There is another term that he did not use, but which is implied in this comparison and ought to have been made explicit: evolution.

Evolution of Curiosity

Darwin on instinct   Whether he realized it or not, Thomas was channeling Darwin. That Darwin spoke of this disposition for discovery as an instinct implies that he believed it had an evolutionary basis. Except for a couple of pregnant lines—"Psychology will be based on a new foundation, that of the necessary acquirement of each mental power and capacity by grada-tion. Light will be thrown on the origin of man and his history."[9]—Darwin mostly omitted discussion of human beings and their mental traits from *On the Origin of Species*, postponing this for later. In *The Descent of Man* he began to lay out his ideas about such matters, including several places where he explicitly mentioned wonder and curiosity as part of his discussion on the evolution of instincts.

In the chapter on instinct in the *Origin*, Darwin covered a wide range of kinds of examples, including pointing behavior of pointer dogs, parasitic egg-laying of cuckoos, slave-making in ants, and cell-making in hive bees. In the *Descent*, he noted that the instincts in humans and other higher ani-mals are fewer and simpler than in the lower animals, but the ones we do share are critical, including self-preservation, sex drive, maternal affection, and infant sucking.[10] Why don't humans have complex instincts? He spec-ulated that complex instincts and high intelligence "interfere with each other's development."[11] Instincts are "definite and uniform" responses to "particular sensations or associations."[12] They "impel" behaviors and, at least in their basic form, do not require learning from experience.

Of course, to say that instincts do not require learning from experience does not mean that this can make no difference—experience and learning may subsequently shape instincts. Even the characteristic, instinctual songs of birds are modified into variant renditions depending upon what dialects the infant birds hear around them. We will later look at learned habits and their relation to instincts, but for the moment let us keep the focus on instincts as evolved traits.

Presumably, the complex environments and complex social relationships that human beings have to deal with made the flexibility of high intelligence and learning more valuable to us in the long run, but Darwin's point is that all mental traits have to be acquired in gradations. Our higher intellectual powers must have evolved on a foundation of simpler instincts. It would be a mistake for us to think that our intelligence has removed us from the realm of instinct.

So how do instincts themselves evolve? They evolve the same way as physical traits. Let us remind ourselves of how this process works.

The Darwinian process The first core element is a mechanism of inheritance. Organisms must be able to pass on their traits to offspring. The relationship of descent is central to the process. Indeed, Darwin himself initially did not use the term evolution for his discovery but rather *descent with modification*. The key to understanding biological inheritance is, of course, the idea of reproduction. It is through a process of replication that traits—Mom's earlobes, say, or Dad's schnoz—are passed on to the next generation.

The second core element is the fact that such traits typically vary and that new variations can continue to arise at random. Earlobes may be attached or detached with all sorts of possible intermediates, and noses come in all shapes and sizes. So, too, do their abilities vary. Most of us have a serviceable sense of smell, though women are generally more accurate than men at identifying odors. We know of some individuals who are born with a poor sense of smell and others whose noses are exquisitely sensitive. The nose of wine critic Robert Parker was famously insured for one million dollars, though even its hypersensitivity paled in comparison with the abilities of the average hunting hound.

In competition for resources, animals with a better sense of smell would be better at tracking prey, or, to take the defender's point of view, faster in noticing and avoiding an approaching predator. Such individuals would accordingly have a better chance of surviving and reproducing. In this way, the automatic process of natural selection shapes a population. This is the third element of Darwin's law. If this process of descent with modification is repeated over time, generation after generation, then such new and improved adaptations will become prevalent.

Instincts are behavioral rather than physical traits, but they can evolve in just the same manner. It may be easiest to see the link between the physical and the behavioral with extremely simple organisms. Paramecia move, gather food, reproduce, and may even be capable of some primitive kinds of learned behavior, but most of their behaviors are inherited. For example, they use their cilia to sweep in bacteria, algae, and other microorganisms they feed upon. Presumably this sweeping behavior gets them more food than if their cilia just twitched slightly, though even slight twitching would be better than cilia that did not move at all. Paramecia do not have a nervous system, so any such behavioral variations would have to be an effect of their physical structure. To move up a step in complexity, it would even be possible for these to respond differently depending upon environmental conditions; perhaps warmer temperatures caused them to move faster, with the behavioral repertoire being built into their physical structure. This sort of phenotypic plasticity of behavior could thus evolve as a function of the evolution of their basic physical traits. Things get even more interesting once a simple nervous system (or functionally equivalent way of storing and processing information) evolves, but at the basic level it should be clear that evolution can shape instinctual behaviors in essentially the same way that it shapes other traits.

Bacteriobot   New techniques in experimental evolution allow us to test some of these ideas directly. We are now able to study the evolution of behavioral instincts using computer models. In my own research lab, for instance, we have observed not only the evolution of phenotypic plasticity, but also the evolution of rudimentary "brains" that can control behavior. In one experiment, we wanted to investigate what steps evolution might take to produce a controller for an organism like a bacterium with a flagellum that had only two basic motion options—move forward or move in a random direction. We also gave it a simple sensor that could measure the strength of some given factor in its environment—the equivalent, say, of the scent of a resource that is stronger as one gets closer. The details are not important here, but suffice it to say that we were pleased to observe the evolution of a software controller that could successfully navigate to find the resource.

We downloaded the evolved controller into a mobile robot, which then performed the rather clever, evolved behavior.

What would happen is this: The robot would take a sensor reading and then move and sense again. If the second input was higher than the first, it would keep moving straight forward. If the second input was lower, indicating that it was moving away rather than toward the resource, then it would switch to its second motion, which took it in some other random direction. Repeating this sense-compare-move sequence would take it up a gradient to the resource.

Imagine a swarm of organisms that have evolved such a capacity when a new kind of resource becomes available elsewhere and its novel scent fills the environment. If resources become depleted along the standard path, an organism that picks up the new scent and is enticed to venture along a road less travelled to follow it will have an advantage.

There is a big difference between a Parker and a paramecium, but such intermediates show how it is possible for evolution to work its wonders. Perhaps sniffing out an interesting new scent is not really different in kind from more advanced forms of inquisitive behavior. It is, thus, at least a reasonable hypothesis to think that the first twinkle of curiosity might have evolved out of this kind of "try-a-new-path" behavior. Does it make sense to talk about bacterial precursors of curiosity? How low can we go? This is a matter for future research. Whatever its original source, curiosity has all the characteristics that make it evolvable, like other instincts.

Curiosity as a Darwinian predisposition Darwin began with examples that show heritability of instincts, particularly in the way they are exhibited even in the young before any experience. No one who has interacted with young animals—think of your pets or your children—doubts the evidence of their innate interest in exploring novelties in their environment. Curiosity extends to novel stimuli across sensory modes—an unusual sound or feeling can stimulate their interest as well as some new sight or smell—and it is also variable. Some individuals are eager to seek out any novelty, while others stick to the familiar if they can, with various grades in between. It is also, of course, subject to natural selection: if curiosity always or even usually killed the cat, evolution would have made a quick end of that disposition. But that is not what we observe, and it is not hard to see how curiosity would provide a selective advantage in many environments.

In particular, curiosity is an *epistemically* advantageous trait for an organism—it is useful for acquiring knowledge. Curiosity and other traits that we will encounter later are valuable general dispositions for figuring out our only partially predictable world. If the world were completely regular in all the ways that mattered to organisms, they would not need to be curious.

Consider an idealized environment with unlimited resources; a feeding organism in such a paradise might need only sit in one place and open and close its mouth. It could move continually in a straight line or move anywhere at random. However, in a world where resources are limited and irregularly located in space or time, its chance of surviving and reproducing would be greatly increased by attending to differences in its environment, deviating from the beaten path, or trying something new.

In such worlds, a predisposition to investigate novelties and to move off of the standard path to check out something that is different can have real utility, giving organisms a competitive advantage. Without such an instinct, discovery of new resources would be less likely. Even elementary mechanisms for investigation could provide an advantage. For example, all an organism might need at first is a simple system of differential behavioral response to perceived environmental differences: if input is the same (constant), do A; if input is different (greater or lesser), do something else. This is a very basic form of neophilia. As evolution hones this disposition, its behavioral forms can become more and more sophisticated. Honeybees, for example, will pause to investigate new elements in their environment. When they find a valuable new food source, they will perform learning flights (doing a "turn back and look" flight) to familiarize themselves with the new landmarks. They stop these learning flights after they are familiar with the area, but they redo learning flights if a landmark changes.

These examples of neophilia relate to the discovery and maintenance of access to new resources, but curiosity is more broadly beneficial than that. For instance, Freud was certainly right that it is also useful for sex. Even well-fed organisms will not do well in the evolutionary competition if they cannot find a mate or whatever else is needed to reproduce.

Furthermore, when considering the value of novelty-seeking, it is important to consider that, while the new may be useful, it may also be dangerous. That rustle in the grass could be new prey, but it could also be a predator. The ability to identify new dangers and learn to avoid them is also adaptively valuable. We have to admit that curiosity does sometimes kill the cat. Organisms that always rushed in quickly where others feared to tread will sometimes step right into danger. A little caution is in order, which perhaps explains why curiosity is experienced initially as a kind of disquiet (a mixture of excitement and apprehension)…a feeling that something is not quite right or as it should be.

There are two possible responses to perceived anomalies. The conservative, incurious response is to ignore the differences and stick with established generalizations. The progressive, curious response is to investigate differences with the hope of discovering a broader or better generalization. The latter approach has the adaptive advantage of self-correction and discovery of new or broader patterns. When one is confused, an incurious response might initially be safer, but a curious response provides an opportunity for learning. An organism's first job is to discover the regularities and irregularities of its world, and this is certainly true for humans. Freud thought that human curiosity was always ultimately sexual, but while we, of course, recognize the importance of reproduction on an evolutionary account, there is a more general advantage that is prerequisite even to this. Infants must be able to figure out the causal structure of the world, and this requires attending to patterns and noticing violations of them. In particular, infants need to be able to do inductive cognition, which is the ability to learn and generalize from new experiences. Novel items disrupt familiar regularities, sometimes by showing they were incorrect, sometimes by revealing more general regularities. Once something has been checked out and is familiar, then things become predictable again. Predictability is useful, and it usually takes curiosity to discover it.

To summarize, curiosity is a beneficial intralife disposition, leading organisms to follow up on deviations from the expected in the hope of discovering some new or improved generalization. Again, this will not be true everywhere for every kind of organism, but it is critical for organisms in environments that change on intralife time scales. Inflexible, hardwired behavioral patterns will not let an organism adjust to changing conditions. Even if following curious deviations from the norm does not lead to new resources, following up on differences may be useful for maintaining one's current resources; deviations from expectations may be a signal that the environment has changed, or that signs one has previously relied upon have changed, or even that one's sensors are malfunctioning. Which takes us back to confusion.

From confusion to contentment   When an organism departs from a regular path, how can we know that it is driven by curiosity rather than confusion? At first glance, the one seems positive, active (one *pursues* curiosity), and adaptive while the other seems negative, passive (one *becomes* confused), and maladaptive. However, in another sense, the two are closely related in that both involve a response to a deviation from an expected pattern. Both are reactions to something in the environment that differs from one's conceptual map (or from some internal conflict within one's conceptual map) and indicate a mismatch with expectations. Both involve a feeling of puzzlement. We are puzzled when we discover an anomaly in what we had previously seen as a regularity. It is as though one's stream of perception goes *Same-Same-Same-Same-Different-Wait... What?... Why?* The deviation from expectation is confusing. Anomalies of any sort are perceived as puzzling and also as potentially interesting.

We pay attention to patterns of similarities and differences in our environment. Pure randomness is not interesting; if we play white noise, after a few minutes we no longer even hear it. Simple repetition can also be boring. We are programmed by evolution to notice the regularities and the irregularities and to try to figure out how to resolve any anomalies. This interest is mingled with excitement, perhaps related to the sensation produced when the body prepares itself for possible danger, or otherwise anticipates action.

Furthermore, inquisitiveness, like all other traits, is variable among individuals. Some respond only to extreme novelty while others feel the itch of curiosity more keenly. In the end, the emotional feeling that comes from the resolution of inquisitive tension is a special sort of pleasure, and for people who are deeply curious, the feeling can be profound. As any cat dying with curiosity will attest, the resolution of some puzzle is a feeling of utmost satisfaction.

Curiosity as a Virtue

Darwin's Moral Sense

So far, I have been focusing on Darwin's extraordinary suggestion that we have an instinct for truth, knowledge, or discovery, and I have argued that it is indeed reasonable to think of curiosity as an evolved instinct. But now I want to turn to the even more astonishing suggestion he made, namely, that it is "something of the same nature as the instinct of virtue."[13] What could it mean to say that curiosity is akin to virtue? And why speak of virtue itself as an instinct?

To help tell the story, let us look first at Darwin's own views about the evolution of what he called the moral sense. Morality, he hypothesized, became possible when animals that had evolved social instincts achieved a sufficient degree of intellectual power. By social instincts, Darwin had in mind those sentiments that lead animals to take pleasure in each other's company or feel some bond of sympathy, such as parental or filial affections. He explained how such instinctive helping behaviors could provide animals with a competitive advantage in various circumstances, improving the chance of survival in a group of related individuals. Such behaviors might be essential parental care of young or even mutual grooming, which usefully removes parasites. In the short run, such instincts might not be particularly powerful, but in combination with improved memory and imagination, they can develop into stronger feelings of dissatisfaction associated with unfulfilled instincts. These feelings, he suggested, are the incipient voice of conscience. Being able, with increased intelligence, to remember such feelings from the past and imagine the same effect in the future, if the helping behaviors were neglected, provides an indication of what they should or should not do—which is the moral sense. Darwin did not think that we could properly attribute morality to lower animals in the same way we do to human beings but hypothesized that any animals with the instinctive ability to feel empathy, together with a sufficient degree of intelligence, would come to feel a moral imperative to behave accordingly.

Beware the naturalistic fallacy   We will examine Darwin's account of the evolution of the moral sense in more detail in a later chapter, but an important word of caution is necessary here; the fact that we require evolutionary science to help explain our moral capacities does not imply that we can read ethical rules directly from biology. Survival of the fittest is a powerful explanatory principle for understanding the biological world but it makes for lousy ethical theory. Consider this thought experiment that Darwin proposed in *The Descent of Man*: If humans were reared in exactly the same conditions as hive bees, then unmarried females would, "like worker-bees, think it their sacred duty to kill their brothers, and mothers would strive to kill their fertile daughters; and no one would think of interfering."[14] If an evolved moral sense worked in this way, the problem would be obvious—it would be to commit a variation of the fallacy that the eighteenth-century Scottish philosopher David Hume identified of trying to derive a moral *"ought"* from a mere *"is."* The fallacy is obvious once one points it out: that something *is* the case does not by itself imply that it *should* be so. It will take a bit more philosophical reasoning to find some moral premise that will link factual claims to moral conclusions. Darwin himself took a few tentative steps along these lines by drawing connections to both Kantian and utilitarian ethical theory.

I have described this story in detail elsewhere,[15] but this short summary is sufficient to see how Darwin thought about the evolution of the instinct of virtue and a few of the complexities that will need to be resolved as we move forward. Some ethicists do not think that the naturalistic fallacy is a fallacy at all; Alistair McIntyre, for example, thought that statements about what is good are just a kind of factual statement.[16] Setting aside that issue, as with the moral sense, one should not expect to derive a robust ethical notion of normativity from biological considerations alone. However, an evolutionary account of the origin of curiosity as an instinct can be given normative import when put in a larger philosophical context. My own approach here will be to do this through an evolutionary variation of Aristotelian virtue theory. Both Darwin and Hume would have found this appealing. The character traits such as curiosity that are conducive to exemplary science are explainable, at least in their source, as evolved, adaptive dispositions, but they also fit well within a virtue framework that connects such mental dispositions to the realm of norms and values. To see this, we need to first get a handle on the nature of virtue.

The Nature of Virtue

As a preliminary point, it should go without saying that virtue is not limited to sexual morality, which is perhaps what first comes to mind for those who think of the prudish Victorian notion of protecting one's virtue. We should also not think of virtue as passive, which is a connotation that the term sometimes carries today. Virtue is not a resting state of being. I want to recall a more traditional element of the concept of virtue, namely, virtue as a power. We retain a remnant of this meaning in a phrase like "by virtue of which" that, like "on the strength of," conveys the idea of a dispositional property, power, or ability that makes something possible or causes it to be so. Molière had in mind this active notion of virtue as a power when he mocked trivial explanations by purported virtues: "Why does opium make one sleepy? Because it has a dormative virtue?"[17] But such glib usage does not undermine the general concept. One may reasonably say, for example, that a material or molecule is strong or fragile by or in virtue of its physical or chemical structure.

We should think of the character virtues that are our subject in a similar way: they are mental traits that dispose us to behave reliably in appropriate ways. They are action oriented, giving us the power and motivation to act as the situation warrants. In the mental realm, virtue does have the connotation of referring not to any power or disposition, but primarily to good or useful qualities. Indeed, the classical Aristotelian term *aretê* that we translate as "virtue" is also often translated as "excellence," and one may accordingly speak equivalently of cultivating a life of excellence or a life of virtue.

As Aristotle defined it, excellence or virtue is "a settled disposition of the mind that determines our choice of actions and emotions." He went on to say that it "consists essentially in observing the mean relative to us...a mean between two vices, that which depends on excess and that which depends on defect."[18] This is Aristotle's thumbnail analysis of virtue, which he took to be a balanced middle state—the golden mean—between vicious extremes. Finding the right balance is rarely a straightforward matter of following a rule; it typically must be determined by a rational principle and the exercise of judgment, which is why he spoke of the idea of inquiring how a person with "practical wisdom" would judge where the balance falls in given situations.

This is a reasonable general notion, but specific details of the account will be of less direct use for the obvious reason that it aimed to understand virtue for us as human beings, whereas I am interested primarily in understanding what virtue means for scientists in their vocational practice. An account of virtue in science, or any vocation for that matter, accordingly needs a narrower and more specialized concept. For these contexts, I want to introduce the idea of vocational virtue as a *practiced disposition*.

Virtue as Practiced Disposition

Although I have highlighted the evolutionary basis of curiosity and other scientific virtues, one does not achieve them by nature alone but must cultivate and train the instinctual traits one starts with. Aristotle was right that this is done by habituation, and we will need to say more about habits later, but there is more to it than that. I use the term practiced here, because it refers both to the fact that one must practice them to do them well and make them habitual (in a similar way to practicing to acquire a skill), but also to the idea of a disciplinary practice (which involves those skills relevant to the practice of a discipline).

That curiosity has a basis in instinct does not mean that it requires no effort to master. We are bipedal, but we still have to practice before we learn to walk, and practice a good deal more to run, skip, jump, or dance.

As we will see, scientific curiosity is more demanding than the ordinary sort of instinctive inquisitiveness upon which it is based. It will involve distinctive methods and specialized skills. It will involve a suite of other character traits. Learning to do it well takes years of training. Standards will seem impossibly high. It will take much practice, in both senses of the term, before scientists feel that they have really answered a question to their satisfaction.

Because it has both causal and evaluative aspects, virtue may be thought of as a bridge concept between the descriptive and the normative. Perhaps this is why Hume thought it could serve well as the core concept for his ethical theory. According to Hume, the fundamental feature of virtue involves "the possession of mental qualities, *useful* or *agreeable* to the *person himself* or to *others*."[19] Again, our interest here is not virtue in general, but rather the vocational virtues of the scientist. However, though our focus will be primarily on values that are useful for the practice of science, our evolutionary perspective will help us periodically reconnect these to the broader notion of human virtue in ways that Hume would have found helpful. Philosophical reasoning is required to reveal the contours of those values for science, but their roots can be traced back to the biological. Hume wrote that it is probable that the fact that character traits and the actions that flow from them are deemed "amiable or odious, praise-worthy or blamable…depends on some internal sense or feeling, which nature has made universal in the whole species."[20]

Hume did not have an explanation of how such internal feelings might come to be, but Darwin's explanation of how instincts evolve show how character traits like instinctual curiosity can be universal, or at least sufficiently so for our purposes. Again, to avoid the naturalistic fallacy, we must recognize that virtues are not derived from biology alone but are normative and intentional—by which we mean that they involve values and choices. Scientific curiosity is rooted in our evolved instinctual curiosity, but, as in Darwin's account of the moral sense, it requires intellectual processing to grow into the space of reasons and justification, and to acquire robust moral content. Attempting to do this is the work of philosophy. To start the process, we begin with an explication of these values in terms of science's goals and methods.

The *Telos* of Science

To understand the normative element in scientific practice, we need to look at what science aims for—what Aristotle called a thing's *telos*, or end. The general idea is that a thing's essential nature and purpose gives a point of reference against which one may judge the value of possible ways of being and behaving. Aristotle was interested in understanding the essential nature and proper purpose for human beings, and in elucidating what that implied about how to foster human flourishing. Our interest for the moment is in scientists and scientific flourishing, and we may begin to trace the moral structure of this more limited field in much the same way, starting with the *telos* of science.

There are, of course, many particular things that individual scientists aim to accomplish. One wants to discover the biochemical causes of cancer. Another aims to explore Mars. A third dreams of a final theory of physics. But these are all just specific examples of a common, central purpose—different ways of getting at the same kind of thing. The quote from Darwin already articulated the more general goal. Nor is Darwin's view idiosyncratic. Different scientists may phrase the idea differently but one finds near-universal agreement. What is the central, guiding purpose of science? In the end, it is very simple: science aims to discover truths about the natural world.

At base, scientists are truth-seekers. Not Truth-with-a-Capital-T, of course, but truth of the ordinary, mundane variety. They are after *empirical* truths. As we saw, these are not absolute truths in some airy metaphysical sense, but down-to-earth, workaday show-me truths that one can test. Scientists want to measure a constant to a new degree of accuracy, isolate a new compound, discover a new species, or uncover a new physical law. Folks with more ordinary interests may find such "geeky" pursuits odd, but these are the sorts of things that make scientists happy.

This is not a superficial kind of happiness, but a kind that is rooted in the core of a scientist's identity and sense of self. Here, too, we may profitably draw from Aristotle's theoretical account, which connects virtue to a deep and robust notion of happiness—what he called *eudaimonia*. Just as we may better approach this essential form of human flourishing as we cultivate the human virtues, so we may better approach the ends of science as we develop excellence in the scientific virtues. To put it succinctly, scientific flourishing involves the excellent practice, to the degree that we are capable, of those forms of being and behaving that are conducive of discovering truths about the natural world. If we must shorten this further to a phrase, we may call it the satisfaction of curiosity.

We should pause here to consider a reasonable objection. One might point out that caring about truth is not exclusive to science. Surely one will not succeed in other endeavors if one does not care about reality. Is not every profession interested in truth? Yes and no. In fact, while truth may be broadly important, it turns out that truth-seeking is not as basic to most vocations as one might think. It typically plays a supporting, not a central, role.

To take a simple example, consider the purposes of business. According to the father of modern management, Peter Drucker, "Because the purpose of business is to create a customer, the business enterprise has two—and only two—basic functions: marketing and innovation. Marketing is the distinguishing, unique function of the business."[21] Drucker intentionally omitted what most consider to be an even more basic purpose of business, which is to earn a profit—that is what "creating a customer" is for, in the end, for a business.[22] Businesses may require scientific truth to the degree that it is useful for innovation and for acquiring customers, but in other cases it could be more in their interest, at least their short-term interest, to deny reality. As a case in point, think of how, for decades, tobacco companies hid, misled, and lied about the known health dangers of their products. Marketing, after all, is less about reality than it is about customer perception of reality. The ideal situation for a marketer is when one's product is in fact superior to the competition, but if the picture is not quite so rosy at some point in time, then it is handy to have a reality distortion field. For most businesspeople, the *telos* of business comes at the end of its balance sheet. The bottom line in business is the bottom line.

What about a vocation like engineering? Though it is closely related to science, the central, guiding purpose of the engineer is not to discover new truths but rather to build new things. An engineering team might discover some new material, but they will care not so much about where it comes from or why it has the properties it does as what they can do with it. Science is an indispensable tool in modern engineering, but engineers are not scientists. They may apply science and sometimes even make real scientific discoveries, but their essential goals and methods are quite distinct from those of scientists.

There is overlap, of course; just as some engineers have a scientific streak and hope to make a discovery that improves their ability to build, so many scientists are also interested in applied research and, in such pursuits, they edge closer to engineering. Many see their work as improving society or bettering the human condition. In their purest form, however, scientists are following their curiosity simply to satisfy it, irrespective of the possible utility of their discoveries. That is, as Darwin put it, "without any practical results ever ensuing from them."

We do not need to multiply examples. Put simply, whereas other vocations may use science to advance their own purposes, science's core aim is to uncover nature's truths.

The scientific virtues   The virtues of science are, accordingly, those traits of character that predispose the scientist to act habitually in such ways that conduce to the ends of scientific investigation. In a sense, we may think of the researcher as an instrument in a study in much the same way that a microscope or a balance pan is. Like any other instrument, researchers can be better or worse at their task, more or less reliable, more or less sensitive. We may think of the scientific virtues as tuning or calibrating the practices of researchers and thereby increasing their sensitivity, reliability, and general excellence. The more virtuous the scientist, in the sense we have been speaking of, the more likely that science will flourish.

This instrumental value is a derivative normative notion, but it is enough to provide a basis for prescription and proscription. That is to say, it provides a basis for judgments of better or worse, and of *shoulds* and *should nots*. If discovering empirical truths is accepted as a worthy end (which we may reasonably take for granted, not only for professional scientists but for other vocations that utilize its discoveries), then becoming a better scientist by cultivating such traits is a morally good thing.

A stronger normative notion may be possible as well. We will look into this in detail in the last chapter, but it is worth mentioning here: if satisfying our instinctual curiosity is necessary not just to scientific flourishing but to human flourishing in general, then the scientific virtues are more than part of a narrow professional ethics. They are part of ethics in as full-blooded a way as any moral duties or any part of a moral calculus.

This is not to say that everyone must become a professional scientist, for there are other aspects of, and thus other paths to, human flourishing. The good life involves health, shelter, beauty, and so on. The healer, builder, artist, and other professionals are alternative roles one may play to promote the good. Moreover, if discovering empirical truth(s) is required for moral action, which also seems incontrovertible, then science is necessary for ethics. It will have to be able to work together with these and other vocations in a unified way for the full completeness of the moral life. Admittedly, this is not easy. However, integrating values is the real meaning of integrity, and that is a worthy goal.

The Moral Character of Scientific Culture

Snow's Two-Cultures Gap

In his influential Rede Lecture at University of Cambridge in 1959, C. P. Snow identified and decried a dangerous gap in communication between scientists and literary intellectuals.[23] As someone who began his career as a scientist and then became a novelist, Snow was unusually well placed to observe this gap. So great is the mutual unintelligibility between these tribes of scholars, he thought, that they might be seen as natives of two distinct cultures, each with language and customs that are essentially foreign to the other. The modern world has little hope of solving the difficult social problems it faces if the sciences and the humanities do not find a way to bridge that gap and form a "third culture" that unites the critical strengths of each. But in over half a century since Snow's clarion call, regrettably little progress has been made toward that goal.

Commentators have offered various explanations for the communication gap between the two cultures, but I find that a philosophical explanation seems to make the most sense. It is not that there is any less intellectual power on one side or the other. Rather, the two groups seem, at least on a certain level, not only to talk past one another but to engage the world in terms of very different kinds of concepts.

Science is more interested in showing than telling, but when it does speak it speaks in the language of facts. Ask scientists to describe their research and you will usually hear about their object of study—a property of matter, a class of chemical reagents, a species of bird—and the factual questions about it they hope to answer. Presentations at scientific conferences are typically short and to the point—a bit of background, a recitation of one's research question, how one's observations or experiments lend increased or decreased credence to one or another of the hypotheses in play, and consideration of possible confounding factors or other issues that point to further research. For many presentations, everything boils down to a single slide with data summarized in a table or graph. In science, our interest is in what is the case about the world and what we know or do not yet know about it.

The humanities, on the other hand, speak more in value terms. In addition to questions about what is (or is not) the case, the focus is on what should (or should not) be our response, attitude, or action. This is true even for fields in the humanities like history that do have a strong empirical component. Historians Jo Guldi and David Armitage write that "The mission of the humanities is to transmit questions about value—and to question values—by testing traditions that build up over centuries and millennia,"[24] and they argue that the role of history is to help guide our future action by lessons learned from the past, thereby providing an antidote to short-term thinking. The notion of "testing" here, which involves evaluation with its emphasis on values, is rather different from the scientific notion of empirical testing. In contrast to science presentations, scholarly talks in the humanities are rarely under thirty minutes and in some cases are much longer to allow detailed articulation of an argument. Philosophy talks commonly run an hour, followed by another hour during which the speaker responds to critical questions from the floor. The overall structure of a talk may be similar to a science presentation—some background to situate the issue, statement of one's thesis, presentation of the reasons for one's conclusion, and responses to possible counterarguments—but the focus is on reasons, values, and recommendations, rather than just the determination of facts. For the humanities, our interest is why we ought to care and what we ought to do.

Hume's point about the differences between *is* and *ought* seems to again apply here. If scientists are alien to values and those in the humanities devalue empirical facts, then it is not surprising that they would fail to connect. This does seem to be a fundamental difference. The methods of science are indeed limited to the testing of factual matters; one cannot perform a scientific experiment to test a moral law. Neither does morality have any authority in determining empirical facts; that the world would be better off if some fact were not the case does not make it so. However, that is not the end of the story if we can find other ways for the two cultures to understand each other's basic concepts. I will not discuss why scholars seeking answers to ethical problems cannot ignore empirical facts if their judgments are to make a connection to—and be relevant to—real agents in the real world, but surely that can be taken for granted. This book is concerned mostly with the other direction, namely that science, too, *is* and *should* be concerned with values.

Science's Basic Cultural Values

Indeed, my central thesis is that, contrary to the just-the-facts stereotype, science has a deep moral structure, and that what I call the "scientific virtues" are a useful way to connect its epistemic and ethical values.

We have already seen how curiosity is a core virtue in science and how it relates to truth-seeking, which is the central guiding purpose of science. Together, these imply and justify a suite of other values and virtues. In the chapters that follow we will examine these in detail, using their links to curiosity as an organizing principle, but here I want to briefly mention the most important of these by way of introduction.

Since the central goal of scientific inquiry is to discover truths about the empirical world, the key scientific virtues are necessarily curiosity and veracity (honesty). Many other scientific virtues, such as objectivity, will flow from this center. Viewing science and the scientific character in the way I propose will seem paradoxical to those scientists and others who are used to thinking of science as amoral and scientific objectivity as meaning free from values. This common view seriously misunderstands the character of scientific practice, which is full of values. To be objective, for instance, is not to stand apart from values. Objectivity is itself a special kind of virtue—one that accentuates a special set of values. It involves a kind of impartiality that has its basis in the epistemic requirements for scientific testing, but also functions in a manner that is closely related to what ethicist Kurt Baier called "the moral point of view."[25]

One of the most significant features of scientific method is that its conclusions must be understood nondogmatically, as they are always open to revision on the basis of new evidence. The well-supported scientific conclusions of one era may have to yield, should countervailing evidence subsequently be found. One must be prepared to bravely abandon even one's cherished beliefs if the evidence goes against them. This epistemological feature of scientific method is one reason that courageous humility is particularly important for scientists. Much will need to be said about the subtleties of such virtues as we examine them throughout this book. To recommend intellectual humility, for instance, is not to say that one must forswear all claims of knowledge. This virtue requires an open-mindedness and a cautious skepticism, but because it arises out of an understanding of the nature of scientific evidence (that is, it is not a general humility that is important in science, but specifically humility to evidence), it is of a sort that should not give way to epistemic relativism or nihilism.

A few of the virtues—not only curiosity but also attentiveness (being observant) and skepticism—may be idiosyncratic to, or at least distinctively emphasized, in science. Others, such as perseverance or patience and meticulousness, which are required because of the slow and deliberate nature of scientific inquiry, are not unique to science but simply take on a special character in the scientific context.

Caveats We will be looking in detail at these and related virtues (and vices) in subsequent chapters, but it is important to highlight a few caveats now.

First of all, this is not an exhaustive list. Why not also include courage, for example, as an independent virtue? Historically, when it may have been dangerous to advance a scientific truth that went against the religious authority of the day, a scientist needed a kind of courage that is not as pertinent today. Although scientific whistleblowers, such as the scientist who challenged tobacco companies, have had to face intimidation, smear campaigns, or loss of an industry job, it is unlikely that a present-day Galileo would have to face the threat of torture from the Inquisition. As suggested above, I will bring in courage in relation to intellectual humility, but one could reasonably argue, especially on Aristotelian grounds, that it deserves separate treatment. In particular subfields, or for particular research projects, other virtues may also become salient. However, this is a core set that arises from general philosophical considerations about the practice of natural science and also that finds support empirically from a survey of practicing scientists.

Second, as mentioned previously, neither are virtues on this list exclusive to science. Perseverance could be as valuable a character trait in the humanities or business as it is in science. And is not honesty taken to be "the best policy" generally? My claim is not that these are unique to science, but that in science they are given special import and are understood and expressed in distinctive ways.

To put this another way, the claim is not that these are exhaustive or exclusive to science, but rather that they have a special centrality and weight in science that is different from similar or overlapping sets of virtues in other vocations or systematic cultural practices. Curiosity may be a useful trait for an engineer or a physician, but we would not expect it to be regarded as of central, defining importance as it is in science. Similarly, creativity may be important for certain scientific tasks but not in the way that it is central in the arts and literature.[26]

Finally, these virtues are not offered as necessary conditions for science in a deductive sense. It would be easy to find scientists who lacked one or another of these virtues who nevertheless succeeded in making discoveries. While I will sometimes speak of these virtues as being "essential" to science, this should not be taken in a classically Aristotelian sense in which entities are taken to have a fixed essence by necessity. My own notion incorporates an evolutionary perspective, which recognizes not only that kinds are fuzzy (e.g., basic science blends in to applied science, just as species and varieties blend at the edges in nature), but also that the utility of traits varies depending upon the environment (including, again as in nature, the social environment). The scientific virtues are essential for science in the sense that they characterize what it means to be excellent in scientific practice, making it more likely that scientists will achieve their goal of discovering truths about the natural world.

Third Culture

I have suggested that the two-cultures gap Snow identified is partially the result of the *is-ought* gap that David Hume identified, and, accordingly, I think that a Humean virtue approach that connects facts and values may provide part of a solution that Snow would have endorsed. The virtues that fall within the scope of this project are primarily those that arise out of the goals of scientific investigation. Moreover, the fact that it establishes science as having a value structure sets the foundation on one side of Snow's gap for a possible bridge.

We may hope that a similar philosophical pylon can be built on the other side: the humanities need to recognize, for example, that proper assessment and attribution of value judgments regarding oppression and justice necessarily involve matters of fact. A scientific virtues account will involve narratives and stories, but they must be ones based in fact rather than fantasy.

Solving the wicked problems facing the world will require what Snow called a "third culture." To the degree that this virtue-based approach can help scientists and scholars communicate and move more easily between their two cultures, then a third culture becomes a more likely possibility. With the explicit recognition of such shared values, new ways may open not only for respectful and fruitful dialogue between the two, but also for common action.

Wonder Room

Entering the main gallery of Oxford University's Museum of Natural History is a stirring experience. Cast iron arched columns soar to five stories, supporting a vast lattice of glass panels that allow the sky to fill the exhibit space with the light. Architecturally, the interior references the elements, scale, and feel of a great cathedral, eliciting awe and wonder. Yet the differences are also clear. Rather than statues of saints, the court is ringed by statues and busts of important scientific pioneers—from Hippocrates and Aristotle to Galileo, Newton, and Darwin. The stone columns that surround the court are not capped by cherubs or mythical beasts, but rather with carvings of plants representing the botanical orders. The objects of wonder on display are not depictions of religious miracles, but rather minerals, fossils, insects, and the remains of the now-extinct Dodo.

At the back of the main gallery is a set of doors to a smaller but equally awe-inspiring museum within the museum. The Pitt Rivers museum focuses not on the natural sciences but on the human science of cultural anthropology. Here the light levels are dimmer to protect the more delicate artifacts that are displayed in dozens of glass cabinets, freestanding, or lining the walls. Items on the cabinet shelves or in drawers are arranged not primarily by chronology or geographical origin, but instead by theme or utility. Cabinets contain hundreds of curious objects collected from cultures around the world. Some are related to methods of making light and fire, others to basketry, rope and string, or writing and communication. Some focus on treatment of the dead, others on magic, ritual, religion, and belief. These dense, eclectic collections are like time capsules that allow us to peer back into an era that predated and gave rise to the modern science museum. Specifically, these displays have roots in the so-called cabinets of curiosities—collections of natural "objects of wonder" that became popular in the seventeenth century. Though we now think of cabinets as pieces of furniture, the term cabinet originally referred to a room, and wealthy collectors indeed outfitted what were also known as Wunderkammern, or rooms of wonder. Prints and paintings of these early wonder rooms are indeed awe inspiring, if a bit garish, with profusions of shells, starfish, other mounted sea creatures, reptiles, birds, and more, covering shelves, walls, and even the ceiling. In many cases, private collections were donated to institutions and became the core of science museums to be used for research and education.

One intent of such exhibits was that such curious, amazing objects might edify and stimulate wonder in viewers, young and old. Contemporary science centers take a somewhat different approach, eschewing "dusty collections" in favor of interactive exhibits that aim to go beyond static displays of the discoveries of science, and try to model its processes of discovery. Setting aside the question of which is more effective, both approaches have a similar goal of inspiring viewers to ask their own questions and perhaps to begin their own investigations.

Most of us see this awakening of curiosity as a good thing, but some critics might say that although science begins in wonder, its real aim and ultimate effect is to dispel wonder. A few even say that science stomps out wonder, brutishly removing all enchantment from the world. Is not that exactly what putting magic, religion, and belief under glass is all about? I would argue that this misunderstands the scientific mindset and the virtue of curiosity. For those who understand its aims and methods, science does not close down but provides ever-expanding room for wonder. Science is the "endless frontier."[27] To pursue our scientific curiosity is to roam freely across the universe. It is an exploration where every turn reveals enchanting new wonders, where the wending path of research may lead from the sparkle in a grain of sand to the starry heavens themselves and back again. This is not the superficial wonder of ignorance but the deep appreciation that comes from hard-earned understanding—knowing the reason why. It is a wonder that brings us back satisfied and alive…whiskers aquiver…ready to pursue the next question.

2 A Kind of Utter Honesty

Plato is my friend, Aristotle is my friend, but my greatest friend is truth.
—Isaac Newton[1]

Explicating Cultural Values

Cultural Sinews

What does it take for science to uncover the hidden structures of the natural world? By what means may we have hope of discovering empirical truths? As we saw, it takes a certain sort of philosophy. Scientific knowledge requires that questions be asked and answered in a particular way. The specialized technical methods of science—observation and experiment, data gathering and statistical analysis, correlation and causation, description and explanation, among other methods—are the easiest part of scientific practice to study, as they are often directly visible. Not all components of scientific practice are so easily seen. Some critical aspects of the process of science are hidden deep within its culture. My thesis is that science requires more than just a set of methods; it also takes the right sort of mindset. Technical methods alone are insufficient. Methods must be embodied in the bodies and brains of scientists, in minds that are readied to use them.

What brings all this together in the culture of science and makes it work? The same thing that defines and unites any culture—values. Values are the sinews that hold the flesh and bones of a culture together and give it integrity. Cultural norms involve overt practices, symbolic forms, and so on, but these are unified by underlying values. Indeed, this unification of being and practice through a structure of values is, I would argue, the very meaning of integrity.

To give a general example of how this works, consider what it is to be culturally American. There are many diverse and distinctive features of American culture, but two core values stand out: equality and freedom. The Declaration of Independence makes these explicit among the truths it takes to be fundamental and self-evident. It is agnostic regarding whether the "Laws of Nature" or "Nature's God" ground them, but the assumption is that all human beings are entitled to a full measure of respect with regard to these values. These two values are manifested in various forms and symbols. Equality is expressed, for example, through the rule of law and symbolized, for instance, by the blindfolded goddess Justice. Liberty is expressed, for example, in the national anthem's phrase "land of the free" and symbolized, for instance, by the mythic cowboy on the open range. Of course, the claim is not that we descriptively achieve these. Indeed, much of our all-too-common partisan antagonism may be understood in terms of a polarizing dispute about which value should dominate. We swing from one emphasis to the other. We always fall short. Nevertheless, prescriptively, we aim for both; we strive for a more perfect union.

Norms as normal vs. ideal(s)   The important point for our purpose is that we must think of cultural norms not in terms of what is normal, but rather what is ideal. This is different from the usual way of thinking about social norms, if they are taken just to be what is normal in a descriptive sense. We are interested in normative structure, that is, the values and ethical ideals that members of a culture or vocation take to be worth striving to achieve. By way of caution, we should not get hung up on the term "ideal," as that may sound too lofty, especially if it is thought to be singular. It makes more sense when pluralized. Ideals, in the way that we are interested in them here, are aspirational values.

Of course, we recognize that we rarely live up to our ideals. Descriptively, we usually lag behind the prescriptions. One response to this is to become cynical. A cynic may have (or have had) ideals but is jaded, pessimistic, or defeatist and so ridicules and dismisses them as a defensive measure. The idealist, on the other hand, takes a youthful, optimistic view and refuses to be discouraged by the gap between the real and the ideal. Perhaps the realist stands in the middle, acknowledging that ideals are never achievable in perfect form, but are nevertheless important as aspirational, regulative principles that guide action in the right direction, leading us toward rather than away from a flourishing life.

In any case, it is important to be clear that we may not base morality merely on a descriptive account of cultural norms. This is implied by Hume's dictum that one cannot derive an *ought* just from an *is*. Even though it is a useful heuristic that when one is in Rome one should do as the Romans do, this does not mean that the Romans are always right. Presumably, Romans themselves aspire to continue to improve their culture.

The same should be true of scientific culture. Thus, our task here is different from that of cultural anthropology. In exercising a scientific study of culture, anthropologists properly avoid moral judgment of what they aim to describe. But our goal here is not to eschew but to exercise moral judgment—to consider how we should be and how we should behave. Science has come a long way in developing practices that are conducive to its goals, but we do not assume that it is anywhere near perfect. How can we improve the practice of science so that more researchers are more likely to achieve their *telos* as scientists so that science may flourish?

If we wish science to flourish, we must go beyond a sociological account of scientific practice, which is also intentionally limited to description. To move from the *is* to the *ought*, we will need to bridge the gap between sociology and philosophy. This means that we need to look not just at what scientists do, as though these were tribal customs observed from afar. If we are to improve science, we need to understand what it aims to accomplish and then do the further work of evaluating its practices. The term here— "evaluating"—marks the key element of our task, namely, an appraisal of the values of a practice. We must uncover what the core values of science are and assess how well they fit together with one another and with the methods that are meant to help achieve them.

Philosophically, we must examine the component parts of this hidden value structure and then consider how to systematize them in a rational manner that provides not only an explanatory understanding of how they work together now, but also the opportunity and framework to think of how they might be adjusted or modified to work better in the future. This is similar to what the philosopher of science Rudolph Carnap called the process of "explication," which involves the attempt to first reveal and then provide a "rational reconstruction" of some feature of scientific reasoning.[2] Carnap focused primarily on constructing a logic-based analysis of science's evidential procedures—a language-first approach that failed—but the general notion and terms can be redeployed in more naturalistic framework to incorporate the broader issues of the values that make up the scientific mindset. Think of it as a rational normative reconstruction.

Unity of virtue? We also must acknowledge the additional complexity that comes with pluralized ideals, and how this will affect the way that a rational normative reconstruction is done. Having a single, ultimate value is always easier both theoretically and practically, as it allows for optimization. Once one allows the possibility of multiple values, balancing becomes required. We should not expect an ultimate resolution of whether there are irreducibly multiple values required for human flourishing, or whether there is some single *telos* to guide them all. This is connected to the question of whether there are many virtues or just one. Plato and Aristotle both made arguments in favor of the "unity of virtues" for human flourishing, but for the time being I will set aside that larger question. With a tip of the hat to Sir Isaac Newton, quoted in the epigraph, I will focus on the more limited question of scientific flourishing where the prima facie *telos* is clearer.

As we discussed previously, science is a truth-seeking activity that aims at acquiring empirical knowledge. The goal of science is to discover things about the natural world. Furthermore, veracity in this scientific sense—the practiced valuing of empirical truth—is a concept that links curiosity and honesty. These are the two virtues that scientists judge to be most important to the exemplary practice of science. If there is unity to be found, it is through this nexus of values that we will come to know the meaning of scientific integrity.

Commencing a Study of Scientific Integrity

In his famous commencement address in 1974 at the California Institute of Technology, physicist Richard Feynman gave an engaging talk that tried to express his understanding of this concept of integrity for science. He began by recounting several amusing stories of pseudoscientific beliefs, including the curious Cargo Cult religions of Melanesia that arose following World War II, wherein tribespeople imitated the American soldiers they had encountered whose planes had brought valuable supplies, in the hope of once again receiving their bounty. For instance, they made headphones out of wood and bamboo to wear while sitting in imitation control towers by fire-lit runways they built, all with the goal of causing the return of the planes. Feynman used this and a couple of similar cases of what he called "cargo cult science," such as reflexology and telekinetic spoon-bending—goofy, unscientific ideas that he encountered among Californians—to illustrate how easy it is for human beings to fool themselves. He noted that the difficulty extends to pedagogy, where he suggested that teachers are often no better than "witch doctors" when it comes to validating their techniques. We may talk about educational science and such, but need to admit that these are not (or at least not yet) scientific, because we had not yet tested our pedagogical practices.

With this kind of difficulty in mind, he offered would-be science students some advice that gets to the essence of what it means to be a scientist. He admonished students to cultivate:

> a kind of scientific integrity, a principle of scientific thought that corresponds to a kind of utter honesty—a kind of leaning over backwards. For example, if you're doing an experiment, you should report everything that you think might make it invalid—not only what you think is right about it: other causes that could possibly explain your results; and things you thought of that you've eliminated by some other experiment, and how they worked—to make sure the other fellow can tell they have been eliminated... In summary, the idea is to try to give all the information to help others to judge the value of your contribution; not just the information that leads to judgment in one particular direction or another.[3]

Notice that Feynman was pointing out just what we have been talking about. He first cited a core value—honesty—which is a central scientific character virtue, and then went on to show an example of what this means for behavior. In saying that this requires a kind of "leaning over backwards," Feynman clearly recognized that this prescription goes well beyond what is normally done or expected. It is an ideal. It may not be impossible to achieve, but certainly it will be very difficult.

Feynman's judgment about such matters is not to be dismissed lightly. He may not be as familiar a name to the general public as some, but among physicists he figured among the elite in the professional pantheon. He was called a "wizard" and "the best mind since Einstein."[4] Feynman was a member of the Manhattan Project team that developed the atomic bomb and he went on to do fundamental work in quantum electrodynamics, which won him the Nobel Prize in 1965. He played an invaluable role on the commission that investigated the causes of the *Challenger* space shuttle disaster. In many ways, he was also an eccentric. A prankster, he figured out how to break into safes to play tricks on colleagues. He played bongo drums at academic parties. He took up drawing and hung out at strip clubs to practice sketching nudes. The punning subtitle of his best-selling first book of anecdotes about his life—*Surely You're Joking, Mr. Feynman? (Adventures of a Curious Character)*[5]—proudly labeled him as a strange sort of person, but it also captured the defining feature of his character, which was his inexhaustible curiosity. More than any other scientist I know of, he not only exemplified this and other core scientific virtues but also was conscious of their import for scientific practice and had an unusual knack for expressing them in an engaging manner. Feynman was a natural storyteller and we are fortunate that many of his tales are recorded in interviews and books. We will discuss some of his other recommendations as we continue, but his Caltech commencement address is an enlightening example, as it occurred at a culturally important occasion.

Cultural bedrock   Culture is what we take for granted in our relationships with one another. In part, this is because within a culture we share identities through our shared values. We do not often discuss the cultural values that define who we are, but just take them as given. In daily conversation, we usually express them only tacitly. However, values are often communicated explicitly at certain culturally important occasions such as rites of passage, funerals, award ceremonies, and the like. In the scholarly community, commencement ceremonies are one such occasion when aspirational values are spoken of explicitly. For these events, the expectation is that a person of stature in the relevant academic discipline will be invited to provide the new graduates with words of practical wisdom and to speak of the community's hopes for their future. It is in such a way that Feynman's commencement speech should be interpreted. It is an attempt to succinctly state some of the values of the scientific community.

Much can be gleaned from these and other sources where the deep values of scientific culture may be seen, like an outcropping of rock. By taking samples and tracing the inclines, one can get a fair sense of what lies beneath the surface. However, to get an even better sense of the hidden structures below the level of the land, it would be useful to have a systematic survey. To that end, our national study of exemplary scientists was the equivalent of a thousand core borings that let us drill through the horizons of soil to extract cores from the very bedrock of values.

Again, the goal was to identify scientific values in the normative sense—science's aspirational ideals. In our study, we asked scientists not what traits they themselves had, but rather what traits they value and recognize as most important for the practice of science. Before we delve further into the specific results of these core drillings, it is worth taking a moment to provide a brief overview of a couple of the theoretical concepts that undergird the project in general, beginning with an account of what is meant by scientific practice.

Scientific Practice

Virtue Theoretic Notion of Practice

In the previous chapter, I introduced the idea of scientific virtues as practiced dispositions. I focused on the point that these dispositions have a basis in evolved instincts but that practice is required to do them well and make them habitual. Now I am going to elaborate on the other aspect of this concept, which is how these dispositions acquire specialized meaning and content through their role in scientific practice.

Plato and Aristotle both articulated a theoretical account of virtue in terms of the dispositions and skills we should develop for human well-being. Excellence of the appropriate kind provides the basis for an ethics of happiness for us as human beings. Aristotle mostly discussed how to understand this idea of virtues as conducive of general human flourishing in terms of essential human functions and purpose (our *telos*), but at various points he turned to the more specific virtues and associated duties of a ship's captain or a farmer. He did so to help illuminate his general idea, I suspect, because it is easier to understand the value of specific virtues that are appropriate to such practical vocations, if only because it is clearer what these practices aim at. Aristotle did not go into much detail about these cases, taking them to be self-evident, but I would argue that the virtues of these vocations could readily be interpreted in terms of practiced dispositions. It is relatively straightforward to identify the aims and distinctive methods of their practices and to see how having the appropriate virtues makes it more likely that one will achieve excellence in those pursuits. Identifying vocational goals involves specification of what things are of value for each practice, that is, the relevant notion of the good (or goods) that each seeks. For the farmer, this involves a plentiful harvest. Similarly, identifying distinctive methods requires coming to understand the specialized techniques and skills that are relevant to the practice. For the sea captain, this involves everything from determining one's position in the ocean to trimming of the sails to best catch the wind and move the ship in the desired direction.

Some vocational practices become professionalized to a point that their goals and methods can be formalized and institutionalized. However, other vocational practices have a coherent set of goals and methods that may be said to form a discipline (a disciplinary practice) without necessarily having been institutionalized into a profession. Some social practices may seem ad hoc but still have implicit normative import in a culture or subculture. There can even be idiosyncratic practices of an individual that may or may not ever become shared by others. Whatever their degree of development, practices have a normative element that distinguishes them from other behaviors.

MacIntyre's definition of a practice   Alasdair MacIntyre, one of the philosophers who deserve much of the credit for the contemporary revival of virtue ethics, provided a technical virtue-theoretic definition of a social practice that highlights the value aspects of this notion in a nicely compact way. We will not be able to adopt MacIntyre's definition entirely and will take a slightly different approach to generalize the concept, but it provides a good place to start. On his account, a practice is "any coherent and complex form of socially established cooperative human activity through which goods internal to that form of activity are realized in the course of trying to achieve those standards of excellence which are appropriate to, and partially definitive of, that form of activity."[6] This identifies practices as more than just activities; they may not involve any explicit agreement but do require some fairly complex, cooperative behaviors. According to MacIntyre, they must be socially sanctioned in some way.[7] His definition also makes reference to the two elements we identified earlier—namely, the values that practitioners aim to achieve (here the "goods") and their specialized techniques and methods (here the "standards of excellence"). We will need to look into these further. Like Aristotle, MacIntyre's goal was to provide a general account of virtue for human beings, so he mostly used this discussion as a scaffold to build that larger account and did not go much into the details of this concept for the sake of analyzing vocational virtues, which is our purpose here.

With regard to the first element, MacIntyre's definition couched the Aristotelian concept of the aim or goal of a practice in terms of "goods," including the useful distinction between internal and external goods. *External goods* are rewards that may attach to a practice from contingent social circumstances that lie outside the practice, whereas *internal goods* may only be achieved through proper participation in the practice itself. MacIntyre explained this in terms of bribing a child with the external reward of candy while teaching her the rules of chess until she has internalized those rules and comes to see playing the game well as its own reward.[8] Standards of excellence involve the norms that govern assessment of value. At the beginning, these may be no more than following the rules of the game and seeing to it that one does not make invalid moves. As one progresses, higher standards involve looking ahead a move or more, protecting critical pieces, controlling areas of the board, and so on. Grandmasters recognize thousands of patterns, know classic opening gambits and endgame theory, and plan many moves ahead. At some point our developing chess player will come to appreciate and find pleasure in the structure and strategy of the game itself and will thrill to the joy of a brilliant match, irrespective of who wins. The winner of a major tournament may take home a monetary prize, but for enthusiasts of the game—those who are possessed by its spirit—that is a lesser, external good.

The notion of a disciplinary, methodological practice   It is only through an understanding of the internal rewards of a practice that one can understand what is essential to it and what provides its normative force. Only when one authentically adopts those goods as guiding for oneself may one may be identified as a true practitioner. Culturally, this is the most important part of what distinguishes an insider from an outsider. For professional practices, committing oneself to the goals and internal goods of the profession is the most important element to being recognized as a professional.

Learning the methods of a vocation is the other critical element, but values are primary, for they are the measure against which other practitioners may judge whether particular methods are adequate or how they may be modified to improve practice. The same is true of disciplinary practices; members of a discipline are partly defined by their mastery of its methods, and those methods are always in principle open to revision (in a principled way) in relation to how well they work for disciplinary ends. Because the term discipline is typically applied to specific sciences (e.g., biology, chemistry, physics) rather than to science as a whole, I will sometimes use the term methodological practice as a more generic category that can apply at either level, depending on whether one is speaking about methodologies that are characteristic of a specific discipline or about scientific methodology generally.

Science certainly qualifies as a practice in MacIntyre's sense of the term, but it will be useful to consider it in these other senses as well. For example, is science a professional practice? It does have prescribed pathways for credentialing and an institutional structure that resembles the traditional professional degrees of law and medicine. However, the kind of training required for a PhD in science, especially with its requirement for original research, is sufficiently different that for many purposes it is best to see the categories as distinct.

We will not quibble here about that terminology, but we must be sure not to be too limiting in that respect, as there is a long tradition in science of recognizing the findings of amateur scientists. Indeed, as we shall see, the core epistemology of science has always been to accept discoveries on the basis of empirical evidence, not authority, even the authority of a professional title. Admittedly, as science has progressed, it has become harder and harder to learn the necessary skills and background knowledge to do original research without extensive academic training, as any dogged science student working on perhaps their second postdoctoral position can attest to. However, if someone has managed to bootstrap themselves independently to a point that they can do original research, then more power to them. An academic degree can serve as a proxy indicator for knowledge and skill, but the evidence itself should always be primary. If an amateur claims to have made a scientific discovery, and offers it for publication, the decision whether to publish will properly be made on the basis of the evidence they offer, irrespective of their academic degree. This is a critical difference from professional degrees, which bar one from practicing law or medicine without the requisite professional license. No one needs a license to practice and publish in science.

The fact that empirical evidence, collected and analyzed with proper scientific methods, is the standard, irrespective of the authority of the presenter, is the reason that science is most usefully understood as a methodological practice.

This also helps one appreciate why it is not correct to analyze science simply as a social practice. In its institutionalized form, the social aspects of science are, of course, important, and there are interesting aspects of the practice to investigate at that level, but the methodological constraints are more salient. For this reason, I want to relax MacIntyre's requirement that the idea of practice (in its generic definition) be sanctioned socially. Practices may be individual and may be justified instrumentally. For instance, P could be my practice and not necessarily anyone else's. Furthermore, P may be justified not by social sanction, but rather because it helps me accomplish my own goals. It will not do to make the strong claim that individuals cannot do science. One does not require the imprimatur of "the scientific community" to make a discovery. That would be a return to an authority model, which science rejects. If I have done my research properly, I can make a discovery on my own, which other scientists should then be able to replicate. This is the important conceptual role for the scientific community— it must be possible in principle for other experts to repeat my work and get the same result. However, it is not anyone's word that licenses a discovery, but the evidential nature of the practice.

This broader concept of practice allows us to consider individual practices separately from social practices. It then allows us to ask how the latter may have arisen out of the former, and how they acquire additional normative content in that transition. It also more readily links practices to habits, which may include naive forms of behavior before they are put into the space of reasons. Finally, allowing this connection in terms of habits makes it easier to see how there might be a traceable and analyzable connection between scientific practice and the epistemic instincts that gave rise to it.[9]

An instrumental notion of justification of this sort fits with the virtue-theoretic approach that judges the effectiveness of virtues and methods in relation to the goals of a practice. For our purposes, this helps us see how the achievement of the internal goods of science depends on scientific methods. The central, guiding purpose of science is to discover empirical truths about the world, but there are severe constraints in science regarding what it takes before scientists may legitimately recognize something as a discovery. The constraints are the methods of science. This is where philosophy of science comes in, which traditionally has focused on explicating such methods. The achievement of epistemic rewards depends upon having followed appropriate methods. One may be identified as a scientist to the degree that one has adopted science's core values, acquired the requisite methodological skills, and embodied the appropriate virtues as part of one's mindset and character.

A possible objection considered   Before going on, we should consider a possible objection to this emphasis on virtues and method. If the ultimate goal is to discover some new thing about the world, and someone does that successfully, why should we care what kind of character they had or what method they used? This suggests that method and virtue are actually not that important. However, it does not take much digging to see the problem with this challenge. One part of the answer is relevant to the context of discovery, and a second to the context of justification.

First of all, we should admit that it is possible that someone could make an important discovery by chance. This has indeed happened—consider the farmer who uncovered a fossilized bone of a dinosaur while plowing his field, or the sea captain who was the first to see a new species of whale. However, this is not problematic for the virtue-theoretic account, which does not claim that any particular virtue is necessary or sufficient for making a scientific discovery, but rather that there is a suite of virtues that are conducive to that end, making it more likely. A person who embodies the character traits most important for science will have a better chance of making scientific discoveries than a person who has the virtues of some other practice. Similarly, even someone who does have the requisite virtues will not thereby be guaranteed to make an interesting discovery. Luck can play a role. However, luck favors the prepared mind. Thus, having the requisite mindset does make a difference for the exemplary practice of science.

Secondly, someone who makes a discovery by luck alone does not thereby qualify as a scientist. Why not? Because one cannot say that someone really has made a discovery (at least not a scientific discovery) if it has not been tested and confirmed by scientific methods. This is why it is sometimes reasonable to distinguish the discovery of an amateur from that of an expert. The farmer, by virtue of farming expertise, is not in a position to know, in the relevant epistemic sense, what he found; he found a bone, but he did not discover a triceratops. Thus, being trained in the relevant methods (such as, in this case, paleontological taxonomy and classification methods) also is important for the exemplary practice of science. Note that this is an issue about what qualifies as science in a rigorous sense. Philosophy of science is about attempting to explicate such elements of scientific practice precisely and systematically.

Again, this is not to dismiss the more informal sense in which people discover things about their world all the time. Indeed, that is what every infant must do. As previously discussed, our thesis is that key elements of the scientific mindset have roots in the evolutionary advantage that they provided before language arose, and even before we evolved to be human. Animals do not do science in the robust sense of a systematic methodological practice, but they do make discoveries in a precursor sense.

Internal Good of Science

For evolution, the epistemic value of curiosity is, initially, the service it provides for the increase in organisms' fitness, which involves their ability to survive to reproduce. Later, culturally, it may become, in a certain sense, detached from fitness and valued for itself. I say "in a certain sense" because one could, of course, argue that the complexities of culture are but rococo embellishments on the same underlying drive to survive and reproduce. But we do not need to get into that debate here. An explanation at that lower level does not negate the reality of the higher-order processes that may have emerged from it. However, there will remain connections, and in either case, as we will see later, there is an emotional element that has an intimate connection with this process.[10]

Thus, while the evolutionary effect and ultimate reward of the instinct for discovery is the increase in fitness, the direct effects are also important in the causal chain. The meal one gets from a ripe fruit or captured prey increases one's chance of living and eventually reproducing, but the immediate reward is the satisfaction of a full belly. It makes sense that the evolved epistemic traits that contributed to finding and acquiring such resources would come with a similar immediate reward that reinforces the salient dispositions. In this case, there is a direct emotional reward that comes with a discovery—the satisfaction of curiosity is its own reward. For the scientist, this is especially potent. It is what Feynman referred to as "the pleasure of finding things out."[11]

Again, the robust sense of science (with regard to this notion of the pleasure of finding things out and the satisfaction of curiosity) includes more than the biological emotion; it also has cognitive content that comes from its satisfaction in the space of reasons. What might have been good enough from an evolutionary point of view (i.e., being better than one's rivals at surviving and reproducing) will not necessarily be enough for the higher epistemic requirements of the methodological practitioner, where the bar has been raised and standards are stricter. It is important from a theoretical point of view that we be able to ground scientific research in a biological instinct, such as Darwin identified, but that will not be the end of the story. In the same way that Darwin noted that there is more to morality than the moral sense that arises out of the social instincts, there will be complexities to bear in mind for epistemic norms of science when we analyze science as a methodological and professional practice.[12] The distinction between internal and external goods is useful for starting to sort out some of these complexities.

Complexities of internal and external goods in science There are various external goods that may attach to scientific practice, from earning a salary up to receiving a Nobel Prize. The internal goods of science include rewards such as making a scientific discovery and the satisfaction of curiosity. Like the child who has become a chess player, no longer needing outside incentives after having internalized the intrinsic value of the game, the scientist primarily values internal over external rewards.

However, to say that external goods are not valued as an intrinsic part of science is not to say they are not important in other ways. One requires sufficient funds to do the science itself, such as materials for experiments. Moreover, Maslow's hierarchy of needs cannot be entirely ignored. In most social situations, one needs a job that earns money to provide food and shelter. Without these, one loses the freedom to pursue the answer to a scientific question because other interests become more pressing. However, one may not need much, and so long as one has sufficient financial and social support, one can live and happily pursue one's research. In any case, science is not a vocation to enter with the idea of making a lot of money. As one wry scientific wag put it, becoming a scientist to get rich is like going into the priesthood to increase one's chance of marriage. For a scientist, freedom to follow one's curiosity trumps money and many other values. Indeed, the drive to discover can at times overpower even basic nutritional and other needs. The stereotype of scientists who lose themselves in their work, and who forget or cannot be bothered to stop to eat, has a kernel of truth. Think of the astronomer William Herschel, who was sometimes spoon-fed dinner by his younger sister Caroline, who became an accomplished astronomer herself, so he could remain for hours at his telescope doing astronomical observations.

Looking at the ways in which the line between internal and external goods changes in relationship to the *telos* of a practice can help one explain how members of a practice think and behave. It can also help one define the nature of one practice in relation to other distant or neighboring practices. Gastronomic satisfaction is an internal good for the chef but is merely an instrumental value for someone focused on scientific discovery. Engineers use scientific discoveries instrumentally as they build machines, whereas scientists use some apparatus built by engineers to do experiments to make discoveries. Of course, there are border cases. In basic science, the satisfaction of curiosity has intrinsic value; the discovery of truths about the world is, as Darwin said, reason enough for scientific research. However, in applied sciences, discovered truths are also valued for their utility. The goal of a biomedical researcher, for instance, may be to cure a disease, which provides a target that is more specific than the general satisfaction of curiosity. Both require the same basic commitment to scientific honesty as both seek truth, but the applied scientist will likely not be satisfied with discoveries that do not move in the direction of the desired application. Being precise about the central guiding purpose of a practice, as well as its internal goods, provides a principled way to see such differences and how they relate to one another.

Such relationships between internal and external goods, and between the values and methods of a practice, become important as we think about what integrity and honesty mean in the practice of science, and what happens when these values break down.

Honor in Science

Honor vs. Honors

Sigma Xi, the Scientific Research Honor Society, is an organization that honors exemplary scientists and promotes scientific investigation. It typically inducts early career scientists, sometimes when they are still postdocs or even graduate students, so the honor is not necessarily based on having demonstrated a long record of discoveries, though many inductees do have impressive publications. One cannot apply for membership but must be nominated by a member who vouches for one's accomplishment in, or potential for, noteworthy, original scientific investigation. Like commencement addresses, such nomination letters (and letters of recommendation in general) are normative outcroppings where bedrock values of scientific practice may be identified.[13] Nomination letters that take the process seriously do not stop with a reference to publications, but also speak of the nominee's qualities—the character virtues, on our account. Both are ways to demonstrate someone's worthiness as a scientist.

To be honored is to be recognized by one's peers for one's worthy embodiment of the intrinsic values of the practice. It is to be held up as a model. Again, what one is modeling are the intrinsic values of the vocation—one is displaying what ideals look like in practice. To be honorable is to embody those values in such a way that one is worthy of being honored. More exclusive honors are reserved for those who have demonstrated this at higher levels. Of course, as noted above, there are elements of luck in science, so one may be a model scientist in the sense that is important, and yet never make a breakthrough discovery.

Looking at this distinction in a different context may help clarify the difference. Consider, for example, the first knight, Sir Lancelot, who provides an archetype for the knightly virtues, which include not just the bravery that is requisite of the warrior, but also constancy, piety, generosity, and chivalry. Together these constitute a particular kind of honor and nobility. Presumably Lancelot had these noble qualities before King Arthur knighted him in recognition of them. (Lancelot certainly thought he did; humility was not one of his virtues!) In knighting him, Arthur also held Lancelot up as a model so that others might emulate him and work to develop those qualities in themselves. However, everyone recognizes that one might have the requisite qualities and be knightly, even if one never has the opportunity to be dubbed by the king. Someone is noble who has and displays exemplary personal qualities or high moral principles and ideas. In science, this means that one embodies the virtues and strives for the goals of science. Yet, one may be scientifically noble and never win a Nobel.

Is seeking honors dishonorable?   As an aside, it is interesting to consider whether seeking honors is dishonorable. The answer is not as straightforward as one might initially think. Einstein was passed over for the Nobel Prize at first in part because his fame was interpreted as the result of self-promotion. He received the 1921 Nobel Prize retroactively in 1922 for his research on the photoelectric effect rather than his research on relativity. According to Einstein's biographer, this delay in receiving the prize was due in part to conflicts over theory vs. experiment, in part to anti-Semitism, and in part to a negative view of Einstein's celebrity status. Frowned upon as unbecoming to a scientist, self-promotion was felt to take away from worthiness for a Nobel Prize.[14] The point here is that being seen as motivated primarily by external goods, rather than by the internal goods of science, was a disvalue even for someone of Einstein's status. Isaacson wrote that eventually this "lack of a prize had begun to reflect more negatively on the Nobel than on Einstein" so he ended up eventually being awarded it.[15] One does sometimes hear of scientists who promote themselves for a Nobel, but it does not reflect well upon them. One person I interviewed told of his professor who related a story about another scientist who changed his research focus after someone else won a Nobel Prize in the area he had been working in (so that he would have a better chance of winning a Nobel himself) as an example of someone who was not well thought of.

Feynman thought that there was something questionable and even distasteful about scientific honors and awards. He had a long-standing dislike of honors of any kind, which he considered to be superfluous ornaments. He often told the story of how his father, who had been in the uniform business, impressed upon him how everyone was a human being who eats dinner, goes to the bathroom, and has the same human problems no matter what hat, costume, or epaulet they wore.[16] Feynman famously hung up on the caller who woke him up at 3:30 in the morning to inform him that he had won the Nobel Prize, and he considered simply declining it, until a reporter convinced him that it would be more of a fuss to decline than to just accept. But he agonized over how to give a gracious acceptance speech without being dishonest about his feelings. In the end, he explained how he had already received his prize in the pleasure he got in discovering what he did and from the fact that others used his work. His Nobel Prize Banquet Speech eloquently expressed this sentiment:

> Imagination reaches out repeatedly trying to achieve some higher level of understanding, until suddenly I find myself momentarily alone before one new corner of nature's pattern of beauty and true majesty revealed. That was my reward.
>
> Then, having fashioned tools to make access easier to the new level, I see these tools used by other men straining their imaginations against further mysteries beyond. There, are my votes of recognition.[17]

He went on to say that one special value of receiving the Nobel Prize was that it provided an occasion for his friends to express their joy and for him to feel the warmth of their affection. He valued these but made it clear that they were external to the primary scientific rewards of the discovery itself. To make them primary, and to pursue them as one's primary end, would be to subvert the essential interests of science.

Viewing our original question in terms of conflicts of interest may be helpful. The core interest of scientists—their central, guiding purpose—is to discover new empirical facts about the world. Scientists may be competitive, and there may be rivalries between labs that work in the same area to be the first to answer a current question and make a key discovery. There may even be a prize or monetary reward that awaits the first discoverer. I do not agree with Feynman that it is necessarily dishonorable for a scientist to desire recognition. Indeed, the drive to be the first to make a discovery can helpfully provide some extra motivation that speeds science along. In such circumstances, interests may coincide and be pursued without a conflict. However, to value recognition *over* discovery would be to pursue a very different interest, and Feynman was right to see that misplaced priority as foreign to the character of science. Questions of conflict of interest become important for understanding scientific misconduct, which we will examine in a later chapter, but for the moment the key point is that scientific honors are secondary, external rewards that have scientific meaning only if they are awarded for the achievement of internal goods, or in recognition of intrinsic virtues. One must keep one's eyes on the right prize.

Exemplary Scientist Survey

Such considerations played a role in our selection of respondents and in the way we posed questions of them for our study of the scientific character virtues. We could not begin by selecting scientists who had particular traits that we expected would be important, as that would have biased our sample. On the other hand, we could not sample scientists indiscriminately; that would be like averaging-in the views of the many novice chess players about what is needed to excel at chess. One wants the considered views of experienced, highly qualified players, not just beginners or average ones. This is Aristotle's point about the need to learn virtue from those who have acquired practical wisdom. We needed to determine the views of exemplary scientists, and we wanted that group to be peer-identified. For that reason, we drew our sample from scientists who had been honored by other scientists. It is not, as discussed above, that others who had not been so honored were not honorable. However, by randomly sampling from lists of Nobel Laureates, members of the National Academy of Sciences, Fellows of professional scientific societies, and similar scientific honors, we knew that we would get a group that the scientific community itself would take to be representative of their best. (Budget constraints meant that we could not do a worldwide study and had to limit our study to American scientists, but we did capture some international scientists who had emigrated to the United States in the mix.) We also drew a separate random sample of early career scientists with similar training, so we could see whether there were any marked differences in attitudes between them and the exemplary, senior group.

Second, we intentionally did not ask scientists what character virtues they themselves had, or thought they had. Instead, we asked them simply to give their views about what values and character traits they judged to be most important to the exemplary practice of science in general. We asked them to give numeric scores, so we could quantify the relative importance of different virtues, and we also asked them to reflect upon their reasons for their judgment and to give examples to illustrate the significance of the virtue in practice. In some cases, scientists described examples from their own experience, but they often spoke of other scientists whom they saw as exemplifying some trait. In other cases, they could not come up with a particular example on the spot but indicated that values such as curiosity and honesty were so basic that science could not really be practiced without them.

Honorable Honors

Feynman was correct to emphasize the internal goods of science and the intrinsic value of discovery for the scientist. An award or honor, even one as prestigious as the Nobel Prize, is but an epaulet, and not the thing itself. Someone who mistakes the one for the other is getting things backward. In this sense, Feynman had it exactly right. However, there is another perspective that Feynman missed in his views about honor and honors, which is the significance of the latter as a public affirmation of community values. When taken seriously and done right, the conferring of an honor is more than a recognition of individuals and their worthiness—it is a recognition of the community and what it means in that community to be worthy. An epaulet is not the good itself, but it is a symbol of those internal goods. The conferring of a scientific honor, such as induction as a Fellow of some scientific society, is an occasion for the aspirations of the practice to be named and affirmed. Occasions and mechanisms for such public affirmation of the values of a practice help to solidify and proclaim what it stands for. In science, these are occasions to affirm the value of curiosity and its fruits, and to publicly share and celebrate the pleasure of discovery.

Honor Guards

The Methods of Utter Honesty

Feynman's Caltech commencement address was a beautiful example of this sort of public celebration and affirmation of scientific values. Looking again at his speech, we may see how science's methods (the second element of a practice we discussed above) are connected to its values.

As we noted, Feynman introduced his notion of scientific integrity as a kind of utter honesty in terms of how we should go about separating real science from what he called cargo cult science. Feynman uses the term to describe activities that have the superficial appearance of science but in one way or another fail to capture the actual requirements of science. At one extreme, this would include forms of counterfeit science such as various forms of creationism; these are deceptive in very fundamental ways. At the other end are cases that, while not intentionally deceptive, involve self-deception, whereby the science is done so poorly that one cannot hope to make a real discovery—the "results" simply cannot be trusted.

All these cases involve aspects of scientific methodology. Creationism and other supernatural views violate fundamental aspects of scientific methodology such as the requirement of methodological naturalism, which limits science to testable, natural factors in its explanations.[18] Feynman's example of telekinetic spoon-bending most likely is of this variety if it is thought of as some sort of occult power that violates the laws of nature. Some might try to claim that telekinetic spoon-bending is a new natural law but such cases are most likely examples of self-deception. One is reminded of the person who inadvertently convinced himself that he could read palms. It turns out to be surprisingly easy to say something that an individual will confirm as true, whether or not it was based on a "true" palm reading, and it was only when the person intentionally made up readings and yet still got confirmation that he realized his error. Many superstitious beliefs arise because one reads in causation to what was only a coincidence. Without proper controls, one does not have a real test.

Scientific Methods

Determining cause and effect relationships is one of the main ways to make sense of the world. Randomized controlled experiments are the gold standard for this practice in science. To test whether C causes E, one cannot just look at cases where C occurs; one must check it against cases where it does not occur. The experiment done without C is called a control. If E keeps occurring whether or not C occurred, then one should not conclude that C is the cause. In a drug trial, for instance, if the group that gets the compound being tested does no better than the control group receiving a sugar pill placebo, then one cannot conclude that the compound works. This seems simple, but experiments may rapidly get complicated as one also has to take care that there were no other factors that may have confounded the result.

Perhaps the group receiving the compound did do better, but it was not because of C but because it was correlated with some other factor that actually was doing the work. However, it is often difficult to control for other factors. Random assignment to the experimental group or a control group can help deal with this, as it presumably distributes other possible factors equivalently to both groups. But what if the subject improves because of belief in C rather than because C actually did anything? To help avoid such a placebo effect, subjects must be blind to which group they are in. But what if the experimenters, hoping that the drug is a success, unconsciously ascribe improvement to those who got C? To avoid that possible bias, the trial should be "double blind" so the experimenters do not know which group is which until after the data are collected. But couldn't some differences occur purely by chance? Here, one should perform careful statistical analysis to determine that any differences are statistically significant. One could go on. Are your instruments properly calibrated? Are your reagents pure? Are your samples uncontaminated? Were there bugs in your computer algorithm? Did you double-check the math? Feynman gave a few examples of the kind of thing that utmost honesty requires methodologically in science, but a science student finds out that there are many more that one must learn to follow if one is to make and confirm a real scientific discovery.

Again, it is important to keep in mind what is going on with regard to this insistence on the importance of following proper methods in scientific practice—it involves, as Feynman rightly points out, paying attention to the conditions that will allow for a proper judgment of the value of a scientist's work. How good is the evidence for the conclusion? How trustworthy is it? What degree of confidence may we properly have in the purported discovery? All of these are normative notions.

We may think of scientific methods as not just ways to better get at the truths of nature but also as ways of keeping ourselves honest and objective in that quest. As Feynman explains, the care that we take with regard to providing complete information about alternative hypotheses, and possible sources of error, is meant to help us keep from inadvertently fooling ourselves, both individually and as a community.

They also help with honest communication. For example, the tight constraints of scientific methods help account for what appears to outsiders as the seemingly odd way of speaking that is characteristic of scientists. Scientists are constantly adding qualifiers of various sorts—"It may be," "It appears to be the case," "The evidence suggests"—that display the strength of evidence attached to their statements. Scientists' speech is infused with these indicators of degree of evidential support. To an outsider, these qualifiers may make scientists seem indecisive or wishy-washy. A nonscientist may wonder why scientists cannot just tell us the answer clearly, without all this epistemological hedging. Such locutions have a specific purpose in context in that they indicate the appropriate degree of belief in the statement. However, they also have a general purpose in that they tacitly reaffirm a core value of science, which is to base conclusions not on authority but on the evidence.

The Practical Limits of Honesty?

One might reasonably object that utter honesty and full reporting cannot always coexist in science, at least not in all the detailed ways that Feynman recommends. For instance, while it might indeed be a helpful contribution to the progress of science to be able to access all the data from unsuccessful studies, this may not be feasible or desirable in practice. Who would publish those data or make them available, and how, if not through publication? Every working scientist knows that for any successful bit of research that leads to discovery, and a published paper, there were many failed experiments and rejected hypotheses. It would seem far *too* easy to publish scientific papers detailing such negative results. It is sometime said that the secret to science is to make mistakes quickly, but is a scientist who has more failures more productive? Would it be economically viable for journals to publish what would be a massive literature of false starts and blind alleys? Moreover, would anyone really care to subscribe to, let alone publish in, the *American Journal of Discarded Hypotheses*, the *Annals of Failed Experiments*, or *PLOS Dumpster*? This seems absurd on its face. Indeed, the idea is sufficiently amusing to scientists that there is a real *Journal of Irreproducible Results* that is devoted to scientific satire and humor.

That said, Feynman is surely right that in many cases such information would be useful for other scientists and important to include for the reasons he listed, namely, that it helps them judge the value of whatever positive result you are presenting. Indeed, most of the specific examples Feynman gave involve tests of alternative hypotheses or possible confounding factors that a good experimenter should check as part of the process of hypothesis testing. It would indeed be dishonest to neglect to report the results of such tests, but notice that Feynman was not *just* saying that it would be wrong to be dishonest. Rather, he was arguing for a *positive* standard of honesty that sets a higher bar and requires scientists to actively commit to that more demanding standard as a matter of scientific integrity. Integrity is the right word here, for the kind of utter honesty that Feynman is talking about involves the integration of values and methods in just the way that we have seen is required for the exemplary practice of science. Given that the goal of science is to answer empirical questions to satisfy our curiosity about the world, it is only by striving for the highest level of rigor and care in our methods and practices that a scientist, and the scientific community, can be confident that a question has been satisfactorily answered, and we have indeed made a real discovery.

However, this is not equivalent to publishing every failed experiment. There are many hypotheses that seem reasonable, or even likely, given the current state of understanding, that scientists would be very excited to know did not stand up to empirical test—some failures are *interesting* failures and worth publishing. If someone were to found the *International Journal of Interesting Negative Results*, it would surely find a receptive scientific audience. But other failures are uninteresting and of no particular value for the scientific community. Not publishing such failed studies is no more dishonest than not publishing uninteresting successful studies. Of course, there will be many borderline cases where the degree of interest is a judgment call. We might even agree that publishers should take a broader perspective and publish more than they have done in the past, especially now that electronic publishing has greatly mitigated the previous economic constraints of journal publication. There are some thought-provoking questions to consider and trade-offs to weigh regarding social and professional policy that arise here, but in general this sort of issue is not really a moral counterexample to the idea of utter honesty. However, other issues are more challenging.

Harder cases   Should utter honesty, which some think implies completely open communication of scientific results, be upheld in cases, for example, where there is a reasonable fear that significant social harm would result from dangerous scientific knowledge becoming known? Many of the Manhattan Project physicists who solved a sweet theoretical and technical problem, including Feynman, regretted their roles in releasing the nuclear genie. Leaving aside the question of whether they should have pursued this research in the first place, surely everyone would agree that honesty does not compel full publication of the details that would allow others to replicate that work. In 2001, two British journalists reported that among materials found in the rubble of a Taliban military building after the fall of Kabul were not only weapons but also instructions for how to build an atom bomb. The fact that the document was later shown to be a version of a satirical piece from none other than the aforementioned *Journal of Irreproducible Results* is amusing, but does not negate what is a legitimate and serious concern.[19] Biological research may also carry a high risk of potential significant threat, and the U.S. government issued policy guidelines requiring oversight and limits on what is termed Dual Use Research of Concern (DURC).[20] It has at times even restricted funding of certain types of scientific investigations, such as so-called gain-of-function research in virology that investigated ways to increase the pathogenicity and/or transmissibility of viruses like influenza and SARS.[21] A critic might point to such examples where science must be kept secret as a way of questioning whether honesty is always the best policy for scientists.

There are several things we should say in response to this criticism of the virtue of honesty in science. The first thing is to distinguish the question of whether dangerous research should be pursued at all from the question of whether, if the research is done, it should be published openly and honestly. The first question is important, but it is not a question about honesty. Rather, it is about the ethical limits of curiosity. Perhaps Feynman and the other Manhattan Project scientists were right that they should not have released the nuclear genie. We will take up that kind of problem later.[22] For the current case, the presumption is that, for whatever reason, the research has been done, and the issue is whether it violates scientific honesty to not publish the findings or perhaps even to publish misleading results to throw others off the track.

A virtue-theoretic approach helps one begin to think through such cases. Honesty is a virtue in science because it is essential for the satisfaction of curiosity; it is a practiced disposition that is important for discovering truths about the natural world. In this sense, one must assume honesty as a core virtue even if we conclude that there are instances where the possibility of severe public harm requires secrecy, in that it was involved in discovering the danger in the first place. What is really going on in such cases is that scientific honesty is taken for granted but must be weighed against other more general social interests that come into play and ought also be taken into account. The discovery of empirical truths is one important end for human beings, but it is not the only one. To say that veracity is a core value in science is not to say that scientific findings should never be kept hidden. Sometimes doing the right thing may mean not being completely honest in a larger social setting in order to prevent a great harm. The scientist who recognizes the possibility of such cases is not denigrating or undermining this core value but rather is properly accepting that professional scientific values may sometimes need to be overridden if they come into conflict with broader human values.

We will take up this issue in greater detail in the final chapter, but for the moment we may just note that while all these cases show the importance of developing scientific and ethical judgment, they also affirm the importance of veracity as a core scientific value. A more difficult immediate question is how to understand cases where veracity breaks down within science itself. It is to that issue that we must now turn.

When Honesty Breaks Down

In the early 2000s, Jan Hendrik Schön was a rising star in physics. His research was at the intersection of condensed matter physics, nanotechnology, and materials science. While working at Bell Labs he published a series of papers—an incredible sixteen articles as first author in *Science* and *Nature*, the top science journals in the world, over a two-year period—giving results of work using organic molecules as transistors. One of his papers was recognized by *Science* in 2001 as a "breakthrough of the year." He received several prestigious prizes, including the Outstanding Young Investigator Award by the Materials Research Society. His research looked like it would revolutionize the semiconductor industry, allowing computer chips to continue to shrink in size beyond the limits of traditional materials. The only problem was that none of it was true.[23]

Other researchers tried to build on Schön's findings but were unable to replicate his results. One scientist then noticed that two of Schön's papers included graphs with the same noise in the reported data, which was extraordinarily unlikely to happen by chance. This was suspicious, but Schön explained it away as an accidental inclusion of the same graph. But then other scientists noticed similar duplications in other papers. It began to appear that at least some of the data were fraudulent. Bell Labs launched a formal investigation and discovered multiple cases of misconduct, including outright fabrication of data.[24] All of Schön's *Nature* and *Science* publications had to be withdrawn, plus a dozen or so more in other journals. Schön was fired from his job, his prizes were rescinded, he was banned from receiving research grants, and in 2004 the University of Konstanz stripped him of his PhD because of his dishonorable conduct.

Schön's fraud was an especially egregious example of scientific misconduct, but unfortunately it is not unique. Every few years there is some case of scientific misconduct that makes headline news. Are these exceptions? It is hard to get good data, but a 2009 analysis of published studies on the issue found that an average of 1 percent of scientists admitted to having fabricated or falsified research data at least once in their career, and about 12 percent reported knowing of a colleague who had.[25] What is one to make of such cases where honesty breaks down in science?

Self-correction and trust The Schön case is demoralizing as a breach of scientific ethics, but it also leads one to question how sixteen fraudulent papers could have made it through the peer review process of the two premier science journals in the world, and twelve more in other top journals.[26] Peer review is an imperfect process. It generally weeds out papers where the presented evidence is inadequate to support the offered conclusion, though, shockingly, there have even been a few papers with creationist elements that somehow made it through the peer-review filter when reviewers were not paying attention.[27] However, even a careful reviewer is likely to miss individual cases where someone fabricated the presented data. The naturalist methodology of science is predicated on the idea that findings are replicable—that is implied by the idea of a world governed by causal laws—but for obvious practical reasons it is impossible for a journal reviewer to actually repeat an experiment to check the accuracy of the data. Journals very rarely require that a lab demonstrate its data-production process; mostly they trust researchers to have collected and reported their data accurately. Trust is the operative term here.

There are various circumstances that can justify trust. One is history—trust can be earned through experience. Another is trust based on interests—knowing that someone shares your interests means that you can count on their actions. A third is character—knowing that someone has certain character traits means you can count on their intentions. In general, scientists assume these common values with other researchers. In the vast majority of cases, it is completely reasonable to do so. Of course, this kind of prima facie trust provides an opening for the unscrupulous, which is why someone like Schön could infiltrate science the way he did. In the short term, it is hard to prevent such intentional deceptions, but science has a built-in safeguard. Reality is the secret sauce.

Adhering to scientific methods can help one avoid or correct one's own errors. But even if one fails to self-correct in an individual case, science has a way of self-correcting such deceptions, intentional or unintentional, in the aggregate. This is because true scientific discoveries make a difference. The first difference is the difference one sees in a controlled experiment, which is a basic test for the truth of some individual causal hypothesis. They also make a difference to the fabric of science as a whole, adding an interlinking strand to the network of confirmed hypotheses. Because of the interdependencies within science, any discovery will be connected to many others that came before and others yet to come. That means that fraudulent claims will not remain unnoticed. Other scientists pursuing investigations in the same field will at some point try to use that finding as part of some other study. At first, they may be puzzled by unexpected anomalies or inconsistent results in their experiment and recheck their measurements, assuming that their own work was wrong. However, if the problem persists, they will start checking factors that they had initially taken for granted. The test of reality will reveal that the original "finding" cannot be trusted. Even if investigators do not go on to uncover the original fraud, they will note the discrepancy, and future researchers will be able to avoid the error. Indeed, the more significant the purported finding, the more likely it is that an error, fraudulent or accidental, will be discovered, for it affects the work and stimulates the interest of all the more researchers.

From this point of view, Schön's case is heartening in that it shows how the scientific process does work to self-correct. His meteoric success was short-lived; within two years, scientists discovered the deception as his fraudulent findings failed the reality test. A second heartening point is that in the aftermath of such cases, the scientific community typically reassesses its practices and attempts to improve its methods. The Schön case led the scientific community to discuss ways that peer review could be improved to better detect fraud prior to publication. Although we typically think of scientific methodology in terms of experimental protocols, statistical methods, and so on, procedures of community assessment such as peer review are also an important aspect of its methodological practice. Science never guarantees absolute truth, but it aims to seek better ways to assess empirical claims and to attain higher degrees of certainty and trust in scientific conclusions.

Curiosity implies honesty   Dishonesty turns scientific methods on their head. Fraud and fabrication are dishonest not just in the ordinary sense, but also in a particular scientific sense. They invert scientific norms by putting the conclusion before the evidence. They undermine the central, guiding purpose of science of discovering truths about the world. Scientific methods are developed and improved to make it more likely that we can avoid being deceived (inadvertently by ourselves, or intentionally by others), so that nature can be revealed. Fraud and fabrication contradict the basic scientific goal of finding things out.

We do not know what led Schön to commit scientific fraud. Perhaps he began by fudging data to make them look a little better than they really were and only later slid into outright fabrication. Perhaps he was lured by thrill of publication in top journals and the awards that came with that success. Perhaps he had always been a cheat. We may not know his specific psychological motivations, but we can say some general things about why honesty may break down.

In general, dishonesty arises when internal goods of science come into conflict with external goods. Competition for limited resources can be a threat to honesty, as may be desire for scientific fame. As noted previously, scientists are rewarded by recognition of their peers. Today, one way that is measured is the H-index of the number of papers that cite one's work. There are also scientific prizes and awards. It is even more of an honor to have a law, or principle, or plant named after oneself. No doubt some individuals are tempted to fabricate results to get such rewards.

However, the conflict of interest is obvious in such cases. To achieve such rewards for fraudulent results mistakes the source of the admiration. Knowing the purpose of science—the discovery of empirical truths—makes it clear why such misconduct is professionally deadly. Scientists immediately fall from grace when fraud is discovered. They can never again be trusted within the scientific community, for they have betrayed the most fundamental values of scientific practice. A truly curious person should not consider making up data because it is inimical to the very concept of curiosity. What can one say of a person who would fabricate data or submit a fraudulent paper? That person profoundly misunderstands the nature of science. That person is no scientist.

Scientific Integrity

The Most Important Thing

The kind of scientific integrity that is fundamental in science involves more than behaviors that investigators should not do. It is a mistake to think of ethics as a list of thou-shalt-nots. Feynman's description of what he meant by "utter honesty" in science, remember, was couched entirely in positive terms. Scientific integrity involves "leaning over backward" to provide a full and honest picture of the evidence that will allow others to judge the value of one's scientific contribution. For the most part, this sort of behavior is part of the ordinary practice of science and occurs without special notice. Occasionally, however, circumstances arise where this bedrock of scientific integrity is made visible in a striking way. Think of scientist whistleblowers in industry who risked their jobs to bring to light evidence about product dangers that their company publicly denied. There are many such examples, but I want to mention a different but equally telling case involving a scientific error and retraction.

In 1991, astronomer Andrew Lyne of the University of Manchester published a paper in *Nature* together with two colleagues that reported the first direct evidence of an exoplanet—a planet orbiting a star other than the sun.[28] Lyne's research was on pulsars, neutron stars that rotate with precise regularity, some at intervals of seconds, others of milliseconds. Lyne had discovered a pulsar—PSR1829–10—whose rotational period was irregular up to a couple of milliseconds over a six-month periodic cycle, a wobble that seemed to be strong evidence of an orbiting planet. Preparing to present a paper on the discovery at the upcoming meeting of the American Astronomical Society, Lyne rechecked his data and realized an error in his original calculations. He had an insight that the six-month wobble cycle was not caused by a planet orbiting the pulsar but rather was an artifact of the Earth orbiting the sun, a movement that his calculation had failed to correct for. In an interview later, he described his reaction:

> I was just completely numb for half an hour or an hour, I just sat there going through everything that I said over the last six months to lots of people…realizing that it was mostly complete rubbish and that I had really made a really fundamental error. I had to let everyone know as soon as possible, so that they would not dwell on an object which did not exist and would not tax their ingenuity [trying to] understanding it.[29]

At the conference, 500 other astronomers had come to his talk to hear the official announcement of the discovery of the first planet outside the solar system and the technical details of its confirmation. Instead, they heard Lyne explain that the result had been a mistake. One member of that audience explained his own feelings about the incident: "This poor scientist had to get up there and tell the world that he had been wrong, and publicly wrong. And I felt very proud, frankly, to be member of a group in which in the end honesty was the most important thing."[30]

Ironically, another presentation at the same conference by radio astronomers Aleksander Wolszczan and Dale Frail announced the discovery of two planets orbiting a different pulsar, and that result turned out to be the first confirmed detection of exoplanets. Lyne's team published a retraction of its original paper, and that disconfirmed result is now but a footnote in scientific history. And yet the Lyne case is not an example of scientific failure, but of scientific success; it was a telling affirmation of the basic values of science. When Lyne announced and explained his error at the conference, the audience of his scientific peers gave him a standing ovation.

Feynman's Closing Wish

We began this chapter by talking about the scientific mindset—the cultural ideals that define the practice. As a practice that aims to satisfy our curiosity about the natural world, veracity is a core value. Scientific integrity is the integration of all the values involved in truth seeking, with intellectual honesty standing at the center. Feynman's notion of "utter honesty" is one expression of this. As we continue our rational normative reconstruction of science's moral structure, we will need to identify and integrate other virtues that orbit the bright lodestar that guides scientific practice.

Scientists do not always live up to these ideals, but the scientific community recognizes them as aspirational values that define what it means to be a member of the practice. Those who flout them do not deserve to be called scientists. Those who exemplify them with excellence are properly honored as exemplars. Andrew Lyne embodied the scientific mindset, and his model demonstrates the kind of standards that scientists recognize they should strive for, resisting conflicts of interest.

Feynman closed his commencement address with a wish for his listeners, the graduating science students: "So I have just one wish for you—the good luck to be somewhere where you are free to maintain the kind of integrity I have described, and where you do not feel forced by a need to maintain your position in the organization, or financial support, or so on, to lose your integrity. May you have that freedom."[31] Fulfilling this wish requires more than individual virtue; it requires unified, vigilant support from the scientific community. Integrity in science involves a community of practice, unified by its shared values.

3 Systematic Curiosity

Scientists do not believe; they check.

—John W. Cornforth[1]

Bridges to Trust

Crossing Bridges

What does it take for you to trust a bridge? Your goal is to cross it safely, so you must to be able to count on its structural integrity. Its foundation must be sound enough. Its beams and rivets must be strong enough, with sufficient tensile, compression, and shear strength. Its design must balance the forces at play, so they can support the intended load. But there are social forces at play as well that affect whether these physical forces will or will not be properly actualized. Was the steel poured with care? Were the riveters rushed? Were there sloppy errors in a structural calculation? Did the company cut corners to win the bid? Were governmentally certified standards adequate? To trust a bridge, you need to trust the construction workers, the engineers, the project managers, and so on, for the final physical form depends as much upon them as the materials. All these pieces, and more, must come together if a bridge is to be trustworthy. The structural integrity of a bridge thus involves a unity of values—not just the calculated values of weights and angles, but the epistemic and ethical values of its builders.

Trust in a practice such as science is much the same. We are justified in trusting science to the degree that it is structured with integrity. What this comes to is that its various parts and participants must work together in an integrated way to achieve their common end. We have already seen how curiosity is a virtue that is intimately linked to science's central, guiding purpose of discovering empirical truths about the natural world. We have also seen how this requires a kind of utter honesty. These are the central character virtues for exemplary science, but other values are also needed for science to flourish. Scientists require a suite of virtues that are well integrated if they are to function well, and other components of science itself, including its methods and practices, must be similarly integrated to that end. None of these work as a pick-and-choose grab bag. To function for the pursuit of scientific excellence, they must be systematically structured in relation to the central, guiding purpose of the practice and to each other. I suggest that the satisfaction of curiosity functions as the characteristic, organizing virtue for science. Put simply, science is systematic curiosity.

Curiosity as a concept works as a systematizing virtue because, as we have seen, it captures within it the idea of the search for truth, knowledge, and discovery. Curiosity has value for other people and other vocations, of course, but for scientists and for science it is an essential, defining element. As we saw in the previous chapter, it also explains why veracity, or "utter honesty," is of such importance for science. Now we want to look at how it pulls other special values into its orbit as well, including objectivity and skepticism. Examining the values that structure science and their interrelationships will help us better understand the ways science is, or is not, trustworthy, and why. In this chapter, we will pay particular attention to the structural aspect of these relationships—not just how the concepts themselves are structured but also how they, in turn, help to regulate scientific methods and practices, and organize the community of scientific practitioners. Looking at these two levels of organization together will help us better understand the notion of scientific integrity and how it helps answer the question about why, and in what ways, we are justified in trusting science.

Mertonian Norms of Science

One important way to think about what makes science trustworthy involves its structure of social controls. This is a level of analysis that was pioneered by sociologist Robert Merton, starting in the late 1930s. Merton was a towering figure in sociology and many of the ideas he introduced and terms he coined—role models, self-fulfilling prophesies, unintended consequences, to name a few—have become part of our common conceptual toolkit. Merton is also properly credited as the founder of the field of sociology of science, and when one thinks of the norms of science, it is initially Merton's list of norms that comes to mind. Merton discussed these under the heading of "the ethos of science," which is what he called "that affectively toned complex of values and norms which is held to be binding on the man of science."[2]

He listed four elements that he took to comprise the scientific ethos, namely, "communism" (he put the term in quotes), universalism, disinterestedness, and organized skepticism. They are sometimes abbreviated by an acronym, as Merton's CUDOS norms. The most important articulation of these occurred in a 1942 article that is now known under the title "The Normative Structure of Science,"[3] though several of them were prefigured in his 1938 piece "Science and the Social Order."[4]

Put briefly, the norm of "communism," which he put in scare quotes to indicate that it was not a political system, was meant to capture the idea that the scientific community does not recognize proprietary ownership of scientific discoveries. The norm of universalism involves the institutional requirement that claims to truth should be evaluated in terms of impersonal criteria, and not on the basis of a scientist's nationality, race, class, gender, and so on. Disinterestedness in science is also to be understood at the institutional level, in the usual sense of impartiality with regard to possible outcomes. Organized skepticism involves a group suspension of judgment and the detached evaluation of claims. We will examine these in more detail shortly, but it is worth noting that Merton explained each of these concepts in just a sentence or so and spent only a couple of pages discussing each. Merton saw himself as a theoretician and had little interest in testing his ideas empirically himself. He did not arrive at his conclusions about the ethos of science from an empirical study of members of the scientific community but drew from his own understanding of community expectations. It is a testament to his powers of analysis that the community found his framework to be stimulating and insightful. Later scholars found themselves needing to correct some of his particular claims, of course,[5] and others have argued that they needed to be extended,[6] but Merton's norms are still a good starting point for thinking about institutional norms that help maintain trust. However, they will need to be supplemented by bringing in other values that are equally important for the integrity and trustworthiness of science.

Epistemic Values

In a previous era, after completing their initial experiments and tests to confirm some discovery, scientists often demonstrated their results in public presentations. Scientific demonstrations drew large audiences and some scientists became minor celebrities. Considered from an externalist historical perspective, one could argue that such demonstrations fulfilled a social function by conferring prestige to scientists. While this may have been one effect, for our purposes their epistemic role is more basic. By showing, rather than just talking, scientists were making it clear that it is the observable evidence, open for all to see, and not their own authority that was basic.

This is not to say that science is nothing more than direct observation. Reasoning is involved as well, of course, and in a good demonstration the presenter should take the audience through the supporting reasoning. We also know that it takes training to acquire the level of education to fully appreciate such demonstrations, so it is not the case that every member of the audience is equally capable of grasping the significance of a demonstration. Scammers took advantage of this, and the demonstration circuit was contaminated with counterfeit scientists and counterfeit science. However, that there are science counterfeiters does not mean that there are not real scientists and real science. Indeed, the fact that scammers could deceive using faked demonstrations supports the central point that the form in general is appropriate for its epistemic purpose. That being said, one might still object that even good demonstrations done by real scientists were misleading in that they presented a false image of how science worked; staged demonstrations were far neater than the messy process of experimentation and discovery. However, this objection is also misplaced, as the point of a demonstration, much like a journal publication, is not to recreate the process of discovery (from the false starts and the first glimmers of light, through the experimental refinements, double-checking of calculations, and so on), but rather to display the discovery—whether it be a new fossil or a new phenomenon—in its clearest light so that it is visible to all.

In cases where a public demonstration was not possible, perhaps because of the complexity of an experiment or the delicacy of required equipment, demonstrations were made to smaller groups of witnesses who could serve as proxies and testify to what they observed. There are other epistemic issues about what makes for a trustworthy proxy, but we do not need to get into that topic here. Today, it is most common to move straight to publication, with anonymous reviewers playing the role of independent observers who check and attest to the adequacy of the methods used to determine the reported findings. The point is that such proxies are not the same as taking someone's conclusion on faith, because it remains possible, in principle, to check oneself. More importantly, the scientific community provides a collective check on the credibility of findings. It is the structure of these epistemic and social relationships, including the character virtues of the practitioners, that give us confidence in scientific conclusions—they are what we might call value bridges to trust.

In this chapter, we will make our way to a consideration of community by way of a discussion of two important virtues—skepticism and objectivity—that are closely tied to scientific curiosity. Our goal is to examine the inter-relationships among values at different levels of organization. My claim is that science may be thought of as systematic curiosity, and here we will pay special attention to the first part of that phrase as it relates to how the scientific community is organized. Community norms, in the sense that Merton examined them, are social norms of organizational control. I hold that these do not exist primarily for the sake of control but rather because of a deeper level of systematic organization, namely, the structure of values and virtues themselves. The social forms of control, I contend, are deriva-tive of these and should be thought of as structures put in place to help maintain those values in the community of practitioners so that the prac-tice is more likely to be trustworthy. Scientific integrity consists of all these elements working together toward the common end of satisfying our hon-est curiosity.

Skeptical Inquirers

Skepticism in Science
In the popular mind, skepticism is a common part of the stereotype of sci-ence and the scientist. Witness the many science clubs that go by the name of "skeptic groups" with their associated magazines like *Skeptical Inquirer* and *The Skeptic*. A characteristic feature of these is the goal of debunking false claims; think back to the popular show *MythBusters* as another example. Unfortunately, this notion of skepticism sometimes gives the impression that science is the obnoxious kid just out to prove others wrong. However, that misses the positive side of science. The *MythBusters* hosts were gleeful

when some popular myth they tested was busted, but they were equally enthused to discover that something they expected to fail was confirmed. The core goal of a scientific test is not to debunk but rather to check, in order to find out what is the fact of the matter. Curiosity is satisfied either way. A skeptical attitude in science should not be valued for its own sake but rather for its instrumental value in making it more likely that we will uncover the honest truth we seek. As we shall see, skepticism is the guard that curiosity sets against attractive falsehoods.

Skepticism has roots in ancient Greek philosophy, especially in the philosopher Pyrrho of Elis. Pyrrho objected to the dogmatic philosophies of his day, each of which claimed to possess certain knowledge. He questioned whether it was possible for human beings to determine any knowledge with certainty and concluded that we must accept that knowledge always came with uncertainty. To express this consistently, Pyrrho would always preface any claims he made with qualifying phrases like "it seems," "it appears to me," or "perhaps." He taught that this view was also a good way to help humans achieve a sense of inner peace. The citizens of Elis honored his wisdom and made him their head priest. As a bonus, they exempted all philosophers from taxation.

Scientists today follow a version of Pyrrho's linguistic practice in the way they, too, use care when making empirical claims by couching them with qualifying phrases to mark their degree of evidential warrant. These may range from verbal indicators of relatively weak evidence that is "suggestive," to indicators of better evidence that makes a conclusion "seem likely," to markers of very strong evidence that allows a conclusion to be drawn with "a high degree of confidence." In any scientific investigation, there are typically several alternative hypotheses that may be in play at any given point in time, and the goal is to sort out their degree of plausibility, eventually eliminating weaker ones, so we may move forward with only the most highly confirmed. The fact that conclusions may achieve higher and higher probability, but never quite reach absolute certainty, is a hallmark of inductive reasoning upon which science is based.

Skepticism: The salt of science   As we have noted, science is not, as commonly thought, about absolute truth but rather about sufficiently confirmed truths—findings that are good enough for present purposes. Science's methods involve empirical testability; hypotheses may not be accepted merely on the basis of authority or personal preference but must be confirmed in terms of observational evidence that, in principle, is public and repeatable. Given that science is an inductive enterprise, one may never take a conclusion as final, because there is always the logical possibility that new evidence may force reconsideration. In this sense, scientists should always be intentionally and self-consciously in doubt. Science requires a comfort with uncertainty. Satisfaction may be temporary. An answer that is satisfactory for present purposes may be insufficient in the future when needs change and standards rise. A question may lead to an answer, but every answer leads to another question. Science is a form of restless questioning, so having a skeptical attitude is an essential tool.

In recommending a skeptical or critical attitude toward something, one often says that it should be taken "with a grain of salt." The expression goes back to the ancient Greeks, specifically to Pliny the Elder's *Naturalis Historia*, in which he describes a traditional recipe for an antidote for poison. Take a couple of dried walnuts, a couple of figs, twenty rue leaves and pound them together, adding the proverbial grain of salt. Take a dose of this mixture, and one will be "proof against all poisons for the day."[7] The figurative meaning developed later, such as applied to certain biblical claims by Scriptural commentators when skepticism was warranted to avoid error. A pinch of skeptical salt helps one avoid taking in dangerous, false claims. Skeptical tests are the proof against such poisons. To test something and to taste something are similar, in that the basic appeal is to the evidence of experience. In this sense, skepticism is the salt of science.

However, to say that scientists should have a skeptical attitude does not mean that they should be equally skeptical of all claims, or even all kinds of claims. A rational scientist pays attention to the possible objects of skepticism and judges them appropriately. For instance, we know that our own memories are sometimes not trustworthy. Even rare individuals with eidetic memory, who can recall images or sensory impressions with great accuracy after short exposures, still find that memories can fade, and most of us are all too familiar with the ways in which our "memories" may be partly or largely fictions. A healthy skepticism about our own memories might help us keep a critical edge that puts us in a better position to discriminate between true and false information in the many kinds of situations where this matters.[8] Furthermore, as there are different possible objects of skepticism, there are also different degrees of skepticism. Should I regard a claim with a grain of salt, or does it require a shaker? Appropriate degrees of skepticism are warranted in different situations.

Appropriate degrees of skepticism　Someone who claims to have discovered something fairly ordinary (e.g., a new species of beetle) should not encounter much skeptical opposition, but someone who claims to have found something earthshaking (e.g., cold fusion) should expect a higher degree of critical scrutiny. Extraordinary claims require extraordinary evidence. This evidential heuristic was popularized by Carl Sagan, but goes back farther to sociologist Marcello Truzzi, who was one of the founders of the Committee for the Scientific Investigation of Claims of the Paranormal (CSICOP) and who was called "the skeptics' skeptic." Discussing the evidential standard that science requires when evaluating claims someone might make, Truzzi wrote "[W]hen such claims are extraordinary, that is, revolutionary in their implications for established scientific generalizations already accumulated and verified, we must demand extraordinary proof."[9] This is a reasonable standard of judgment: adjust one's degree of skepticism toward a claim depending on the accumulated weight of evidence for associated generalizations that would be affected if it turned out to be true.

Science is a web of interlinking theories containing observations, explanations, models, and hypotheses, some highly confirmed, some more tentative. It is not possible to introduce a new idea in science that would not have dependency relationships of various sorts with the existing body of scientific knowledge. Part of what it means to be trained in science is to acquire an understanding of one's specialized area of this web. This involves developing a good sense of where the open questions are, what parts are most secure, and what parts might be open to refinement, revision, or even rejection. Developing good judgment about these interconnections, and the weight of evidence for each, allows one to assess the prior probability of a claim, whether it be ordinary or revolutionary, and so adjust one's degree of skepticism accordingly.

One effect of Thomas Kuhn's influential historical and philosophical research on the structure of scientific revolutions is that scientists love the idea of being involved in a paradigm shift. Scientists, therefore, feel a mixture of frustration and excitement if an experiment, or some new observation, goes against expectations. While the most likely explanation is that one made a mistake in a procedure or calculation, there is always the chance that it represents an opening through which new light will be shown. A general skeptical attitude allows one to be open to this possibility. This is not to say that one must be perpetually open-minded for, as the philosopher Bertrand Russell pointed out, that would be a mind that is vacant. However, scientists need to be ready to question even long-held assumptions in the field if the evidence begins to go against them. One should develop the habits of mind that allow for a properly balanced and calibrated judgment.

Huxley on Active Doubt

In his review of Darwin's *On the Origin of Species*, Thomas Huxley referred to Goethe's notion of "active doubt" (*thätige skepsis*) to explain the kind of attitude that a scientist, or any scientific reader, should have when approaching a hypothesis. This is, he explained, an attitude that "so loves truth that it neither dares rest in doubting, nor extinguish itself by unjustified belief."[10] Actively working to hold the center between relativism and dogmatism is a good way to understand the notion of scientific skepticism. Finding and maintaining this sort of balance is a practiced virtue. Darwin's explanation of the origin of species by natural selection was indeed revolutionary, so initial skepticism was certainly warranted. It is testament to Darwin's exemplary scientific character that he not only was able to follow the evidence to the truth himself but also was able to present it in a way that allowed other scientists to do the same.

Huxley's view about the value of this kind of skepticism derived not just from Goethe but also from his reading of Rene Descartes, the philosopher who is most known for setting the foundation of knowledge in what he called the Method of Doubt. The idea behind this method was that one would identify a secure foundation by systematically examining things one thought one knew, setting them aside if one found any reason to doubt them. Descartes directed his doubt-ray against progressively more basic claims of knowledge, finding that he could not even accept basic sense impressions with certainty. For instance, we all know of cases where persons cannot trust what they see, perhaps because of optical illusions, hallucinations, or the like, so how can we be sure that we do not ourselves suffer from some similar interference in our own sensory experiences? Following this skeptical method, Descartes was left with nothing but his own thoughts, before recognizing that this was something which was not open to further doubt. *I think!* That, at least, was clear, distinct, and certain. And if

I think, that means that I exist. *I think, therefore, I am.* With this as his base, Descartes rebuilt the knowledge that he had originally been led to doubt.

Descartes aimed for knowledge that would be like mathematics in its clarity and security. With such a secure foundation, additional knowledge might be deduced in much the same way that one proved theorems from axioms in geometry. It was by way of skepticism that one was led to such a place from which deductive reasoning could take over. Huxley was calling attention to the role of this doubting attitude. That is our own focus as well, though we do not adopt Descartes's particular epistemology, such as his assumption that the only possible foundation for knowledge claims is absolute certainty. Huxley's point that truth-seeking in science involves finding the balance between bottomless doubt and unjustified belief is apt for science's inductive epistemology. It is perhaps even more in line with Aristotle than with Descartes, as the virtue-theoretic approach holds that virtue is to be found in the mean between the extremes of excess and deficiency.[11]

Of course, understanding what moderation in skepticism (or any other virtue) comes to from a theoretical perspective, and developing the judgment to ascertain its application in practice, is not easy. To see how one would begin to do this, it is worth briefly exploring some of the complexities. One must consider the appropriate objects as well as the appropriate degree of skepticism. As we will see, scientific skepticism requires a doubting attitude not only toward claims (to the degree that they may lack sufficient evidence), but also sometimes toward others (as authorities per se, as opposed to proxies for the evidence), and also toward oneself (lest personal biases or insufficiencies obscure the evidence). Before looking into these, it is worth beginning with a short inquiry into a very basic object of scientific skepticism, namely dogma, and the difference between skepticism and iconoclasm.

Skepticism of Dogma

Merton introduced his notion of organized skepticism in terms of what he identified as the "iconoclasm" of the scientific attitude.[12] The "detached scrutiny" of science is viewed as a challenge to what others view as the "comfortable power assumptions" of other institutions. "Organized scepticism," Merton wrote, "involves a latent questioning of certain bases of established routine, authority, vested procedures and the realm of the 'sacred' generally."[13] Other institutions demand faith in some areas that they take to be sacred, but "the institution of science makes skepticism a virtue"[14] and declines to adhere to any "sacred sphere," whether religious, political, or economic.

Iconoclasm does seem to stand out as a distinctive feature of science, especially if one looks to significant episodes in its history. The notion seems to be inherent in the very idea of the birth of science as a "revolution," which implies the overturning of a previously established order. That the classic example of this involved a dispute over the revolution of the planets and our natural place in the heavens gives the notion itself an iconic feel. The idea that the earth was the unmoving center of the universe, with the moon, planets, sun, and stars revolving around us was an established view that not only had the advantage of common sense, but the imprimatur of the Church. One cannot underestimate the cultural significance of the change to a sun-centric view with the earth moving in an orbit no more special than those of the other planets. For our present purposes, it does not matter whether one focuses on Copernicus, Kepler, or Galileo as the scientist who is most deserving of credit for this revolution. All made contributions that are rightly viewed as revolutionary, and scientists speak of them with the same sort of respect that Americans speak of the nation's founders who broke with the British king to establish a constitutional democracy. If Galileo stands out as a giant among giants in stories of the scientific revolution, it is not just because of the breadth and depth of his discoveries, but also because he suffered for his science. Daring to overturn Church dogma, and facing the Inquisition for his heresy, he was not only an iconoclast, but an especially courageous one.

Because revolutions always have political dimensions, it is tempting to analyze the scientific revolution in terms of political power dynamics, with scientist-iconoclasts simply gaming to topple the established power structure and make themselves the new elite. This is too superficial a view. We will need to look at this in more detail later, but here we may just note that it fails to appreciate what the virtue of skepticism means in scientific practice and the sorts of situations when iconoclasm becomes valuable. Scientific iconoclasm is not confined to revolutions against the Church or other external institutions. The Newtonian and Einsteinian revolutions, which were internal, are equally celebrated. So, it is not correct to view this scientific attitude as simply a clash of institutions vying for power or an assertion of dominance over other cultural institutions. Science is an equal-opportunity iconoclast.

Historians of science have highlighted ways in which various forms of iconoclasm—methodological, conceptual, experimental—have been instrumental for scientists who moved a field forward by overturning central organizing assumptions that for a period seemed to define a discipline.[15] In fact, it is too weak to say that science is skeptical of dogma; it rejects dogma (as such) outright. However, one must be wary of letting the thrill of iconoclasm go to one's head.

Scientists themselves do sometimes emphasize this iconoclastic attitude with an unbecoming, cocky air of superiority. Perhaps there is a bit too much pride in emphasizing that Galileo was a heretic and that scientists are mavericks and rebels. Seen in this way, part of their role is to be transgressors and boundary crossers, standing up to the man, and violating his sacred spaces. Edwin C. Kemble, an American theoretical physicist, observed that most scientists had little enthusiasm for history because "the job of the scientist is to make discoveries and develop new ideas."[16] New ideas that have proven to be effective, he pointed out, tend to make old ones obsolete, and in this sense "it is not altogether unfair to say that the typical scientist is in the business of creating the future and destroying the past."[17] Kemble was known to be self-deprecating and personally humble, but this way of making the point again carries a self-satisfied whiff of superiority, positing the scientist as Nietzsche's Ubermensch in his powers of creation and destruction.

While there does seem to be something of a macho "bad boy" element in some of this, it is not restricted to male scientists. Biologist Barbara McClintock, who won a Nobel Prize for her work on transposable genes, was self-consciously iconoclastic. Lynn Margulis, another biologist who bucked the establishment to promote her idea of endosymbiosis, could be described in much the same way. Her radical idea that the evolution of eukaryotic cells (that is, cells with nuclei) involved the permanent, symbiotic merging of previously independent nonnucleated bacteria, eventually won the day over the initial resistance of the discipline, and her election to the National Academy of Sciences was well deserved. However, for both these scientists, it is fair to say that their iconoclastic, contrarian approach did not always serve them well; in her later years McClintock began to investigate the paranormal. Margulis not only flirted with HIV denialism, but even became a "9/11 Truther," claiming that the collapse of the World Trade Center was the result of a controlled demolition and was a false flag operation to justify the wars on Afghanistan and Iraq. As Aristotle explained, any virtue can become a vice when taken to an extreme, and these are clear examples of skepticism taken too far.

In thinking about what it means for skepticism to stray too far into iconoclasm, it is worth considering the views of philosopher of science Paul Feyerabend, which may be mentioned as an example of this kind of extremism. In his characteristically provocative way, Feyerabend argued for the value of "anarchism" in science, which is iconoclasm taken to the limit. Feyerabend thought that no methodological rule was universally applicable in science and that none should be endorsed. One can find, he claimed, an exception to any proposed rule, and he thought that new theories were accepted as much because of rhetorical tricks as because of any particular method. Thus, he argued, the only thing one can say about science practice and progress is "anything goes." To give just one example of his contrarian viewpoint, Feyerabend argued that the Church was actually more on the side of reason at the time than Galileo, who had to make use of ad hoc hypotheses to oppose the established idea of impulse and motion in order to get his ideas about the moving earth moving, so to speak—hypotheses that were supported only later.

We should note in passing that there is at least some reason to think that Feyerabend was exaggerating his case as a devil's advocate. Our concern here is not to offer a judgment on that question but simply to note the position itself as an example of one way that the virtue of skepticism may, when taken to the extreme, become a vice. Feyerabend's arguments against method began as a reaction against Logical Positivism but that view was already collapsing by that time, and his skepticism of method was made, provocatively, against *any* methodology. However, to point out that there seem to be exceptions to proposed methodological rules in science is hardly a sound reason to say that there are no legitimate methods in science. Nor should one disparage scientists who are good rhetoricians. It is not rhetoric itself, but *empty* rhetoric that is problematic. Feyerabend stacked the deck by talking about rhetorical "tricks" but gave no examples in his review of scientific revolutions that were based on trickery rather than reasonable judgments of evidence. Of course, scientists make use of rhetoric, in the sense of persuasive argument and well-worded presentation of the evidence, and so they should.

What about the Galileo affair? We can agree that, like most affairs, it was a messy business. Galileo himself, despite his real courage and heroism, was not without fault. Moreover, the evidence for the movement of the earth was, admittedly, not conclusive at the time. On the other hand, it would be a mistake to get lost in the complexities of the episode and not just miss, but dismiss, the forest for the trees, as Feyerabend did. Just as in evolutionary biology, where species events always have fuzzy boundaries and it may not be possible to identify a critical turning point until one can regard the episode in hindsight, so it is reasonable to recognize the significance of the Galilean revolution in hindsight, even granting that the evidence that contributed to the revolution was not clear and fully in place until later. Even revolutions have a transitional period. Feyerabend's criticism that there is not a clear rule that determined a sharp break is not a reasonable objection to the basic claim that the Galilean revolution was in the end justified by evidence and evidential methods. Furthermore, Galileo's skepticism (and his other virtues) may also reasonably be credited for this revolutionary scientific advance.

Although it is right to praise Galileo and other scientists who overturned unjustified orthodoxies, whether religious or scientific, it is wrong to think that scientists are morally superior because of such critical powers. Moreover, it is misleading to view this as essentially a destructive instinct. There is no honor in destruction for its own sake. Skepticism of dogma is one thing, but indiscriminate skepticism is no virtue in science; some established truths deserve their position. Indeed, iconoclasm for its own sake makes no sense in science, for that would have the effect of undermining its central, guiding purpose of discovering truths about the natural world. If iconoclasm were really the method of science, then there would be no reason to take any of its conclusions seriously. Methodological iconoclasm would be as deleterious to science's goals as unquestioned dogma. If evidence accrues to the degree that a conclusion is strongly confirmed, one has good reason to hold on to it until new evidence seriously calls the old evidence into question. Properly understood, scientific reasoning may support conservatism as well as revolution, each as the evidence demands.

Skepticism as willingness to overturn that which is taken for granted is a virtue, but only if it remains in the service of honest curiosity. Revolutionary, iconoclastic episodes should be hailed not because they are iconoclastic per se, but because they are places where scientists escaped from unwarranted assumptions in a way that moved science forward. They are places where scientists took a stand for their basic values and demonstrated their commitment to the search for empirical truth on evidential grounds, even in the face of powerful opposition.

Skepticism of Others

There are, of course, various forms of opposition that a scientist should be ready to face, and we now turn to sorting out another form that gets to the heart of what is distinctive about scientific skepticism. More than anything else, scientists are skeptical not only of dogma but also of any individual who claims some special privilege to determine facts by virtue of their authority. Huxley identified rejection of authority as critical to understanding why skepticism was so important in science, writing: "The improver of natural science absolutely refuses to acknowledge authority, as such. For him, skepticism is the highest of duties: blind faith the one unpardonable sin."[18]

I suggest that this form of antiauthoritarianism is fundamental to the scientific mindset and is conceptually linked to the methodological principle of determining matters of fact on the basis of empirical evidence. When confirming matters of empirical fact, the scientist respects neither priest nor president. If our goal is to discover what is true about the world, we ought to consult the world itself. As Einstein put it, "Unthinking respect for authority is the greatest enemy of truth."[19]

It is this element, more than iconoclasm per se, that was salient in Galileo's revolutionary work. Einstein identified this as a character trait that he most resonated with, writing in his foreword to a reprint of Galileo's classic *Dialogue concerning the Two Chief World Systems*, "The *leitmotif* which I recognize in Galileo's work is the passionate fight against any kind of dogma based on authority. Only experience and careful reflection are accepted by him as criteria of truth. Nowadays it is hard for us to grasp how sinister and revolutionary such an attitude appeared at Galileo's time, when merely to doubt the truth of opinions which had no basis but authority was considered a capital crime and punished accordingly."[20] As we will see in more detail in a later chapter, the notion of humility at play here has a special scientific sense; it is a particular kind of intellectual humility, namely, humility to the evidence.

The turn to empirical evidence as the sole acceptable basis for conclusions about the natural world is the single most revolutionary idea of the scientific revolution. The Royal Society adopted the motto *Nullius in verba* to explicitly distance empirical justification in the new modern science from the ancient method of justification by reference to an authority. "Upon the words of no one" was to stand in bold, defiant contrast to the idea of submission to authority, referencing Horace's line *Nullius addictus iurare in verba magistri*, that is, being "not bound to swear allegiance to the words of any master."

When it comes to matters of empirical fact, scientists should feel free to ignore what any "authority" says about the world and focus only on trying to discover what the world itself may be coaxed into showing. One might be tempted to say that science should let the facts speak for themselves. However, this is not quite right, for facts do not "speak" at all. Nor do scientists simply "read the book of Nature," lovely though that image may be. Scientists do, of course, need to describe their discoveries in words, but that comes later; scientific findings come neither from creative combinations nor analytic unpacking of words but rather are uncovered by special observations and operations. It is partly for this reason that descriptions of scientific discoveries commonly must use technical vocabulary that requires specialized, often hands-on training, to truly understand. Discoveries themselves, and the processes of discovery, involve a more direct communion with nature. This harkens back to the point above about the epistemic priority of demonstrations. Blaise Pascal, the seventeenth-century mathematician, physicist, and philosopher, put the point bluntly: "On subjects in [the physical domain] we do not in the least rely on authorities—when we cite authors, we cite their demonstrations, not their names."[21]

This is a good place to reiterate the nature of our project and the important difference between a descriptive and a normative account. Although Pascal seemed to be speaking descriptively about what scientists do, I see his comment as what might be thought of as a prescriptive description, a statement of what it would look like if scientists did as they should. I do not claim that all, or most, scientists always meet the ideal. Even the Royal Society failed to live up to its own motto on occasion. It is fitting to quote Alasdair MacIntyre's version of such a case: "Charles II once invited the members of the Royal Society to explain to him why a dead fish weighs more than the same fish alive; a number of subtle explanations were offered to him. He then pointed out that it does not."[22] Here is one example where scientists fell into a trap by accepting the word of an authority and failing to check the facts themselves.

Trusting experts vs. authorities   At this point, someone might reasonably object that such an ideal is not realistic, given that one cannot always check everything oneself. In practice, scientists have no choice but to rely on the testimony of other scientists, and they do so all the time. Surely, that is a perfectly justifiable practice. Is it not, therefore, disingenuous to say that science rejects, or should reject, authority?

To address this objection, we will need to draw some distinctions. To help with the discussion, I will suggest some vocabulary to separate different kinds of bases of justification. However, I do not want to place much emphasis on the terms themselves, so it may be helpful to think of them as "technical" terms, defined for the sake of the discussion. We will then be able to better assess the objection in light of relevant conceptual differences.

First of all, we may acknowledge that rejection of all testimony does not follow directly from curiosity, because there can be reliable individuals whose views about certain questions we should reasonably accept. In that sense, we may agree with the objection without a problem. However, here things become more complicated, for we need to ask on what basis should a researcher determine whose testimony can be trusted? The answer must be made in terms of evidence—both of someone's general character and their specific expertise. Their embodiment of scientific values, together with their application of scientific methods in particular areas, makes them a trustworthy proxy for the evidence itself. Because they are known to embody those values (i.e., their good epistemic reputation precedes them), we may reasonably believe that they will have properly applied appropriate methods in making their observations and conducting their experiments and thereby judge that we would have gotten the same result if we had done this ourselves. We begin with skepticism, but trust may be earned by an evidential record.

Another way we might express this (and here is where I introduce the terminological distinction) is that scientists should not be thought trustworthy because they are authorities, but because they are experts. The two terms are sometimes used interchangeably, so this is not ideal terminology, but there is good reason to distinguish them.

Someone in a position of authority has the power to authorize something. That is to say, they have the power to make something so just by their word, just as an author does. The author gets the first word, the final word, and every word in between. The Latin root of the words "authority" and "author" has the sense of *originating* something, and in Middle English it takes on the sense of being the maker or cause of something. It is in this sense that God is spoken of as the author of the world. It may also be why the Christian idea that in the beginning was "the Word" is taken to be apt, if God can originate and "make it so" as the cause and governor of the world by His say-so. That is not science.

On the other hand, to be expert at something is to be skilled and knowledgeable about it because of experience and practice. The Latin root of expert is *experiri*, which means *to try*. From that same root stems not only the word experience but also *experiment*. To try something is not just to attempt it, but to subject it to a trial—to test it—and that is the basis of all scientific findings.

The knowledge someone has by virtue of being an expert is unlike that of an authority. These have different areas of application and completely different kinds of justification. Here we are primarily interested in the latter issue.

In the case of an authority, something one claims to be so, is so, because one says it is—saying it makes it so. Authority must be granted. The judgments and pronouncements of a person in a position of authority are justified by virtue of their position, which gives them the relevant legislative power. That power is not absolute, however, but is limited to the scope of their authority.

In the case of an expert, however, something one says is so, is so because one has the requisite expertise. Expertise must be earned. The judgments and pronouncements of a person with expertise are justified by virtue of their experience, which gives them the relevant technical knowledge. The knowledge of experts is also not absolute but is limited to the scope of their experience.

Authorities are justified because of the power they hold, while experts are justified because of the knowledge and skill they have. Put another way, authorities may pronounce because of who they are and how they were granted their authority, while experts may pronounce because of what they know and how they came by that expertise. As an illustration, consider the difference between a law professor and a judge. Professor Peters may be more knowledgeable about the law and its history than Judge James and so have greater expertise, but Judge James's opinions from the bench become precedents that have the force of law because of Judge James's legal authority, whereas Professor Peters's opinions do not. On the other hand, if one requires facts about the history of previous rulings, Judge James's pronouncements carry no special weight; in that case one should rather turn to Professor Peters, whose scholarly experience provides a basis for a knowledgeable answer. In ideal cases, there is a Judge Jones who has the requisite expertise as well as the authority, and a nation can count itself fortunate when this ideal for judgment is met.[23]

There are a few cases in science where authority and expertise meet in this way, such as when a society grants a science-based profession special privileges that require licensing. In such circumstances, professionals do stand in a position of authority with regard to certain matters taken to be under their legitimate purview. This is more applicable in applied sciences like medicine (e.g., physicians are granted authority to pronounce the time of death) rather than in basic research science, but the distinction can sometimes be fuzzy. However, even in such cases the presumption ought to be that one's authority does not go beyond one's expertise.

This difference between authority and expertise is the basis for the standard challenge regarding matters of authority: "Says who?" That may be impertinent, but it is appropriate, for it is asking after a person, questioning whether the speaker does indeed have the requisite authority to decide and to lay down the law. One answers such a challenge by, for example, showing one's badge to prove one's identity as an authority. Religious authority works similarly; when Moses asks to know God's name to be able to tell the Israelites who sent him, the enigmatic answer is that of primal authority—"I am who I am."

Scientists are seen as impious because they do not accept such an answer, but the fact of the matter is that they would not even care to ask for such. The scientific challenge is not "Says who?" but rather "Where's the evidence?" Facts demand acquiescence and must be obeyed, but a scientist accepts statements of fact only if they are supported by evidence. And when it comes to ascertaining whether they should or should not accept some purported fact, scientists are not impressed by badges or titles, as these are irrelevant. (Thinking back to our discussion in chapter 2, this is likely the basis for Feynman's disdain of honors.) When researchers report back to their lab colleagues about something they learned at a professional conference, they do not cite the presenter's name as justification; they cite the data. That is why

"Where's the evidence?" is the appropriate challenge to a scientific expert. Furthermore, in responding to such a challenge, the final appeal in science is not to persons, even divine ones, but to impersonal evidence—not to "I am who I am," but "It is what it is."

This is a basic difference between science and many forms of religion. An appeal to someone's title except as a proxy for evidential warrant is a sign of unscientific deference to authority. At scientific meetings, it is rare to hear anyone's title or degree mentioned, and informality of attire and address is the norm. It is not uncommon for scientists to look somewhat disheveled. Scientists do not care about researchers' appearance or formal titles; we just want to see their graphs.

Skepticism of Self

Scientific skepticism applies internally as well. A skeptical attitude toward oneself is a way to help avoid confirmation bias, which is the all-too-human tendency to interpret information so that it confirms one's preexisting beliefs. Giving preference to one's preferred hypotheses, and less consideration to alternative hypotheses, is hardly a good method to follow if one is truly curious to discover which is true. Francis Bacon, the seventeenth-century polymath whose 1620 book *Novum Organum* articulated the "New Method" of inductive science, expressed this concern and proposed skepticism as a remedy: "Let every student of nature take this as his rule,—that whatever the mind seizes upon with particular satisfaction is to be held in suspicion, and that so much the more care is to be taken in dealing with such questions to keep the understanding even and clear."[24]

Bacon's warning remains as pertinent for scientists as it ever was. Today we sometimes hear the same warning made with different terminology, as in cautions to avoid "motivated reasoning," which is not meant to imply conscious desire but is defined as an unconscious, emotional tendency to see, interpret, and reason about things in ways that support a prior belief, need, or goal. If we are considering some particular claim we *want* to believe, we ask "Can-I-Believe-It?" with a weak standard of evidence. If we *don't* want to, we ask, "Must-I-Believe-It?" and look for any reason to reject the evidence. This is not good evidential practice.

Knowing that we have such a psychological tendency, as scientists we must find ways to guard against it. Bacon's recommendation is that we be careful whenever we think that we know something: "If a man will begin with certainties, he will end in doubts; but if he will be content to begin with doubts, he will end in certainties."[25] The zoologist Konrad Lorenz put the point more whimsically: "It is a good morning exercise for a research scientist to discard a pet hypothesis every day before breakfast," he wrote. "It keeps him young."[26]

Connecting Skepticism to Curiosity

Merton's notion of organized skepticism as one of the norms of science highlights how the scientific community suspends judgment until there has been a detached evaluation of claims. Merton devoted only a few paragraphs to the idea, mostly focusing on its purported iconoclastic element, which refused to "preserve the cleavage between the sacred and the profane";[27] it was something that he took for granted. He mentioned that it had a methodological element but deemphasized it. From our perspective, however, it is the way that the methods, values, and norms of social control are interconnected and mutually supporting that is of interest. Skepticism would not be a reasonable norm of social control for science if it were not conducive to the central scientific goal of discovering natural truths by empirical tests. It is a social norm for science because it does serve scientists' ultimate goal.

Skepticism, by its nature, is negative, critical, and cautious. It is related to the methodological rule about withholding of assent until the matter of interest has been thoroughly checked. It is also relevant to the doubting attitude of scientists, especially with regard to authority. However, it is important to see the norm, value, and practice of scientific skepticism not as negative for the sake of negativity, but as being in the service of the prime mover in science, which is positive: curiosity is the driver. Being skeptical of a hypothesis until the tests have been done and the evidence is in, is a corollary, because one wants to satisfy one's curiosity, not fool it.

This critical feature of science explains other distinctive elements of scientific practice, such as why its conclusions should be based on intersubjectively testable, public evidence. This feature is illustrated in the general expectation that data should be open. One ought to never have to accept something in science simply on the scientist's authority; one should, in principle, always be able to check the evidence oneself. This, in turn, helps define responsible conduct in large and small ways. For instance, a principle of open evidence informs the structure of scientific publications, which should not simply give results, but also describe methods and data. Moreover, it is not only in publications and conferences that scientists promulgate and check their results; the community of scientists is like a nest of social insects in their sharing of information. Darwin was in constant correspondence with other researchers around the world, pulling together information from a wide range of sources as part of his own investigations. Social norms involving the sharing of knowledge are part of the culture of science. Usually this is just taken for granted, but in certain settings it becomes important to make them explicit, and one finds institutional expressions that affirm the community's commitment, such as in university policy statements on openness in research.[28] We will later see more about how a virtue-theoretic approach may shed light on such specific aspects of responsible conduct of research, but for now the example highlights a general lesson that may be drawn from the discussion of this chapter, which is that scientific integrity is an integration of its values in all aspects of scientific practice.

Individual curiosity is the instinctual and normative source of scientific values, and science, as a discipline, aims to systematize and institutionalize it. Skepticism is but one value that fits into this coherent set of epistemic, social, and ethical norms. What I want to do now is explore how this notion of science as systematic curiosity relates to another scientific value, namely objectivity, and how this (and other values) is justified and supported, philosophically and institutionally. The goal is to better connect the normative and sociological analyses as well as the personal and institutional levels of practice.

Seeking Objectivity

Objectivity: Common Misconceptions

In some respects, objectivity is very much like skepticism, especially in its rejection of dogmatism. Taking claims on faith is inimical to the idea of viewing and assessing claims objectively. In many religious traditions, the faithful believer is not supposed to assess received dogma at all but rather is expected to obediently acquiesce to it. Faith has its own reward system, but scientific objectivity cannot abide it. As is the case with being skeptical, being scientifically objective means never believing something without question but always trying to base belief on evidence. The scientific community's rejection of political ideology as a legitimate basis for empirical conclusions is of a kind. A classic example is the Marxist rejection of Darwinian evolution led by Trofim Lysenko in the Soviet Union starting in the 1920s. For ideological reasons, Lysenko promoted a false Lamarckian account of the inheritance of acquired characteristics. Lysenkoism set Soviet biology back for decades, and the term is now used generically to refer to any ideological distortion of objective scientific assessment. However, objectivity has its own conceptual complexities that are different from what we have previously considered with regard to skepticism. Indeed, the concept is sufficiently vexed that we need to set aside a few common worries before we can see how it operates as a virtue in science.

One sense of objectivity is suggested by the root of the word itself—object. A common definition of objective knowledge is that it involves knowledge of objects in the world, especially in contrast to its linguistic opposite of subjective knowledge, which is located in subjects. The grammatical separation of subject and object is easy to export into an unbridgeable separation between ourselves as subjects and the world of objects that we find ourselves in. If this separation is taken as a given, it is not surprising to see why some take objective knowledge to refer to objects in the world as they exist independently of ourselves. Moreover, it is easy to slip from a grammatical distinction to an ontological distinction and thereby see the realms of objects and subjects as potentially unbridgeable. This notion is legitimately criticized as being impossible to obtain. Philosophers often express this by saying that we have no "God's-eye view" of the world; even if God has such a view, we are in no position to see it ourselves. This way of thinking about objects and objectivity is not helpful for understanding science, as it incorporates a metaphysical distinction that science may not assume. Our evolutionary history situates us squarely in among the other objects (and subjects) of the world.

We should be similarly careful to avoid simplistic contrasts such as the stereotypical view that being objective is the opposite of being emotional. This mistaken conception probably arises from a common view that takes emotions to be inherently weak, unstable feelings that are easily swayed. Objectivity does imply a steadiness in one's perspective and judgment, and it is true that being objective means having a clear-eyed view that is not colored by fleeting whims or irrelevant desires. However, this is not the source of the problem; rather it is the simplistic view of emotion that is unreasonable. While certain emotions may be unstable, others are deep-rooted and steady. We already saw that there is some reason to expect that evolutionary history provides part of the explanation of why some emotions are one way or the other, and in the next chapter we will examine the normative role of emotion in science in detail. Here we can just note that it would be a mistake to think of objectivity and emotion as opposites—a passion for truth may be deep and abiding.

Another simple misconception is that objectivity means accuracy. One hears this mistaken view when people say that such and such "provides an objective picture of the world" as a way of saying that it provides an accurate picture and that being more objective means being more accurate. Although related to one another, objectivity and accuracy are distinct concepts. For instance, if I weigh a package on my bathroom scale I have an objective measurement even if it is not especially accurate compared to the scale at the post office—the first is accurate to within a pound, the latter to within a gram. I could then bring the package into my lab and use a scale that is accurate to within a microgram. These measurements are equally objective even though they are not equally accurate, so it should be clear that the concepts are different. They are connected, however, in that objectivity in science involves, among other things, care with regard to ascertaining, calibrating, and reporting the degree of accuracy of one's measurements.

Is objectivity illusory and unachievable?   Most importantly, we should disabuse ourselves of another common view of objectivity, namely that it is an illusion because it is thought to be, for either theoretical or practical reasons, unachievable. We must reject the all-or-nothing fallacy that an assessment is not objective just because some element of subjectivity entered or might have entered in the evaluation. This is of a kind with the way that scientists offer evidence to support their conclusions—objectivity always comes with degree of strength just as statistical and probabilistic reasoning do. Thus, rather than rejecting it because we may never achieve it perfectly, we should think of objectivity as an aspirational value—a bar that may always be raised higher than we can reach but that sets a standard against which we may judge improvement. As evolutionary biologist E. O. Wilson put it, "Although seemingly chimerical at times, no intellectual vision is more important and daunting than that of objective truth based on scientific understanding."[29]

Forms of Objectivity in Science

How should we understand the notion of objectivity in science? This question has been a staple of discussion in philosophy of science with articles and entire books devoted to analysis of the concept, each more detailed than the previous.[30] We do not have the space to review the many distinctions, but we may get a sense of the complexity by looking at the various nouns to which the adjective is applied. A common generic way to speak of objectivity is in terms of objective *knowledge*, but scientific knowledge is not a simple thing and may be further analyzed into various components or viewed from different perspectives. For instance, philosopher of science Israel Scheffler investigated the question of the objectivity of scientific *statements*, arguing that "Scientific statements are objective if they have been tested against impartial and independent criteria."[31] Others ask about how to understand objective *facts* or objective *methods*. The physicist John Ziman, writing on philosophy of science from his own experience, focused on what made scientific *data* or results objective, writing, "Scientific data or results are objective if similarly situated observers using the same techniques would produce the same data or results."[32] The idea behind these formulations, including the notion that scientific knowledge must be public, goes back at least as far as philosopher of science Karl Popper, who argued that "[T]he objectivity of scientific statements lies in the fact that they can be inter-subjectively tested."[33]

However, our primary interest here is not with what it means to say that knowledge, statements, data, or the like are objective, but rather what it means for *scientists* to be objective and why they should try to be. Albert Einstein, drawing on the philosopher Schopenhauer, made reference to objective *perception* and *thought* as a refuge from the crudity and dreariness of everyday life: "A finely tempered nature longs to escape from the personal life into the world of objective perception and thought."[34] Einstein and Schopenhauer's notion, of course, is related to the other senses of objectivity, but this is not the place to delve into the subtleties of those interrelationships. For our purposes, the most relevant notion of the objectivity of scientists has to do with an *attitude*. Specifically, to be objective as a scientist is to have an attitude of impartiality regarding the possible outcomes of fair procedures of inquiry. That is, I ought not let my personal biases for or against some hypothesis override the results of a fair assessment of it. This helps explain why the idea of objective statements and knowledge in science is indeed based on intersubjective testability. Scientists implement this attitude of impartiality by way of scientific methods, which are systematic procedures of bias reduction. Objective knowledge cannot be grounded in idiosyncratic beliefs but must come from a rational synthesis of perspectives based on constraints of shared observations and experiments. My approach here will be to focus on how we should understand this notion of objectivity as a virtue for scientists, especially to highlight connections between the epistemic and ethical values that are part of its moral structure. I hold this view of scientific objectivity to be another expression of science's insistence on drawing conclusions about the world based on evidence rather than authority, which is itself derived from the basic virtue of curiosity.

Objectivity as Impartiality

We can begin to see the moral element at the heart of objectivity if we view it as a kind of epistemic impartiality.[35] It is not that the scientist expects to have some special access to objects in the world as they exist without reference to human observers. Rather, the idea is that one should not be partial to one's own particular viewpoint. Barbara McClintock, the Nobel Prize–winning geneticist, spoke of this in terms of attempting to remove oneself as a separate entity and to get "a feeling for the organism" (or whatever thing one is studying) itself: "As you look at these things, they become part of you. And you forget yourself. The main thing about it is you forget yourself."[36] This is similar to ethicist Kurt Baier's discussion of "the moral point of view," which, he argues, looks at the world and our choices in it not simply in terms of our personal interests, but in terms of moral reasons. Science, too, has its moral reasons that, as we have seen, structure its practices with the aim of satisfying our curious search for empirical truth.

Specifically, objectivity involves a kind of impartiality that is expressed in the epistemic requirements for scientific testing—one should set up experiments that test a hypothesis without prejudice or favor regarding the outcome. This notion of impartiality is not the same as a view from nowhere; it does not mean that one does not have a point of view. Indeed, the point here is that because I know that I may come in with some investment in a particular outcome, or because I have prejudices of which I may not be aware, I need to take steps to overcome these. Science is a method of bias control. We aim to find and follow procedures that reduce the chance that our biases will skew our results.

Similarly, one should judge scientific findings without improper biases such as nationality, race, gender, or other such irrelevant factors. Merton discussed this under the norm of universality. The key element of his notion of universality was that scientists may appeal only to what he calls "preestablished impersonal criteria" when judging truth claims. His slogan for this principle was "Objectivity precludes particularism."[37] Merton also mentioned objectivity as part of his explanation of organized skepticism, noting that science is not deferential to "sacred" authority, which (unlike "profane" matters in the natural world) is not subject to objective analysis.[38]

Merton's norm of universality is a good example of how objectivity is tied to the idea of impartiality and why having an appreciation of this is so important for science. Merton was explicitly contrasting the scientific norm of universality with views that claimed that particular groups had special epistemic viewpoints different from others. This was the argument that Nazis used to disqualify so-called Jewish science. To say that each "race," or other particular group, has a unique viewpoint that is in principle inaccessible to others is not only subjectivist relativism, but it also opens the door to legitimizing the worst kind of prejudice and oppression. Merton was adamant that particularized criteria of this sort were inimical to science. The framework that we have been developing provides an important reason for why this is so, namely, that it improperly asserts authority over evidence. Viewing scientific objectivity as impartiality may also be thought of as an epistemic variation on John Rawls's theory of justice as fairness, where structural justice is judged in terms of what he called the "Original Position"[39]—a theoretical perspective where we are under a "veil of ignorance" about what position we will have in society as we figure out how to structure it, considering alternative models until we come to a reflective equilibrium. This is another sense in which intersubjectivity is important. It should not matter who brings forth the evidence, and their sex, tribe, or skin color is beside the point, because it is the evidence alone that should determine whether a conclusion is warranted.

Correcting failures to live up to the ideal This is not to say that scientists always live up to this ideal. Far from it. Historians of science, feminist philosophers of science, and others have provided a great service in pointing out cases where scientists have failed to be objective and have thereby imported various biases into their conclusions. The point here is that this virtue-based analysis helps explain why such cases ought to be recognized as failures and why steps should be taken to improve scientific practices to make it less likely that conclusions will be contaminated by unrecognized prejudices.

For instance, this explains the value of and limits to what some philosophers have offered as "standpoint" epistemologies.[40] At their best, these strengthen objectivity by giving a voice to the views of people (especially from marginalized groups) whose experience gives them a perspective that may be invisible to people (especially from socially privileged groups) who take their own viewpoint for granted and do not recognize their own biases. For instance, especially in social science and other areas of research where sex or gender is causally relevant, it should not be surprising that women scientists entering a field notice things that a previous generation of male scientists had overlooked. It also explains why it would be problematic to take the idea of an epistemic standpoint to an extreme, positing a special epistemic privilege for such viewpoints, thereby isolating them from the possibility of critique from within or without. It would be bizarre and not helpful to the goal of reducing biases if one thought that this meant that a new idea offered by a woman scientist could not be challenged (or supported, for that matter) by her male colleagues because they, by their biology or their social upbringing, do not share that standpoint. Such separate but equal epistemologies are at best philosophical halfway houses, which may be temporarily useful as a corrective to systematic social biases but which cannot be adopted as basic without inadvertently reintroducing and reinforcing the very same sort of biases in a new form or in an old form of the sort that Merton warned against.

We will want to look carefully at what scientists should do to reduce the effect of personal biases, but for the moment let us keep our focus on the institutional level, to see how objectivity played out in Merton's social control analysis. Objectivity is not one of the basic institutional norms that Merton identified, but it played a role in several of them, such as his norm of disinterestedness.

Disinterestedness

Disinterestedness was the third of Merton's four concepts that he thought comprised the scientific ethos. Merton did not define the term, but emphasized that he viewed it, like the other concepts, as an institutional element rather than a motivational element. This is one key point of contrast between Merton's sociological analysis and our philosophical analysis, which focuses on what justifies their practice and thus should motivate them from within. Merton's interest, as a sociologist, naturally focused on external social factors that structure practices. Indeed, he seemed skeptical that scientists had any distinctive motivations, except perhaps as an indirect effect of social control, such as to avoid community sanctions against someone who violates a norm or (if the norms have been internalized) to avoid psychological conflict. Merton was insightful in recognizing the importance of values in structuring science, but he analyzed them primarily in sociological terms, as elements of social control rather than in terms of their own ethical logic and justification. A philosophical analysis is different from both sociology and psychology. One unintended consequence (to use Merton's own helpful concept) of Merton's sociological analysis is that ethical and epistemic norms become seen as external factors that are imposed and enforced rather than as internal structures that are constitutive of what it means to be a scientist. This comes from Merton's point about scientists having many psychological motivations but conforming to a characteristic set only through a "distinctive pattern of institutional control" by means of the threat of institutional sanctions.[41] However, focusing on such cultural norms without reference to any philosophical justification leaves the position open to the damning charge of ethical relativism. Such an externalist view of norms leads to an image of ethics as little more than a matter of power and enforcement.

Merton goes on to relate the norm of disinterestedness to what he saw as the virtual absence of fraud in science. It is not, he suggested, that scientists are inherently more moral than other people but rather that the institutional emphasis upon verifiability of results means that a scientist's research activities are subject to "rigorous policing" through the "exacting scrutiny of fellow experts," presumably keeping everyone in line.[42] No one should doubt the importance of such social controls for community norms, but it is a mistake to leave the analysis at that level, ignoring the underlying philosophical justification for these norms. Doing so is problematic from both a theoretical and a practical view.[43]

What did Merton mean by disinterestedness? He gave it no technical definition, so presumably he meant it in the ordinary sense of being free of selfish motives or interests. This interpretation is supported by the one thing he did to clarify his meaning, which was to contrast disinterestedness with altruism as a psychological trait. Again, this must be understood in light of Merton's institutional-level analysis, but it is compatible with the vocational virtue approach we have been describing, and it suggests that he would accept our notion of impartiality as part of something like an institutional form of objectivity. This fits with his earlier paper, in which he first discussed the ethos of science. In a brief aside on disinterestedness in that piece, he reflected on it as involving the scientist's "primary goal, the furtherance of knowledge" that is "coupled with a disregard of those consequences which lie outside the area of immediate interest."[44]

Again, because Merton's focus was on institutional values and modes of social control, he did not pursue this point about the methodological function of scientific values.[45] He recognized the furtherance of empirical knowledge as the central, guiding purpose of science but did not analyze it, or its relationship to scientific virtues, in a philosophical sense.[46] However, this is our task, and when we examine the ethos of science in this way, the norm of disinterestedness is seen to have a clear basis in science's core values. The idea is that scientists are, first and foremost, interested in ascertaining truth as best they can—they seek the honest satisfaction of curiosity. The virtue of objectivity encompasses the idea of disinterestedness in the sense that research conclusions should be judged solely on the evidence and should not be swayed by financial or academic reward, politics, religion, or ethical views. Consequences in these areas should not be salient to the assessment of the degree of truth of competing hypotheses; such things, as Merton said, "lie outside the area of immediate interest," which is just to discover the facts of the matter.

Historian and theologian Reinhold Niebuhr saw the link between being disinterested and avoiding bias in examination of nature, writing that "The scientific observer of the realm of nature is in a sense naturally and inevitably disinterested. At least, nothing in the natural scene can arouse his bias. Furthermore, he stands completely outside of the natural so that his mind, whatever his limitations, approximates pure mind."[47] Niebuhr contrasted this with the situation of the historian who, he said, "cannot be disinterested in the same way, for two reasons: first, he must look at history from some locus in history; secondly, he is to a certain degree engaged in its ideological conflicts." Here again, however, we must be careful to distinguish the descriptive and the normative. Niebuhr cannot be speaking merely descriptively, for we know that scientists can never stand outside of nature as pure minds, even approximately, as any evolutionary account recognizes. Moreover, scientists, like historians and everyone else, view the world from a temporal locus, which means that prejudices such as sexism that are pervasive in the wider culture can infect their practice and findings. It is because such cultural prejudices can be so difficult to overcome individually that, to enhance objectivity, science should not only be open to but should actively seek to include a diverse community of researchers, whose perspectives will make it less likely that patterns of bias will be overlooked.

Objectivity as Fairness

One final point to emphasize about how to understand objectivity in science as a normative concept with both epistemic and ethical elements is that, as noted previously, it also involves a notion of fairness. When judging hypotheses, virtuous scientists should be keen to make a fair assessment of the evidence, whether in or against their favor. The weighing of evidence should be done without putting one's finger on the scale to boost a personally favored conclusion; that would not be fair to competing hypotheses. Again, this is a notion that arises directly from an understanding of science as seeking the honest satisfaction of curiosity.

This has implications not only for how scientists should approach their own research but also for how they should behave with regard to that of others. For instance, it helps us appreciate the significance of the scientific practice of peer review. As we have seen, science demands that conclusions be drawn on the basis of empirical evidence that anyone, at least in principle, could check. It is not just that a finding will be better accepted if it is open to public scrutiny (though that is certainly one useful effect), but that the very idea of basing conclusions on evidence, rather than authority, demands it. If it were not possible to check a conclusion independently, then we would be back to accepting it on the basis of authority. This expectation of fair, objective assessment is the equivalent to the moral point of view for science. It is not meant as a barrier to participation but rather the opposite—it is meant to guarantee the possibility of participation, in the sense that anyone should be able to check and make the same observations. The rationale for this requirement is not primarily social, but epistemic.

What this means for scientific peer review is that it is not simply a matter of social control, though, of course, it does have that useful consequence.

Social control supports the practice and does help to keep it trustworthy, but neither keeping scientists in line nor even keeping science trustworthy should be thought of as the primary justification. The basic justification is that evidence ought to be subject to the test of intersubjective triangulation. This is what a fair assessment of the evidence comes to in this aspect of scientific practice. As a member of the community of science, one has a professional duty to participate in peer review as part of this process of fair assessment. The end result will be that a submitted paper is or is not accepted for publication. However, although the peer review filter has this practical purpose in the context of the institutional procedures for scientific publishing, the fundamental idea is that a purported finding becomes certifiable as scientific knowledge because it can fairly be tested by anyone in the community.

To put this another way, science does not aim for objective tests in order to make its conclusions *seem* more trustworthy but rather because objective tests make it more likely that the conclusions will, in fact, be more trustworthy. That is why we should feel confident in crossing an epistemic bridge. While it is possible for an individual to do science alone, it is far better when done by a diverse, curious community that can validate results impartially. Objectivity as a character virtue is something that individuals should aspire to because it benefits the search for truth and the satisfaction of curiosity, but it is also something that must be advanced by a community that is appropriately structured. It is to the scientific community that we now turn, for that is where individual virtue is made more powerful by becoming organized.

Competitors and Collaborators

Community and Division of Scientific Moral Labor

Community is not a virtue itself, but rather an organized structure of relationships that allows the virtues to be constituted, expressed, or amplified. For example, scientists often point out that science is self-correcting because of community checks and balances; errors due to subjective biases are more likely to be identified because of community procedures of mutual critique. What this means is that not every scientist needs to be equally skeptical, or even equally objective, so long as enough individuals are, and the community is structured in such a way that these qualities can make a real difference in practice. We may think of this as a division of scientific moral labor. Scientists in our survey noticed this as they thought about the relative importance of different virtues. While some virtues, like honesty and curiosity, are so central that it hardly seems possible to be a real scientist without them, other virtues, like meticulousness, could be weaker in some individuals in a lab group, as long as others in the group made up for them.

This leads naturally to the question of what other aspects of science's community structure depend upon its aims and values. In thinking about this, it is important to remember that the scientific community extends to the past and future. Scientists collaborate with their predecessors, and they leave their witness to those scientists to come. Scientific theories are interconnecting webs of observations, models, and previously confirmed hypotheses, and just as no scientific finding stands alone, neither does any scientist.

Merton's institutional norms may be thought of as part of the answer that the community of science has given to the question of how it should structure itself, but they are only one level of analysis. A full account requires looking at science's methods and practices in greater detail to see how the epistemic and ethical norms of the practice, the virtues of the practitioners, and the community norms of social control all fit together. In this chapter, I have been highlighting a few aspects of this, especially the core idea that scientific conclusions are to be made on the basis of evidence rather than authority. That idea finds expression in the scientific mindset, in the scientific character traits that are valued by the community, in the methods of experimental testing, and even, as noted above, in specific practices like peer review.

Peer review   Peer review is neither the first nor the final step in the evaluation of hypotheses, but at least in the present scientific culture, it represents the most critical step in the communal system of checks and balances. This is the stage at which a research team lays out its evidence for scrutiny by the community of peers, which in this circumstance refers to other experts in the field who have the wherewithal to assess its value. Was the question defined with sufficient precision to allow possible hypotheses to be tested? Were the experiments adequate? Are there possible alternative explanations for the observed results that were not considered? Were the right statistical methods used and were there no errors in the calculations? Peer referees consider the proposed findings in light of these and other questions to see whether the evidence adequately supports the conclusions.

A paper may be rejected as fatally flawed or possibly returned for major or minor revision. Or it may be accepted after a round or two of revision, and then finally published. This is the point at which a finding is given its first official community endorsement and enters the stockpile of scientific knowledge for general use. It is also a key point of validation in the assignment of credit. Merton emphasized the importance of peer recognition in science, and it is worth looking more closely at a few of these elements to appreciate the ways in which they are conceptually and practically interconnected.

Communalism and Credit

Recall that the second institutional norm that Merton listed was "communism," referring to "the nontechnical and extended sense of common ownership of goods."[48] As I noted, the fact that he put it in quotation marks indicates that Merton was not entirely happy with the term. He certainly did not mean to say that science endorsed communism as a political ideology. Some later scholars use the term communalism instead, which avoids that problem, while still capturing the idea that scientific discoveries are not to be thought of as any one scientist's personal property, not even that of the discoverer, but are to be recognized as shared resources for the entire community. The only claim to intellectual property that the discoverer may make is that of "recognition and esteem."[49]

One observes the workings of this scientific norm regarding credit and community recognition in the importance that scientists give to the public acknowledgment and recognition of the shoulders on which one stands. The most obvious place one sees this is in scientific publications, which typically begin with a section that reviews previous related research by other investigators and which always provide citations for results used from others. For especially significant findings, the discoverer's name is attached to the finding. Halley's Comet, the Doppler Effect, the Kuiper Belt, and the Heisenberg Uncertainty Principle are just a few examples. However, eponymous discoveries may become problematic when the work was collaborative. It is easy enough to include two discoverers (e.g., Epstein-Barr Virus) and three is possible (e.g., the Einstein-Podolsky-Rosen Paradox), but beyond that it becomes impractical. We remember Peter Higgs as the theoretical physicist credited with the Higgs's Boson, but Higgs's work was done together with five others. Higgs was expressing the usually unstated ethical norm about credit sharing when he pointed out, as he often did, that the boson should not have been named just after him but should have a very long hyphenated name. The scientific norm of crediting discovery is also exhibited in every scientific conference. In presentations, scientists always include an acknowledgment slide that cites the names of their collaborators and will often mention specifically who made what contribution to the study. Such public displays of acknowledgment are one sort of scientific convention that reinforces community relationships and values. Paleontologist and evolutionary biologist Neil Shubin may be the name and face that one associates with the important discovery of *Tiktalik roseae*, a striking transitional fossil between fish and amphibians, but in his talks, Shubin always points out that he did not make the discovery himself but was a member of a big team. "Science," he says, "is a collaborative enterprise."

Collaboration vs. Competition

At this point, someone might object that all this talk of generous credit sharing and collaboration is just a polite veneer and that scientists can be just as stingy with credit and as cutthroat in their competitiveness as anyone. Scientists must compete for grants to conduct their research, and funding is limited, so many will go unfunded. Science students must compete to be accepted by researchers to work in their lab groups, and plum spots are few. Research groups must race to be the first to discover an answer to an outstanding question, and the first to publish gets the glory.

We certainly must acknowledge the tension that can arise because of the importance of reputation in science. Since community recognition typically goes to the first discoverer, it can occasionally lead to rancorous priority disputes. Even Isaac Newton became involved in such a battle, with Gottfried Leibniz, about who deserved credit for developing calculus. In professional settings, this tension can be exacerbated when publications become the coin of the realm for career advancement; if another scientist scoops one's research by publishing first, this may result in a real loss. Merton's norm about the "common sharing" of scientific goods also finds itself in tension with profit-making motives because of the use of science in industry, which expects to assert intellectual property rights. Since the passage of the Bayh-Dole Act in 1980, even universities are on the lookout for possible patentable scientific results. However, in the ideal at least, plaudits, patents, and profits are external goods that should not affect scientific practice. Think back to Feynman's explanation of why he abhorred honors, and why the real reward for the scientist was the pleasure of finding things out and knowing that others could use one's findings in their own research.

Rather than collaboration, one might even argue that competitiveness among scientists is actually good for science and ought to be encouraged. Nobel Prize–winning biologist James Watson thought that high-pressure competition was part of the "formula for breakthroughs" in scientific research: "Take young researchers, put them together in virtual seclusion, give them an unprecedented degree of freedom and turn up the pressure by fostering competitiveness."[50]

However, consider an alternative formula for breakthroughs articulated by computer scientist Jim Warren, who was a leader in the microcomputer revolution of the 1970s. He credited the communal counterculture of 1960s California with its spirit of working together and sharing freely for the technology revolution that was spawned there. As an example, he described the community spirit of sharing that permeated the Homebrew Computer Club, which is now recognized as the epicenter of the personal computer revolution:

> As soon as somebody would solve a problem they would come running down to the Homebrew Computer Club next meeting saying "Hey everybody, you know that problem that all of us have been trying to solve. Here's the solution. Isn't this wonderful? Aren't I a great guy?" And it is my contention that that is a major component of why Silicon Valley was able to develop the technology as rapidly as it did. Because we were all sharing. Everybody won.[51]

So, which is it? Should scientists follow Watson's model or Warren's? Should they emulate competitiveness or collaboration? In our national survey of exemplary scientists, the virtue of being collaborative received only an average rating in the quantitative scale of importance. Some people have found the average value placed on collaboration to be vaguely troubling, indicating a possible weakness in science's value system. It should be noted, however, that our exemplary scientists rarely even mentioned being competitive as a virtue for science at all. Indeed, competitiveness was most often mentioned as being negative. Watson's view seems to be an outlier. Watson was himself very competitive by nature, so perhaps his recommended emphasis on competition was more a matter of his own personality, which certainly stimulated him to great productivity.

It also likely makes a difference what kind of research question one is pursuing. We asked our subjects to explain their ratings, and one who gave collaboration a low score noted that Einstein was not very collaborative, and he did just fine. A theoretical physicist in Einstein's day could still do science on notepaper while on a break in the patent office. However, Einstein's ideas could not be tested without the help of experimental physicists, and in that area, collaboration was essential. Moreover, even with regard to pure theory, Einstein was hardly a lone wolf but participated in the scientific community both historically (e.g., standing on the shoulders of giants) and contemporaneously (e.g., in correspondence, conferences, and through peer review).

Americans tend to think that having people compete for rewards is a good thing in general, but this model has its limits in science. It wrongly makes it seem that one scientist wins and others lose in a zero-sum game. This mistakes the nature of the practice. The central goal is not to win a grant, or prove one's own view, but to discover the truth. This makes science a different kind of race. Thus, it should not be seen as a defeat if my colleague's hypothesis "wins" and mine "loses."[52] For example, in a "race for the cure" for some disease, it is not just the final discovery of a treatment that counts, but also all the prior discoveries that led to it, including lines of inquiry that were discovered to be dead ends. The scientific community as a whole must work together to make progress by testing various hypotheses, discarding ones that fail, and letting others know that they should look elsewhere for the answer they all seek. All scientists may not need to be collaborative, but they all do need to cooperate. The key thing is not that scientists ought to be subject to competition, but that hypotheses should.

Evolutionary Model of Scientific Testing

The philosopher of science Bas Van Fraassen pointed to competition among hypotheses as a simple explanation of why science is so successful: "The success of current scientific theories is no miracle. It is not even surprising to the scientific (Darwinist) mind. For any scientific theory is born into a life of fierce competition, a jungle red in tooth and claw. Only the successful theories survive—the ones which in fact latched on to the actual regularities in nature."[53]

Van Fraassen was making a point that many philosophers of science have made—that one may fruitfully analyze scientific progress in evolutionary terms. It does not matter so much how scientists come up with the hypotheses to consider, as how the hypotheses are tested. For instance, the pioneering organic chemist August Kekule's idea that the benzene molecule had a ring structure came to him in a dream. He explained that he awoke and spent the rest of the night testing the idea. How well does the ring-structure hypothesis work compared with alternatives in accounting for what was known about the properties of benzene? Alternative hypotheses must be competed against one another to see which can provide the best explanation of the phenomena of interest.

Karl Popper, an even more well-known name in philosophy of science, made the same suggestion and put a version of the evolutionary model at the center of his analysis of the logic of scientific discovery, which he called Falsificationism. The way that science should proceed, he argued, is to subject hypotheses to the most rigorous tests one can imagine and see whether they can survive. Consider what your hypothesis implies—some observable prediction—and then run the test. If the predicted result is not borne out, the hypothesis has failed. Falsified hypotheses should be discarded and the processes repeated. Hypotheses that pass the test should be subjected to further crucial tests that can weed out the losers and see which is left standing.

The theoretical framework here is evolutionary, and the language is drawn straight from Darwin. Indeed, Darwin made the point himself about the nature of scientific progress. "False facts," he said, "are highly injurious to the progress of science, for they often endure long; but false views, if supported by some evidence, do little harm, for everyone takes a salutary pleasure in proving their falseness."[54] The phrasing seems a bit odd in that some might think that facts are by definition true, but what Darwin likely had in mind here are mistaken observations that are taken to be facts, to which a hypothesis must conform. Imagine some mistaken observation recorded about something in the past that one cannot now go back in time to check; it might be a long time before such a false "fact" of this sort were dropped. As for false "views," Darwin had in mind hypotheses that scientists put forward. Even if they are supported by some initial evidence, they will quickly succumb if other scientists subject them to more rigorous tests.

My point here is that we should recognize that this insistence upon testing competing hypotheses against the evidence has both an epistemic and a social dimension. The logic of evidential testing is implemented in experimental methodology, embodied in a scientist's character, and reflected in the community's social practices. It is an empirical method, a habit of mind, and a social norm.

This is relevant to our previous discussion of skepticism. Biochemist Daniel Koshland appealed to the evolutionary model in explaining how a bit of nonconformism is needed to advance science: "Non-conformists are necessary for progress in science, just as mutations are necessary for progress in evolution."[55] As we saw, evolution requires random mutations to provide the variations for natural selection to act upon. Skeptically questioning assumptions, and offering alternatives, is one way to fill that necessary role in science. However, skepticism can cut both ways, and scientists may initially be skeptical of "crazy, but correct" ideas that question some established view. Sometimes skepticism of the unconventional causes a stubborn conservatism that blocks progress. On the other hand, as we saw above, extreme skepticism can also be deleterious to scientific progress. While it is true that mutations are necessary to provide the variations that are the grist for natural selection, too high a mutation rate will eliminate beneficial variations. On this particular point, evolutionary theory and virtue theory are again in agreement. Nonconformism, like iconoclasm, is a virtue only in moderation. Perhaps without recognizing it, Koshland was channeling Aristotle's notion in explaining the middle road that science aims to take between conservatism and anarchy. He explained how peer-reviewed journals and grant-making agencies help provide the needed balance and revealed that during his tenure as editor-in-chief of *Science* he kept a friendly eye out for nonconformism to provide "skeptical encouragement."

The Nobel Prize–winning chemist Roald Hoffmann appealed to this as part of the psychology of the scientist, to help explain the self-correcting nature of science even in the face of occasional fraud. There are two facts that help keep the self-correction of science working, he said. The first involves what he called "the psychopathology of fraud," which is that fraud is very likely about something that is interesting. After all, what value is it to the fraudster to put forward something that no one cares about? The second fact is that interesting results will be checked.[56] This is a nice corollary to Darwin's point that scientists are always keen to disprove false views. The more interesting the fact, the more likely it is to be checked.

The evolutionary model is enlightening, but we do need to be careful not to mistake what is doing the work. As noted above, it would be a mistake to think that the appropriate idea of a zero-sum competition among hypotheses is equally appropriate as a model for interactions among researchers. It is not competition per se that is essential in science, but evidence. Moreover, even competition is predicated upon cooperative standards. Science works as well as it does because scientists value evidence over authority and attempt to hone their methods and practices to support that central value for validating discoveries. Only one hypothesis may win a rigorous evidential competition, but in that process all scientists win in the sense that everyone's curiosity is satisfied. This takes us back to the idea of science as systematic curiosity.

A Curious Culture

Scientific Practice: A Culture of Curiosity

Merton's interest as a sociologist was in the social aspects of science, so it is not surprising that his analysis of the normative structure of science focused on institutional norms. As we have seen, however, that is only one part of the picture. Viewed from the point of view of philosophy, one discovers that the normative structure of science goes much deeper. Scientists may not be more moral in a general sense than other human beings, but they do not conform to scientific norms simply because of institutional structures of social control. They do it because they share the goal of discovering empirical truths and the values that go along with that goal. Curiosity is not just a character trait of individual scientists; it permeates the culture. The word and its cognates and synonyms glint and sparkle in nearly every discussion among scientists when they talk about their research. I started a collection of some examples I have heard during various conversations and presentations: "We were curious to find out...," "This is a question I've wondered about for a long time," "We wondered to ourselves why...," "We'd love to know why X." The most common question one hears after a talk at a scientific conference is, "I was curious about what you found about X" or something along similar lines.

In this chapter, we have looked especially at skepticism, which keeps scientists on their guard against accepting some hypothesis because they wish it to be true instead of because of the evidence, and objectivity, which seeks out impartial tests and attempts to eliminate biases that could contaminate judgment. These, and other traits that we shall examine later, are character virtues that stand out as of particular importance in the scientific culture, but all of them arise from the central guiding purpose of science as the curious search for truths about the empirical world.

Cooperation and Trust

Scientists sometimes collaborate and sometimes they compete, but they must always cooperate. It is their shared curiosity that makes scientific cooperation possible, and that structures the community of practice. We began this chapter with a question about what made the practice of science trustworthy, and we now have a better sense of the general answer. Science requires a delicate balance between the freethinking revolutionary who is willing to question dogma and the hard-nosed conservative who insists upon preserving well-confirmed findings in the face of the fickle winds of fashion. Because scientific research relies on prior conclusions, for mine to progress I need to be able to trust the results I make use of, which means that I need to trust that they were based on evidence that was honestly, skeptically, and objectively acquired. That is the bridge that we all need to cross as scientists.

Empirical progress requires that scientists can trust each other's results. We must be able to trust that the methods were sound, that data were collected patiently and accurately, that calculations were done meticulously, and that no evidential corners were cut. The shoulders that we stand on need not be those of giants, but they must be strong. I have argued that this strength arises from the values that structure scientific practice, especially in the honest search for truth and the satisfaction of curiosity. These underlying values are the rationale for the norms of social control that Robert Merton identified and for a wide variety of other aspects of scientific culture. Common goals and values help determine the structure of the practice, from the most basic expectations regarding public evidence to options for the division of epistemic labor. Curiosity—the search for truth—structures scientific practice. The degree to which our shared scientific norms allow us to take this for granted increases the degree to which scientific results may be trusted.

The values of science are built into its DNA. They are expressed in institutional structures, but not only there. They also drive its methods. And they drive scientists themselves. The normative structure of science works because of the way it provides a rational justification that unites social organization with methodology and with a scientist's emotional motivation and habits of mind. We will look more closely at scientific methodology in chapter 6 and at disciplined habits of mind in chapter 5, but first we must delve into the relation of virtue to emotion. As we shall see, curiosity is not a superficial trait; it is a deeply passionate search for truth, both for its own sake and for the benefits that arise from an intimate acquaintance with nature.

4 Ardent Observers

Hardly any faculty is more important for the intellectual progress of man than Attention.

—Charles Darwin[1]

First love is only a little foolishness and a lot of curiosity.

—George Bernard Shaw[2]

Companions in Zealous Research

Vocational Friendship

The Greek initials of Sigma Xi, the Scientific Research Honor Society, stand for *Spoudon xynones*, which means "Companions in zealous research." This motto expresses the connection between the community and the purpose of science that we previously explored, but it also captures two additional elements that deserve further discussion. The day-to-day practice of research is often a solitary activity—a researcher may spend weeks or months at the bench observing bacterial cultures through a microscope, or in the field collecting samples and taking measurements, or in the laboratory attending to experimental protocols—but it is rarely done in isolation. Most scientists do their work as part of a research group, usually a lab headed by a professor plus a small coterie of graduate students and postdocs, a senior scientist overseeing staff scientists and technicians, or some similar arrangement. Typically, such research groups are local, with researchers working in parallel within the same space and coming together for regular lab meetings to discuss progress, though technology now allows collaborations of researchers over global distances. The usual analysis of "the scientific community" focuses on institutional structures, overlooking the important interpersonal relationships within research groups. It also gives short shrift to the nature of researchers' attentiveness, especially its emotional character. Sigma Xi's motto highlights this important quality; scientists pour their hearts into their experiments and observations—they are "zealous" in their research.

Founded in 1886 at Cornell University but now with members all over the world, Sigma Xi's constitution emphasized the importance of scientific friendships for supporting the "highest and truest advances" in science:

> Friendship in Science. While those whose heart and soul is in their work, are coping with the great problems of Nature, let them remember that the ties of friendship cannot be investigated, but only felt. Let them join heart and hand, forming a brotherhood in Science and Engineering; thus promoting and encouraging by those strong, personal attachments of friendship, the highest and the truest advances in the scientific field.[3]

The reference to "brotherhood" reflects the gendered terminology of the day, but this scientific "fraternity" was not limited to men; five women were elected to full membership within the first three years. Today, one often speaks of research "collaborators," though Sigma Xi's notion of research "companions" is certainly fitting, as is the term "colleagues," which is common in academic settings. All of these terms connote closer working relationships than the more abstract notion of being members of the scientific community. The concept of a scientific colleague in this sense has not received the attention that it deserves within philosophy of science. Neither has the conative aspect of science been given fair treatment compared to the cognitive.

Scientists may be more likely than others to be introverted, and they may score higher in rational than emotional intelligence, but it behooves us to look beyond such psychological metrics. These may be valid and reliable for ordinary applications of these traits but miss the distinctive ways in which social bonds and passions play out in the context of scientific research, with its goal of focusing attention on the world in ways that may uncover its secret causal structures. Many aspects of these are worth analyzing, but there are especially good reasons to take seriously the notion of friendship in the Sigma Xi statement of ideals, so let us begin with an analysis in those terms. Friendship figures prominently in both the *Nichomachean Ethics* and the *Eudemian Ethics*, Aristotle's central treatises on morality. In his *Nichomachean Ethics*, friendship is allotted more space than any other virtue; Aristotle devotes books VIII and IX entirely to its consideration and mentions it in several others as well. Its function in the account is signaled by its placement, together with the six other books on the virtues, between the opening and closing books on that "best, noblest, and most pleasant thing"—*eudaimonia*—which is variously translated as happiness or flourishing. According to Aristotle, friendship is a key relationship in fulfilling the *telos* of the human being. In addition to being a support for other virtues, friendship is constitutive of fundamental aspects of our nature as social animals.

The *telos* of the scientist is narrower than that of the human being, but science's distinctive virtues also have key social supports, so it is reasonable to suppose that interpersonal relationships of colleagues may play a similar role. I suggest that there is a parallel concept of "vocational friendship" that is essential in science. In an Aristotelian sense, scientific colleagues may be thought of as vocational friends.

It would be naive to suggest that all research groups maintain collegial relationships—scientists can relate stories of tyrannical research heads, dysfunctional lab groups, and so on—but as has been our approach throughout this book, we are less interested in descriptive issues except as they are revealing of normative ones. We are not asking primarily what is the case, but what ought to be so—what should scientists aspire to and aim for? One long-term research group makes its common goal explicit in its informal motto: have fun by doing good science with friends.

Friendship and fellow-feeling in this sense is relevant to and supportive of a variety of the scientific virtues, but I will use attentiveness as the focal case because it also serves well for explicating the role of emotion in science, which deserves special treatment because of its important role in virtue theory. It will let us more clearly see a distinctive feature of science, which is how the research relationships of scientific friends is structured by their shared passionate curiosity about nature. To ground this discussion, a brief review of Aristotle's account of the role of friendship in virtue will be helpful.

Aristotle on Friendship

Virtue for human beings, according to Aristotle, involves those activities that make for human flourishing, and these are a function of human nature. Because we are rational animals, the highest good for us involves the exercise of our rational faculties—study and contemplation. It is not surprising that a philosopher would emphasize this. Of course, other activities may also come into play because human nature is composite and hierarchical, but for the sake of simplicity let us take the extreme "strict intellectualist" case. There is one passage where Aristotle notes the value of friends even for supreme contemplators. Aristotle argued that the life of study is self-sufficient in a way that many other practices are not, meaning that an ideally wise person can study alone. This would seem to imply that friends are not necessary, but Aristotle went on to note that, though the wise person is able to study in solitude, even such an individual "presumably [studies] better with colleagues."[4] Substitute "research" as the specifically scientific form of study and the point applies directly to science. One can do scientific research alone and make true discoveries, but one's practice is enhanced by having research colleagues. How do friends help? On Aristotle's view, friends enhance our ability to act and our ability to think, which human flourishing requires. These helpful qualities are no less important in science.

Aristotle classifies kinds of friendship along several different axes, the first being their motives. Utility-based friendships arise as a function of reciprocal utility, dissolving once the partnership is no longer useful. Some research collaborations go no further than this, as when a research project requires someone's specialized expertise on just a temporary basis. Pleasure-based friendships arise from emotional connections. One might put many sorts of scientific relationships in this category, based on a shared pleasure of finding things out. However, in scientific friendships that pleasure is in concordance with the relevant scientific virtues, which better places scientific friendships into the category of what Aristotle calls "goodness-based" or "perfect" friendship. He takes this to be the highest and best form of friendship, which holds between good persons who are alike in their virtue. It implies a form of mutual trust and generosity. Aristotle also distinguishes between equal and unequal friendships. In the former, both partners wish and receive the same thing from each other, while in the latter these may differ because the relevant virtues of each partner are different. As examples, Aristotle mentions what he calls "role-based friendships," which combine usefulness, pleasure, and ideally also goodness. In role-based friendships, such as those between parent and offspring, king and subject, or wife and husband, the relevant virtues of each partner are different, depending on their different functional roles. I am inclined to classify scientific friendships as a special form of role-based friendship in which the relevant virtues are usually equal. Research colleagues share in their mutual utility, pleasure, and the unqualified goodness they seek in their practice.

The characteristics of the relationships among ideal scientific research colleagues have many of the same qualities that Aristotle identified as basic to friendship in general, though with a focus on the scientific *telos*. Humans are rational beings but they are also social beings, so our natural condition involves relationships with others. Speaking of the supremely happy man, Aristotle said, "Whatever is the purpose that makes his life desirable, he wishes to pursue it together with his friends."[5] The most basic characteristics of friends are their mutual good will, their wish for the good of the other, and their awareness of the other's good will. In this sense, one may think of a friend as "another self," for they are alike in finding joy in actions that conform to the virtues of their practice. If one has had the good fortune to be a member of a well-functioning research group, one knows how collegial relationships of this sort create an atmosphere of mutual affection that makes science flourish. Aristotle notes that friends spend time together, have the same desires, and share their joys and sorrows. In science, this is the camaraderie of a lab group—colleagues who are united in their research goals and passions.

There are many aspects of the idea of scientific friendship that would be worth exploring, but here I will focus on just one element that links the notion of a research community, considered broadly, to the mindset of individual researchers, and that is important to a virtue-theoretic account of scientific normativity. It stems from the instinct for truth that we have already discussed—the biological drive of organisms to attend to and investigate their environment—which is eventually given more robust cognitive content in what we take to be scientific curiosity. For scientists, a basic impetus for their work is this will to discover. They are, indeed, zealous in this pursuit. This is the motivational side of curiosity. It is to this emotional aspect of the scientific mindset that we now turn.

Darwin on the Social Instincts

Darwin devoted an entire book to emotion and its evolution. *The Expression of the Emotions in Man and Animals*, published in 1872, elaborates on the model of evolution by natural selection that he established in the *Origin* and the *Descent*, applying the same careful reasoning and presentation of evidence now to a variety of mental traits. Even high-order traits like language ability, which is especially relevant to us as social animals, arose out of simpler traits that were different not in kind but in degree. Animals communicate using inarticulate cries and Darwin noted that humans use similar sorts of cries, gestures, and facial expressions when they express simple feelings. This suggests that, for humans, not only is there a cognitive element to emotions but also that there is a concomitant emotional element to higher-level cognition. Darwin's account of the evolution of other higher mental traits follows the same pattern. Significantly, he argues that one may account in this way even for the moral sense.

Darwin did not think that animals have morality in the same sense that humans do, but he argued that the moral sense is an emergent property that was built upon social instincts that came into full bloom once we had evolved to a sufficient level of intelligence. As we briefly explained previously, by the "social instincts" Darwin referred to those instincts that led animals to take "pleasure" in one another's company and to feel "sympathy" with each other. He explained that, for animals that benefited by living in close association with one another, individuals with stronger fellow-feeling of this sort would stand a better chance of escaping various dangers compared to those who cared least for their comrades and lived in isolation. This provides a basis for natural selection to operate on the traits, just as it does for other mental traits, such as memory and imagination, that are relevant for the development of the moral sense.

Darwin's more detailed account of that development continued to highlight emotional aspects of this evolutionary process as he spelled out its hypothesized causal linkages. Animals endowed with the social instincts feel a certain commonality and sympathy with their fellows that leads them to tend to perform various services for each other. These may be definite services, such as mutual grooming, but may become, in the higher social animals, a wish to aid in general ways. Increased intelligence improves memory and imagination, bringing images of past actions and motives constantly to mind, and with them the "dissatisfaction" or even "misery" associated with unsatisfied instincts. Since the social instincts cause the approbation of fellows to be highly valued, and because reason and foresight now allow the weighing of the consequences of possible courses of action, the organism is able to conclude that it ought not be swayed even by strong desire for transitory pleasure or against transitory pain from choosing that action demanded by instinctive sympathy. This feeling of the "ought" grounds the moral imperative. Sympathy is thus the "foundation" not only of the social instincts but also of morality. It is because of instinctive sympathy that we feel another's pleasures and pains as our own and thus are led to feel the moral imperative to try to enhance the former and relieve the latter. With the addition of reason that can then guide and refine our choices and behaviors, it is easy to see how we might eventually express this imperative as a rule that would be phrased in familiar terms; Darwin wrote, with apparently complete confidence,

> the social instincts...with the aid of active intellectual powers and the effects of habit, naturally lead to the golden rule, "As ye would that men should do to you, do ye to them likewise"; and this lies at the foundation of morality.[6]

Darwin went on to try to show how this account was compatible with not only the Golden Rule but also the best ethical theories of his day—those of Immanuel Kant and John Stuart Mill.[7] As we have already seen, an even more convincing case could be made for its relevance for virtue ethics. There is much more that could be said about this, but here I want to simply highlight a few elements of Darwin's account that are especially relevant for our current discussion of scientific virtues.

The first is that both Darwin and Aristotle took for granted that human beings are rational animals. Aristotle took this to be the very definition of what it is to be human; animal is the generic kind of thing we are, but rationality is the trait that distinguishes us from other animals, thereby determining our essence. This makes it central to Aristotle's account of human flourishing. Darwin did not have this sort of analysis in mind, but he too emphasized that rationality was a prerequisite to real moral choice. Philosophers typically focus on rationality, but we should not neglect the fact that we are animals.

Second, Darwin, like Aristotle, took it to be relevant and significant that humans are social animals. The social aspect of morality is obvious given that most ethical decisions involve actions that affect others, but that is not the only way that community comes into play. We saw in the previous chapter how science is systematic curiosity, and Merton's institutional scientific norms are especially important at this level. This is not to say that one could not make scientific discoveries alone, but there are important aspects of scientific practice that could not be done, or at least, not be done as well, without the support of the scientific community in general and one's research colleagues—the vocational friends we spoke of above—in particular. Fellow-feeling is especially important in Darwin's account in the development of the moral sense, because on his account individual moral sense evolves because of group-level behaviors.

Finally, Darwin's account is significant in that it is an evolutionary account of normative feelings generally. As we discussed previously, one may not simply read off ethics from the biology; justification in the realm of reasons is different than explanation in the realm of causes. Nevertheless, as Hume argued, we miss a necessary element of human morality if we ignore the passions. Normativity involves interrelationship of cognitive and emotional elements. With regard to the scientific case, we have seen that curiosity is both an instinct and a virtue, so we should expect that its cognitive elements will also have an emotional basis. And yet, at first glance, this seems to go against the usual view of science and scientists. Even Darwin sometimes said things that seemed to rule out a role for emotion. The issue deserves further investigation.

Romancing the Stone

Heart of Stone?

In his autobiography, Darwin compared himself to a machine "grinding general laws out of large collections of facts."[8] The image of the scientist as an inductive automaton seems intentional; he wrote elsewhere that "a scientific man ought to have no wishes, no affections—a mere heart of stone."[9]

Quotes like this seem to confirm the familiar stereotype of the cold-hearted, emotionless scientist as often portrayed in popular culture. One might even say that this stereotype is not far removed from some of the characteristic virtues of the scientist; do not objectivity and disinterested-ness fit with the view of scientists as being dispassionate and unemotional? Indeed, literary characters—even stock, stereotypical characters—often do function to exhibit the practiced dispositions and recognized norms of professionals, albeit in an exaggerated manner. One of the best examples of a character depiction that puts a positive spin on this combination of traits of the scientist is *Star Trek*'s science officer, Mr. Spock, who regularly extolled the virtues of pure logic and the Vulcan discipline of suppressing emotions. As a half-human, half-alien, Spock was hardly a typical stock character, but in other respects he served the same literary purpose of personifying certain values of the scientific profession, sparring with the starship's physician Dr. McCoy, the character who embodied the value of human emotion.

Spock is an exception in having rationality and emotionless logic portrayed as positive character traits. More often than not, scientists are depicted as emotionally wanting or stunted. Occasionally, one hears the scientist's attitude critiqued as male, controlling, and lustful, lifting the skirts of nature, but this is not representative; the sexually clueless, unfeeling scientist is by far the more prevalent image. The stereotype of the female scientist is consistent with this as well; she is most often portrayed as brainy, but asexual and emotionally cold. Some have even identified this stoneheartedness as a personality disorder that comes with the profession.

Whether or not scientists are more likely to have such a diagnosable disorder is interesting but not especially pertinent to our inquiry, which is not a descriptive thesis about psychology or personality, but rather a normative inquiry into virtue and character. Nevertheless, I want to suggest that both of these stereotypes are mistaken. In a straightforward sense, the descriptive stereotype of the emotionless scientists is, of course, false. As the Nobel Laureate physicist Leon Lederman put it:

> Physicists today feel the same emotions that scientists have felt for centuries. The life of a physicist is filled with anxiety, pain, hardship, tension, attacks of hopelessness, depression, and discouragement. But these are punctuated by flashes of exhilaration, laughter, joy, and exultation. The epiphanies come at unpredictable times. Often they are generated simply by the sudden understanding of something new and important, something beautiful, that someone else has revealed.[10]

Leaving aside the simple descriptive point, I want to argue for a normative place for emotion. Over and above the ordinary human emotions cataloged by Lederman, there are more distinctive sentiments that are attached to, and important for, the characteristic practice of science. Lederman gestured toward this in the last sentence, which relates to the core scientific goal of revealing new discoveries and understandings about the natural world. The proper scientific attitude is not distant and emotionless but rather involves a passionate, even sometimes intimate, desire to see the world for what it is (rather than to impose upon it). One sees this undercurrent of scientific emotion even in the character of Mr. Spock, though it was rarely expressed beyond the raising of a quizzical eyebrow and his characteristically understated expression of curiosity—"Fascinating!"

Science as an Adventure

The idea of science as a fascinating pursuit is far from emotionless. On the contrary, scientific discovery is better articulated as exhilarating. To make a discovery—to see something that no one else has ever seen before—is thrilling. Indeed, scientific exploration itself as a process has all the elements of romantic adventure tale.

Star Trek wonderfully captured this captivating worldview. One must be jaded indeed to not be excited and inspired by the stirring admonition "To boldly go where no one has gone before." *Star Trek* creator Gene Roddenberry pitched the show as a "Wagon Train to the Stars," and the series did draw from a key aspect of the romantic tropes of old Westerns—the idea of a limitless frontier to explore. (Vannevar Bush, who headed the U.S. Office of Scientific Research and Development during World War II, used this metaphor in making his argument for what would become the National Science Foundation, speaking of science as the "endless frontier.")[11] *Star Trek* also drew out an idealistic and optimistic strand in this vision of the future, where the fruits of science included not just cool gadgets but peace, social justice, and a sense of wonder in a beautiful, inviting universe. It should come as no surprise that many scientists cite *Star Trek* as an early stimulus that drew them to science and provided a pervading influence on their outlook.

Darwin's experience as a naturalist on the *Beagle* was a first-rate scientific adventure of just this sort. The five-year mission of *HMS Beagle* was not unlike that of the fictional Starship *Enterprise*: to explore and survey the new world and beyond. Darwin observed flora, fauna, and fossils, making discoveries that would have established his scientific reputation even if he had never gone on to make his grand discovery of the law of evolution by natural selection. Some of his most important initial observations involved the geology of South America. Darwin's training was initially as much as a geologist as a naturalist, though he did not enjoy the boring lectures on the subject by Robert Jameson at the University of Edinburgh and vowed never to read a book on the subject or study it again. However, it was probably Jameson's mode of pedagogy that had been the problem, for Darwin's view completely changed when he transferred to Cambridge University and encountered a more active form of learning.

Field trips with John Henslow, Adam Sedgwick, and other professors, through the English countryside led him to appreciate geological science. He was especially influenced by Charles Lyell's uniformitarianism, a principle of scientific method that sought to explain the geological features of the earth in terms of uniform causal processes rather than catastrophic episodes. Examining the formations of stone that he saw on the *Beagle* voyage, Darwin accrued evidence of the great age of the earth. In one instance, he excitedly wrote about of his observations of land that had risen two to three feet around the Bay of Concepcion following an earthquake. That, together with evidence of previous cases of the same sort, suggested how landforms could change over the decades and centuries. Such observations contributed to the growing body of scientific evidence that revealed for the first time a hitherto unimaginable deep history of the earth. Darwin's first book—*The Voyage of the Beagle*—which described this voyage of exploration and discovery, was read not as an academic treatise but as an exciting, romantic, adventure tale and became a best seller.

Science as a Romance

Adventure and exploration is not the only literary genre that is applicable to science. One could also present science as romantic in a more intimate sense, as in the romance novel. Ask scientists what attracted them to science, and they will often recount stories that sound very much like stories of falling for one's first love. One will sometimes hear of love at first sight or of a fascination that begins with a youthful flirtation and then ignites into a passion. Scientists regularly speak of the beauty of their objects of study; it is not a stretch to see them as attentive lovers, burning with desire to discover Nature. Historically, one finds scientists exhibiting every mood and form of romantic love. Always eager. Often ardent. Occasionally lustful, dominant, or controlling. Sometimes even leading to jealous rivalries.

In his pre-*Origin* study of barnacles, Darwin initially spoke of them as "my beloved Cirripedia,"[12] explaining that it was "heart-breaking" how not even one species had been properly defined.[13] However, by the end of his eight-year-long investigation, he had come to despise them more than "a Sailor on a slow-sailing ship,"[14] as one might swing from love to loathing for someone one came to know much too well. In general, though, Darwin was consistent in his expressed love of his objects of study and of the process of scientific research itself. Darwin's biographers regularly take note of these feelings, such as Janet Browne's representative comment that "his work made him feel alive, helping his mind sing, was the one thing that blotted out his cares."[15] This is nothing if not the language of romantic love.

Wide-Eyed Wonder

While resonating more with the Trekkie side myself, I do not want to advance one genre over the other as a more appropriate way to understand the romance of science. Both capture elements that stand in sharp contrast to the stereotypical picture of scientists as emotionless automatons—a view that ought to be rejected. From both an evolutionary and a virtue-theoretic point of view, it makes sense to recognize the significance of the emotional element in science.

As we have seen, curiosity is the first motion of science. Just as gnawing hunger moves one to search for food, a gnawing question impels the search for an answer. Curiosity leads to observation and attentiveness. Wide-eyed wonder and keen, eager, observation are closely related to it. Moreover, observation is basic to scientific methodology, and attentiveness is part of what makes for exemplary scientific practice. Darwin said that one of his few special virtues as a scientist was his power of observation, and he regularly wrote about this using expressions of pleasure, delight, and love. Attentiveness is, of course, not the only virtue for which there is an emotional component, but it will serve as a good example for us to examine the role of emotion in the scientific character.

Attentiveness

How should we understand and define attentiveness so that it is a useful concept for understanding scientific practice? The first thing to recognize is that attentiveness in its base form is a prerequisite for curiosity. One cannot be curious about something if one is not paying sufficient attention to notice it in the first place. As a character virtue, we may define attentiveness as the dispositional trait of being observant of, and sensitively attuned to, one's external and internal environments and their interrelationships. Attentiveness is a feeling—a mental state—that connects subject and object. This has epistemic import: to the degree that we are paying careful attention (or not), we are more (or less) justified in trusting what we see.

Darwin himself noted its special importance for human beings, writing that "Hardly any faculty is more important for the intellectual progress of man than Attention"[16] Attentiveness, like curiosity, has an instinctual, evolutionary basis. Darwin noted how its strength varied among individuals, and he gave various examples of its selective advantages for many organisms, such as for a cat who watches a mouse hole. Wild predators similarly become absorbed in attending to their prey when hunting, ready to spring forward to catch a meal. Organisms with greater powers of attention would likely do better in the evolutionary competition and not just with regard to getting food. Just as attentiveness is closely related to curiosity, it is similarly useful for learning. Darwin related the case of a monkey trainer he knew, who explained that he was willing to pay double if he was allowed the chance to select more attentive individuals. Inattentive monkeys were easily distracted and were untrainable, but "a monkey which carefully attended to him could always be trained."[17] As with curiosity itself, an attentive interest in patterns, including changes in one's environment, provides a selective value in environments that are variable, as it allows organisms with the ability to learn to be generalists rather than specialists and so adapt to changing conditions.

Of course, our interest here is not just with its general selective advantage, but whether this kind of disposition has a special role in science. Clearly it does. Detecting patterns and regularities in the world is a central, essential part of the practice of the scientist. Thus, attentiveness has a real practical, normative import for scientific practice. Being attentive and observant helps us be better scientists.

The connection of attentiveness to scientific methodology is important for understanding why this should be recognized as a normative virtue. One of the most basic elements of natural philosophy is its focus on observational evidence as the only legitimate basis for drawing conclusions about the natural world. Indeed, this demand for observational evidence, in opposition to appeals to authority, was the rallying, revolutionary call of the scientific revolution. We take this for granted now, but one cannot overemphasize just how revolutionary it was to insist that empirical knowledge claims ought to be restricted to conclusions that could be inferred from observation. Inductive inferences are a topic of their own and understanding their logic has been an important area in philosophy of science. But that is not our issue here. The point about attentiveness is that this disposition is critical, in the sense that careful observation is as important for drawing justified conclusions as the reasoning itself. The computer science adage GIGO ("garbage in, garbage out") applies—it does no good to get the logic right without good observational inputs. Moreover, attentiveness is not an all-or-nothing trait but comes in degrees; being more observant and more attentive can make the difference between good science and excellent science.

British physicist and Nobel Laureate Patrick Blackett admonished students to be ever attentive to the details of their experiments, so as to not miss something that might turn out to be important: "May every young scientist remember and not fail to keep his eyes open for the possibility that an irritating failure of his apparatus to give consistent results may once or twice in a lifetime conceal an important discovery."[18] There are many examples of the importance of attentiveness in the history of science. For instance, it is well known that we owe the discovery of antibiotics to such scientific attention. Alexander Fleming was led to the discovery of penicillin after noticing mold growing on his discarded culture plates and observing that the staphylococcus germs surrounding it had died. This was more than just luck; had he not been paying close attention, the plates and the discovery they contained would have been lost in the trash.

I might also mention an example from the lab of my colleague Richard Lenski, famous for his Long Term Evolution Experiment (LTEE) with *Escherichia coli*. Having started in 1988 with a single strain of bacteria divided into twelve populations in different flasks, Lenski's research group has observed the evolutionary changes over tens of thousands of generations. At one point, in 2003, it was noticed that the liquid medium in one of the flasks was unusually cloudy. What was going on? Could there have been contamination? Rather than just tossing out the flask and retitering the solution, they did further observations, carefully checking to try to determine the cause. This led to discovery of a new kind of *E. coli*—one that was able to metabolize citrate, a previously unused chemical in the growth medium—that had evolved from the original strain.[19] Had they not been paying close attention, they might have overlooked an incipient species that had evolved right in front of them.[20]

The kind of keen attentiveness that makes such discoveries possible is not detached or dispassionate but requires an intimacy with one's study system that involves an emotional bond. Nobel Laureate Barbara McClintock, whose research on the genetics of maize led to the discovery of transposons—so-called jumping genes—explained that she had to develop a close attentiveness and familiarity to be able to truly see a thing for what it was. Her colleague Marcus Rhoades said he would never forget the story of how he once marveled to her about how she could see so much when looking at a cell through a microscope, and she replied "Well, you know, when I look at a cell, I get down in that cell and look around."[21] She called this kind of intimate attentiveness "a feeling for the organism."

What Curiosity Feels Like

We previously investigated some of the elements of this special kind of feeling in our initial discussion of curiosity in chapter 1. It may begin with a feeling of confusion or puzzlement that begs to be resolved, piquing one's interest, and putting one on edge with excitement, perhaps with anticipation of possible danger, or the satisfaction of reward. One's senses are heightened, and the heart beats faster. This is little different from the passionate stirrings of love, whether of high adventure or of romance. The feelings associated with absorbed attentiveness are intimately related. In both cases, individuals in the grip of such a passion gaze with ardent interest in, and fascination with, their object of affection. In both cases, they give rise to similar emotional pleasures.

In his letters, his autobiography, and elsewhere, Darwin regularly expressed the depth of his emotional involvement in his scientific research. He described, for instance, how "my love for science, gradually preponderated over every other taste."[22] It even overshadowed his love of hunting, which had been his dominant pleasure since boyhood. His father had been so frustrated that his son seemed to care for nothing but dogs and shooting that he told him he would be a disgrace to his family and would never amount to anything. Even at university, Darwin admitted that he liked hunting far more than attending classes. Recounting the transformation in his mindset, Darwin explained that on the *Beagle* voyage he initially enjoyed shooting all the birds and animals for his scientific collections himself but that over time "my old passion for shooting" gradually gave way to a new passion of determining the geological structure of the country. He wrote: "I discovered, though unconsciously and insensibly, that the pleasure of observing and reasoning was a much higher one than that of skill and sport. The primeval instincts of the barbarian slowly yielded to the acquired tastes of the civilized man."[23] Darwin's father noted this dramatic change in his son's disposition and, making a phrenology joke, said that "the shape of his head is quite altered."[24]

For Darwin, the "primeval" instinctual pleasure in the attention devoted to hunting for food or sport, and the satisfaction of getting one's prey, gave way to the more "civilized" pleasure of observation itself in the hunt for truth, and for the satisfaction of curiosity. Aristotle would have completely understood this. Human beings, he argued, are best understood as rational animals, and thus contemplation is the most characteristic and proper activity in accord with our *telos*. What Aristotle did not think about, of course, was the way that evolution can help make sense of how we became rational animals in the first place and what that means for our self-understanding. As I have argued, even these "acquired" habits had their roots in instinct.

The Love Life of Scientists

As we saw, there is a reasonable account to be made of how mental traits like curiosity and attentiveness could provide a selective advantage to organisms in the kind of variable environments we evolved in. The salient point for our purpose here is to recognize one important feature of instincts, namely that, rather like hunger, they are felt in the gut as well as the brain. That is to say, they are experienced *emotionally*.

We find this linkage between reason and emotions everywhere in science, such as in the way that Albert Einstein tried to explain the scientist's search for truth: "But the years of searching in the dark for a truth that one feels, but cannot express; the intense desire and the alterations of confidence and misgiving, until one breaks through to clarity and understanding, are only known to him who has himself experienced them."[25] To reiterate, attention is not the only scientific virtue with an emotional component. I focus on it here as an example, especially because of its close connection to curiosity and the striking elements of love one finds associated with it.

Edward Eickstead Lane, a physician friend of Darwin, wrote about their walks, saying, "No object in nature, whether Flower, or Bird, or Insect of any kind, could avoid his loving recognition. He knew about them all…could give you endless information…in a manner so full of point and pith…that you could not but be supremely delighted."[26] Alfred Russell Wallace, who independently discovered the mechanism of evolution by natural selection, attributed their insight in part to the fact that both had been longtime beetle collectors. Even more important may have been how this was but an effect of their passionate delight in their subject matter. Wallace described how he would always take long walks every Sunday through the mountains with his collecting box, recalling that, "At such times I experienced the joy which every discovery of a new form of life gives to the lover of nature."[27]

Love and Curiosity

Is love as a form of curiosity, as Shaw and others have proposed? Or is curiosity a form of love? Perhaps the relationship is symmetrical. Whatever the case, the evidence is suggestive of a deep connection.

Psychologists have studied the physiological signs of love. In the presence of one's object of desire, especially in the early stages of attraction, the pulse rate rises and the pupils of one's eyes widen. With love comes a passionate mindfulness—lovers have a heightened attentiveness to the object of their affection. Whether one thinks of it in terms of erotic or platonic desire, there is a wish to "know," and truly "become one" with one's love.

The curiosity of scientists for their objects of study goes without saying, but this passionate kind of love for their object of study is less recognized, though in many cases its effects are indistinguishable. Scientists get wide-eyed and excited as they describe their research. Sometimes the reaction seems oddly funny—they can have the same response even upon hearing another scientist tell about getting any sort of data. We eagerly describe our evidence, expecting that other scientists will similarly appreciate how fascinating and cool it is to discover some new fact about nature.

A humorist might joke that individuals who fall for science, and develop an intimate relationship with nature, are people who are too introverted, or repressed, or who lack the social skills to relate to other human beings. Even if there is some truth to this, the revised stereotype should make us reassess the old one. A sublimation of desire is still desire. Whether love is a form of curiosity, or vice versa, the emotional connection is vibrant and strong, especially in exemplary scientists. We might again take note of Einstein, who said of himself, "I have no special talents. I am only passionately curious."[28] It is striking how clear an echo this is of Darwin's own way of describing himself. Such statements are clear evidence against the stereotype of scientific objectivity as distant and unemotional. Science is a loving and even passionate practice.

Know Them! Love Them!

Once one is sensitized to this emotional aspect of science, one begins to see and hear it everywhere. In a recent National Public Radio piece about research in the Smoky Mountains, a journalist followed an entomologist as she slogged through stream muck that looked to him like "gunk from your gutter" to examine native bugs. She finds, and excitedly holds up, a mucky insect to observe, saying, "Isn't it pretty!?" The amused reporter noted that that wasn't the word he would have used. One might think that such emotional expressions are anomalous or limited to situations when a scientist is speaking to the press or public. Not so. Even in staid, professional conferences, where there are expectations for efficient presentation of one's data and findings, scientists regularly bubble with giddy expressions of love for their objects of study.

Biologists often organize their research around some particular model organism, and they regularly speak lovingly of their organism—its beauty, elegance, and so on—in their presentations. I recently began noting such examples whenever I attended professional talks and now have a large collection of similar statements from scientists speaking about their study systems. Eminent scientists are as likely, if not more likely, as young scientists to make public pronouncements of this sort. In her presidential address to the Society for the Study of Evolution at one of its annual meetings, evolutionary biologist Kim Hughes spoke about the Trinidad guppies and other poeciliid fishes she studied. "They are not just bags of genes," she insisted (though of course she knew their genetics inside out), "I love these little guys!"

Neither does the object of study need to be an animal to elicit such feelings. In another talk at the same meeting, an evolutionary plant ecologist spoke about how researchers can do better research by picking a plant lineage and learning *everything* about them: their varieties, populations, ecology, genomes. "Know them! Love them!" was the recommendation. The importance of specific case studies was stressed: "This is where we should be focusing our love and attention." Why? Because then "surprising and wonderful things will be found." No biological kingdom is without its ardent admirers; I have heard scientists speak of bacteria as "beautiful" and even "charismatic."

Nor is it only softhearted biologists who are so enamored. Chemists and physicists may not as often have the same sort of attraction to particular molecules or leptons as biologists do about particular organisms (I do not attend those professional conferences, so I lack the base of evidence I have for evolutionary biology), but they express the same feelings at a more general level. Physicist Robert Oppenheimer was representative of that subject: "Science is not everything," he once said, "but science is very beautiful."[29] A student who is not stirred by science in this way or who is embarrassed by such feelings ought to consider a different vocation.

Theoretical Import of Emotion for Excellent Science

One should not think, however, that the emotional account of science is nothing but scientific hormones. Just as we cannot derive ethics from biology, so we must be careful as we move from the biological level to the normative for epistemological purposes. For philosophical tasks, emotions need to be connected to the relevant cognitive content. One such example involves how physicists will often appeal to the beauty or "elegance" of some theory as a reason to think that it is true. Philosophers of science have analyzed this kind of aesthetic factor with regard to what role it might or might not properly play in theory assessment. This is an important and interesting issue, but here our focus is not on beauty as a possible feature of a scientific explanation, but rather on the emotional aspect of scientists' normative practice.

Why does an emotional component to attentiveness and other traits matter in science? The answer takes us briefly back to more general ethical considerations, specifically the issue of ethical motivation. It will have little effect that reason tells one what is the right thing to do, if one lacks the motivational impulse to do it. Virtue theory incorporates the necessary drive (to do the moral thing) that David Hume had recognized as necessary for ethics.

Hume stated the point starkly, noting that "it is not contrary to reason to prefer the destruction of the whole world to the scratching of my finger."[30] Reason can determine logical relationships and figure out what means to use to achieve a given goal, but by itself it cannot fix our goals. It is for this reason that Hume argued that "Reason is, and ought only to be the slave of the passions, and can never pretend to any other office than to serve and obey them."[31] One may put this in a way that meshes with our thesis: without desire, reason is impotent. Rationality alone is not enough. It just gives if-then information, for example, about cause-effect relationships—if I do A, then B will follow. But morality needs passion in addition to rationality so that one then acts to do the right thing.

Hume was writing about ethics in a general sense, but I want to again suggest that the same point holds with regard to normative issues in science, including epistemic norms. If proper scientific conduct is analyzed simply in terms of rules, there may be a disconnect that short-circuits action. Given the goals of science, virtuous actions are more likely to lead to success (including following appropriate methods). If this were no more than a theoretical point, it might not actually lead to the requisite actions. However, dispositions with real emotional content provide motivation to act. As paleontologist Mary Leakey put it, "Basically, I have been *compelled* by curiosity."[32] Nor are the goals of science ungrounded; as we saw, they are rooted in an evolutionary account that provides an objective basis for our core needs and desires.

As a brief aside, I want to point out that one should not think that incorporating the passions reduces morality to subjectivism. Even with regard to aesthetic judgments of beauty, there are standards that the "eye of the beholder" may be trained to recognize. In *Of the Standard of Taste* Hume argued for objective aesthetic standards. Like Aristotle, he argued that our appreciation may be sharpened by learning from ideal critics—the practically wise—who, with thought and experience, have developed a more delicate taste, and an unbiased understanding of the subject.[33] Even in a hard-nosed field like physics, no one blushes when scientists appeal to beauty and elegance, and there is broad agreement in judgment about the application of such standards. There will, of course, still be differences of perspective that may cause conflict, but a virtue-theoretic approach provides a useful framework that may help us better understand differences of judgment and how they might be resolved or avoided.

What I Have Observed of the Pond

Thoreau as Romantic, Literary Scientist

There are many possible ways that one might try to bridge the gap between science and other cultures. For the most part I have been interested in highlighting the view from the tangled scientific bank, but I want now to briefly discuss the view from the side of poetry. To that end, I will focus on a writer whose exquisitely sensitive descriptions of nature often reach toward the attentive sensibilities of the scientist—Henry David Thoreau. We now most often think of Thoreau as a nature writer, and it is fair to say that no writer was more influential in evoking an emotional reverence for the environment as well as a recognition of our place in it. The character of his nature writing was derived in part from his participation in a broader philosophical movement, Transcendentalism, which links to several of the elements we have been considering in this chapter. The Transcendentalist movement arose in Boston in the early nineteenth century with a unique combination of religious, philosophical, and scientific elements. Religiously, it was most associated with Unitarianism, the dominant Christian denomination in Boston at the time, but also drew in spiritual elements from Hinduism. Philosophically it drew from David Hume and Immanuel Kant, especially with regard to their epistemological views. Furthermore, it aimed to incorporate a scientific worldview. Through all of this, it also retained elements of English Romanticism, out of which it had emerged. Because of the way Thoreau attempted to synthesize and express these elements, he provides a useful way to think about the emotional aspect of attentiveness.

Thoreau's published journals and books may be read as an object lesson on the exercise of the power of attentiveness. "The true man of science," he wrote (using, as we noted previously, the gendered phrase that was in common currency before the term scientist was coined), "will have a rare Indian wisdom—and will know nature better by his finer organization. He will smell, taste, see, hear, feel, better than other men. His will be a deeper and finer experience."[34] I am not convinced that this is true of scientists in general, but Thoreau's own descriptions of nature clearly reveal his own highly developed sensitivity. His early attempt to study science was not promising; in his journal, he recalled trying to learn plant names from a botany book but, lacking a systematic method, he quickly forgot them. Persevering in the task, he wrote, "from year to year we look at Nature with new eyes," and he made another attempt at "attending to plants with more method," meeting this time with more success.[35] Thoreau admitted that he never studied botany, as such, but in his two years at Walden he applied himself to the task of learning the plants in the area in order "to know my neighbors, if possible,—to get a little nearer to them." He would take notice of when plants first blossomed or leafed, writing that "I often visited a particular plant four or five miles distant, half a dozen times within a fortnight, that I might know exactly when it opened, beside attending to a great many others in different directions."[36] In time, he learned to become attentive to individual plants and leaves, and his descriptions of Walden's natural environment, local though it was, gave a much broader view than even he recognized. Thoreau was a budding scientist but never a full-blossoming one. He wrote:

> I fear that the character of my knowledge is from year to year becoming more distinct & scientific—That in exchange for views as wide as heaven's cope I am being narrowed down to the field of the microscope—I see details not wholes nor the shadow of the whole. I count some parts, & say "I know."[37]

Thoreau may be said to have leaned toward science and understood some of its aspects, but, not having ever truly studied it, he never completely bridged the gap. In many places, he seemed to retain a nostalgia for a prescientific view, as when he wrote of figuring out the puzzle of the stones that paved the shores of Walden Pond. He recounted a Native American origin legend told by townspeople and another told by an old settler. Then he mentioned the evidence for his own conclusion, noting with some regret that, having detected the paver, the puzzle of the stones "unfortunately ... is no longer a mystery to me."[38]

Beyond knowledge by acquaintance   Science cannot leave mystery alone, but it sees joy rather than regret in its solution. Scientists desire to know nature for what it truly is, so they press on for answers. Thoreau thought that we will learn whatever answers may be found only through direct acquaintance. He wrote, "We do not learn by inference and deduction, and the application of mathematics to philosophy but by direct intercourse. It is with science as with ethics—we cannot know truth by method and contrivance—the Baconian is as false as any other method."[39] Here Thoreau was referring to methods for discovery that had been proposed in the early part of the scientific revolution by Francis Bacon. But Thoreau was only half-right.

We now do recognize that Bacon's early attempts to formulate inductive methods were too primitive. One cannot simply crank out natural laws by a simple automatic method. The grinding and polishing required to confirm a scientific discovery may have a logic to them, but there is no easy logic that will produce an initial insight. Philosophers of science have distinguished what they call the "context of justification" from the "context of discovery" to highlight two linked but importantly different aspects of the scientific process. Whether or not Thoreau was right that there could be no method or contrivance for learning truths, he was certainly right that for human beings our best bet involves close, intimate observation. That, in part, is what the virtue of attention is about—those who are passionately curious and lovingly attentive of nature are more likely to observe something that may lead to a true discovery. Of course, scientists must then go farther than even deep knowledge by acquaintance. They must test their understanding of their loved one to make sure that they are seeing reality, not a fantasy.

Sounding the Bottom of Walden Pond

In fairness to Thoreau, there are cases where he did take that extra step. In one striking example, he described a time he took it upon himself to measure the depth of Walden Pond, which local legends held to be bottomless. He wrote:

> There have been many stories told about the bottom, or rather no bottom, of this pond, which certainly had no foundation for themselves. It is remarkable how long men will believe in the bottomlessness of a pond without taking the trouble to sound it. I have visited two such Bottomless Ponds in one walk in this neighborhood. Many have believed that Walden reached quite through to the other side of the globe. Some who have lain flat on the ice for a long time, looking down through the illusive medium, perchance with watery eyes into the bargain, and driven to hasty conclusions by the fear of catching cold in their breasts, have seen vast holes "into which a load of hay might be driven," if there were anybody to drive it, the undoubted source of the Styx and entrance to the Infernal Regions from these parts.[40]

Thoreau responded scientifically to these mythic stories. Using "compass and chain and sounding line" he took more than a hundred measurements to map the hidden depths of the pond. With dry wit, he assured his readers that "Walden has a reasonably tight bottom at a not unreasonable, though at an unusual, depth. … The greatest depth was exactly one hundred and two feet; to which may be added the five feet which it has risen since, making one hundred and seven."[41] Scientific measurement revealed the answer.

Thoreau went on to point out that although Walden is not bottomless, human imagination will use every inch of depth it does possess for its symbolic value. "I am thankful," he wrote, "that this pond was made deep and pure for a symbol. While men believe in the infinite some ponds will be thought to be bottomless."[42] Such symbolic value is completely compatible with science, which he properly understood to be the search for the truths of the world and, more basically, its natural laws. Indeed, based on his measurements of Walden Pond, Thoreau proposed what he thought could be a lawful relationship between the intersecting point of the width and breadth of the pond, and the location of its deepest point. Impressively, he even went as far as to test it against measurements at another pond.

> If we knew all the laws of Nature, we should need only one fact, or the description of one actual phenomenon, to infer all the particular results at that point. Now we know only a few laws, and our result is vitiated, not, of course, by any confusion or irregularity in Nature, but by our ignorance of essential elements in the calculation. Our notions of law and harmony are commonly confined to those instances which we detect; but the harmony which results from a far greater number of seemingly conflicting, but really concurring, laws, which we have not detected, is still more wonderful.[43]

All of this is of interest both for its own sake and as an illustration of the scientific mindset and methods, but read on and one sees that it also served as Thoreau's setup for an observation about ethics and human character.

> What I have observed of the pond is no less true in ethics. It is the law of average. Such a rule of the two diameters not only guides us toward the sun in the system and the heart in man, but draws lines through the length and breadth of the aggregate of a man's particular daily behaviors and waves of life into his coves and inlets, and where they intersect will be the height or depth of his character. Perhaps we need only to know how his shores trend and his adjacent country or circumstances, to infer his depth and concealed bottom.[44]

This nicely brings us full circle to the task that we have set for ourselves—mapping the hidden depths of the scientific character. We have taken several soundings already, and there will be more to come before we have a handle on its deep normative terrain. However, we are now in a position to draw at least some tentative conclusions about the nature of passionate curiosity and attentiveness, in both their emotional and rational aspects, and how these principles relate to the epistemological revolution that was natural philosophy.

It is a fundamental principle of science that one cannot rely on blind faith. Nor may a scientist have a flash of insight and call it a day. There is a proper emotional, even romantic, aspect to science, but we have to admit that in the end it is a nerdy sort of romanticism. A fuzzy romantic vision or a "watery eye" will not suffice; one needs to put on one's spectacles and always have one's compass, chain, and sounding line at the ready.

An Emotion of the Intellect

Once More, with Feeling

With our new perspective on the proper relevance of emotion and loving attention to the practice of science, we are now also in a better position to reassess Darwin's comment about the need for a scientist to have a heart of stone. The scientist's true love is for the truth itself. Although passionate observation is required in the context of discovery, as one enters the context of justification, one must be wary to not allow infatuation in one's own shiny hypothesis blind one to the real nuggets of truth amid the glinty false. Heated palpitations of excitement in a promising hypothesis must, at least temporarily, be cooled for the proper assessment of the evidence in its favor.

Darwin's scientific virtues were well balanced in this way, and his colleagues noted this with respectful admiration. In John Tyndall's famous 1874 Belfast Address to the British Association for the Advancement of Science, he cited not only Darwin's discovery of evolution but also Darwin's character as exemplars in his clarion call for the revolutionary scientific worldview. He began by highlighting the massive amount of evidence that Darwin had collected in support of evolution and how it readied him to defend it against all challengers, noting that "This largeness of knowledge and readiness of resource render Mr. Darwin the most terrible of antagonists."[45] Even unfair criticisms do not irritate him, said Tyndall, and he "treats every objection with a soberness and thoroughness which even Bishop Butler might be proud to imitate, surrounding each fact with its appropriate detail, placing it in its proper relations, and usually giving it a significance which, as long as it was kept isolated, failed to appear. This is done without a trace of ill-temper."[46]

Darwin, he concluded, "moves over the subject with the passionless strength of a glacier; and the grinding of the rocks is not always without a counterpart in the logical pulverization of the objector."[47] The metaphor of stone grinding is very similar to the one we saw Darwin use in describing his scientific work as grinding general laws from collections of facts. But Tyndall did not stop with the image of the cold, powerful glacier. "All passion has been stilled," he wrote, in the way Darwin presents his evidence and handles his mighty theme; yet there remains shining through all this "an emotion of the intellect incident to the discernment of new truth which often colours and warms the pages of Mr. Darwin."[48]

Darwin's Final Advice to Young Scientists

This "emotion of the intellect," its nature, operation, import, and justification, has been our subject in this chapter. It has its roots in instinct, its expression in curious attention, its import as a drive that moves reason to action, and its justification in connecting ethics and epistemology through the scientific virtues. It is the energy that mines the rocks and that then turns the grinding stone in the quest to discover some new, small nuggets of truth about our natural world.

We give Darwin the last word on the proper balance of emotion, attentiveness, and reason for the practice of science. Indeed, these are among Darwin's final public words, written shortly before his death, in his preface to the English translation of Muller's treatise on the fertilization of flowers. He exhorted the "young and ardent observer" to "observe for himself, giving full play to his imagination, but rigidly checking it by testing each notion experimentally."[49]

5 Scientific Discipline

I have worked as hard and as well as I could, and no man can do more than this.
—Charles Darwin[1]

The longing to behold harmony is the source of the inexhaustible patience and perseverance with which [the physicist] devote[s] himself.
—Albert Einstein[2]

Disciples of Nature

Science as a Vocation

Science should be recognized as a vocation—an occupational calling that requires special dedication for a worthy purpose. Perhaps because the term has become more associated with religious callings, it is somewhat rare today to speak of science as a vocation, but it is worth resuscitating the usage, for it rings with a warmer tone than the somewhat frosty and sterile term "profession." The word "vocation" has its roots in the Latin *vocare*, which means "to call," and the notion of following a call is more appropriate to science than the notion of professing, which also has a religious etymology but stems from the idea of a public declaration or "confession" of beliefs. As we have seen, science aims to follow only the evidence, rejecting appeals to authority, so scientists ought to shy away from being seen as "professors" in that sense. The notion of listening and following where the evidence calls is more fitting. Moreover, the concept of vocation carries the sense of a moral structure that is appropriate and deserves greater emphasis in science. Science is an occupation with a worthy purpose that requires special dedication—the scientist is one who has heeded the call to explore and discover. The passion and ardor that we discussed in the previous chapter are part of the mindset of someone who is called to a vocation and becomes a disciple to its special practices.

Robert Merton did not analyze science as a vocation, but given his influence from Max Weber, it was surely in the back of his mind. Weber's paper "Science as a Vocation," originally delivered as a speech at University of Munich in 1918, began with an overview of external factors about the profession of science, and differences between how it was organized in Germany compared to the United States, but then moved on to his core discussion about "the inward calling for science." Weber spoke of this in emotional, and sometimes quasi-religious terms, suggesting that science required a "passionate devotion" and "enthusiasm." He wrote, "Whoever lacks the capacity to put on blinders, so to speak, and to come up to the idea that the fate of his soul depends upon whether or not he makes the correct conjecture at this passage of this manuscript may as well stay away from science. He will never have what one may call the 'personal experience' of science." Nonscientists may ridicule the scientist's "strange intoxication," he said, but "without this passion, this 'thousands of years must pass before you enter into life and thousands more wait in silence'—according to whether or not you succeed in making this conjecture; without this, you have no calling for science and you should do something else."[3]

Weber's talk highlighted the fact that science had become specialized to the point that it was no longer possible to know much about neighboring areas of research. Science had divided into different disciplines and subdisciplines. Even within these, scientists must direct their focus even more strictly, putting on blinders to extraneous distractions. One's scientific calling must be similarly restricted. Weber wrote: "[I]nwardly, matters stand at a point where the individual can acquire the sure consciousness of achieving something truly perfect in the field of science only in case he is a strict specialist." The only truly definitive and good accomplishment in science any longer was a specialized accomplishment, he said. What this meant was that "Only by strict specialization can the scientific worker become fully conscious, for once and perhaps never again in his lifetime, that he has achieved something that will endure."[4]

What scientists need to reach this state of mind is what we might call *disciplined attentiveness*. It requires the kind of focused devotion to the task of answering some scientific question using specialized tools and methodology, especially mathematics. We have already looked at the virtue of attentiveness, especially its emotional content, but in this chapter, we turn to another aspect of the nature of that trait. I use the term "disciplined" here in two of its senses, the first having to do with the notion of specialization into scientific disciplines, and the second with the mental and sometimes physical discipline that is required for such specialized, technical pursuits. Both have a conceptual root in the notion of being a disciple, someone who trains in the ways of a specialized practice, in the service of some goal. This is the same kind of relation that we have previously seen in the virtue-theoretic account of how the character virtues arise and are justified in relation to the central, guiding purpose, or *telos*, of a practice. Different scientific disciplines specialize in one of the many mansions of the world, but they share a single goal, which is to discover the secret truths about the empirical world; scientists are disciples of Nature.

Einstein on Planck

Albert Einstein spoke of this scientific attitude, also using religious terms, in a speech he gave at a celebration of physicist Max Planck's sixtieth birthday. Beginning with a biblical reference, he noted that in the "temple of science are many mansions."[5] Within these are some scientists who ambitiously seek the expression of their intellectual superiority and others who offer the products of their brains at its altar in the service of some utilitarian purpose, but should an angel drive these individuals out, there would still remain some like Planck, he said. Such individuals seek only to develop the honesty and objectivity needed to ascend "into the silence of high mountains, where the eye ranges freely through the still, pure air and fondly traces out the restful contours apparently built for eternity," and where, through patient and meticulous rigor, they may find a "simplified and intelligible picture of the world." With a reference to the philosopher Gottfried Leibniz, Einstein observed that this is what is required to fulfill their longing, which is to behold nature's "preestablished harmony." It is this lofty goal that provided exemplary scientists like Planck with the "inexhaustible patience and perseverance" required for the difficult scientific task, which "demands the highest possible standard of rigorous precision in the description of relations, such as only the use of mathematical language can give."[6]

In this chapter, we will explore this notion of scientific discipline in the dual sense of the term, focusing on the virtues of meticulousness, patience, and perseverance. These are not the only traits that are involved in the disciplined practice of science, but they are among the ones most commonly cited by exemplary scientists. Our goal will be to see how the notion of a disciplined practice connects the somewhat abstract notion of character virtues with practical action through the development of systematic scientific habits.

Curiosity and Systematic Habits

My Habits Are Methodical

In Darwin's published account of his voyage on the *Beagle*, he mentioned the question of the relationship of species only in passing, writing that there was "not space in this book to enquire into this curious topic."[7] What he did write about was his adventure as a naturalist on that voyage of exploration and discovery. In his diary, he noted that "it is a new & pleasant thing for me to be conscious that naturalizing is doing my duty, & that if I neglected that duty I should at the same time neglect what has for some years given me so much pleasure."[8] The emotional satisfaction that we discussed in the previous chapter that a scientist feels in this kind of activity is displayed clearly in such passages. Darwin was also very cognizant of how his experience shaped his scientific character and developed in him the necessary habits. "The voyage of the *Beagle*," he wrote "has been by far the most important event in my life, and has determined my whole career." At least as much as the collections he made on the voyage, he credited the journey as having engendered in him the "habit of energetic industry and of concentrated attention to whatever I was engaged in.... Everything about which I thought or read was made to bear directly on what I had seen or was likely to see; and this habit of mind was continued during the five years of the voyage. I feel sure that it was this training which has enabled me to do whatever I have done in Science."[9]

Darwin had made reference to these traits he had developed when he was asked, in 1848, to help write the Admiralty Manual's instructions on scientific field work for sailors. In the mid-nineteenth century, expectations for the modern British officer were high, including knowledge of matters mathematical, vegetable, animal, and mineral. Referring to matters mineral, Darwin's chapter explained that geologizing required "slight preparation and little apparatus, in fact only the sort of curiosity and systematic habits that the officers of a man-of-war were expected to possess."[10] By learning to meticulously label and describe specimens, by increasing the clarity and exactness of one's power of observation, and "by acquiring the habit of patiently seeking the cause which meets his eye" one may in short order "infallibly become a good geologist" and certainly come to "enjoy the high satisfaction of contributing to the perfection of the history of this wonderful world."[11]

Darwin was probably giving short shrift to the degree of preparation required, as he had not found his formal education to be very useful. He was surely underestimating the preparation he had received informally as a student at Cambridge, with regular field trips and meetings with some of the greatest geologists of his time. Leaving this aside, we see that, in Darwin's considered opinion, it truly was a set of habits of the mind that stood out for him as the most valuable assets a scientist could possess. He called out traits that we have already seen as important for the exemplary scientist. Although we would not expect these to be as defining a set of virtues for an officer as it is for a scientist (probably the marshal virtues would be), Darwin seemed to think that the Admiralty did value them and that the overlap would suffice. Our current task, however, is to look more closely at the notion of systematic habits. Darwin's reference to these traits in "persons embarked on a naval expedition"[12] provides us with a nice point of contact for thinking about the role that habituation plays within the virtue-theoretic framework that we have been adapting from Aristotle.

Aristotle on Excellence and Habits

As we have noted, because he was interested in human flourishing, Aristotle's inquiry into virtue focused on what this means for us as human beings. However, in trying to explain what he meant by the concept, he occasionally resorted to examples of virtues in people in different occupational roles, including those aboard ship. The task of the ship's pilot is different from that of a rower or a lookout, and each occupation requires its own special virtue in the service of the general task of preserving the ship, which the ship's captain must oversee. The ship captain's virtues are the qualities needed to accomplish that general task with excellence. It is because of these practiced virtues, as we have called them before, that a sea captain is ready to guide and protect the ship and its crew, making appropriate decisions, and issuing needed commands.

Aristotle asked us to consider the case of the captain of a trading ship in a storm who, in order to save the lives of his crew, chooses to throw his cargo overboard.[13] We are not interested here in the question Aristotle used the example for, which involved whether that choice and action was voluntary or involuntary, but rather on the nature of the decision. Aristotle's point was that the captain cannot simply follow some simple moral rule (e.g., whenever lives are threatened by a storm, ditch the cargo!), and that ethical choices involve judgment—practiced judgment, we might say, or practical wisdom, as he put it...judgment that has been gained from experience. In the case of a sea captain, that practical wisdom requires having absorbed the relevant knowledge of the ship and crew, learned the needed skills of navigation and seafaring, all repeated until they become habitual, so that one embodies the virtues requisite to the practice. While not all skills that are useful in a practice would be viewed as virtuous in a general moral sense, they do properly constitute excellence within the more limited scope of a vocation. Excellence is earned by training and habituation; doing a practice well requires practice.

This is an important point, for it illustrates how a person is not born virtuous in the practiced sense that is required for action, even though virtues may have their beginning in instinctual traits. Action flows in part from heritable traits, but character must then be developed. One develops one's sense of justice by performing just acts. One develops self-control by performing acts of self-control. The process by which one acquires or earns excellence in a practice is subtle and complex. Virtue, recall, involves dispositions—powers that enable action—and Aristotle was clear that dispositions are developed by habituation: "[A]ctions determine what kind of characteristics are developed."[14]

Dewey on Habits of Mind

To expand on this idea, it is worth turning briefly to the work of John Dewey, who was important for his pioneering work both in the philosophy of Pragmatism and in functional psychology, as well as in their application for educational reform. Dewey was influenced by Aristotle, and the role of habit played a significant role in his account of moral behavior. Aristotle explained the role of habit in the development and expression of virtue, characterizing habit as the repetition of similar activities. Dewey took a similar approach but gave a more detailed analysis of habit as "an acquired predisposition to ways or modes of response." Habits are like tools in that they provide the means for the accomplishment of intended ends. In this way, he argued for a use of the term that is broader than the way it was used in psychology (to refer to repetition). Moreover, Dewey insisted, habits should not be thought of as passive, but as "energetic" and even "dominating," to the degree that they "rule our thoughts." This is obvious with regard to bad habits, but he argued that it holds generally. All habits, he said, "are demands for certain kinds of activity; and they constitute the self. In any intelligible sense of the word will, they *are* will. They form our effective desires and they furnish us with our working capacities."[15]

In acquiring a set of habits, we are shaping the desires that lead us to action, as well as the kinds of actions that we are capable of performing reliably. It is because these together are so fundamental to who we are as individuals, and because he equated this to our will, that Dewey argued that habits "constitute the self." On Dewey's view, which expanded on Aristotle's, habits are key for understanding how virtuous dispositions are formed and operate. He argued that it was "superstitious" to think that "by feeling strongly enough about something, by wishing hard enough, we can get a desirable result, such as virtuous execution of a good resolve, or peace among nations, or good will in industry. We slur over the necessity of the cooperative action of objective conditions, and the fact that this cooperation is assured only by persistent and close study."[16] We have already discussed the important role of community cooperation and collegial fellow-feeling in science, but Dewey was making a different point here, namely, that a similar kind of relationship holds with regard to individual action. It is not enough to simply want to have good posture, as that leaves out the required means to that end, which is to say that "it leaves out the importance of the intelligently controlled habit."[17]

There is a lot more in these short passages that one would need to unpack to understand Dewey's complex account of habit, but that is not our purpose, and we do not need to adopt the full Deweyan view. Here, I want to highlight an aspect of Dewey's account that is relevant to training to be observant and attentive. Dewey argued that training and habit is involved even with regard to having clear-cut sensory perceptions. He wrote:

> To be able to single out a definitive sensory element in any field is evidence of a high degree of previous training, that is of well-formed habits. A moderate amount of observation of a child will suffice to reveal that even such gross

discriminations as black, white, red, green, are the result of some years of active dealings with things in the course of which habits have been set up. It is not such a simple matter to have a clear-cut sensation. The latter is a sign of training, skill, habit.[18]

This is an important claim, but we must be careful not to interpret Dewey as saying that sensory perception is nothing but a matter of social training. We are equipped by evolution with basic sensory capacities that have already been well adapted to our environment. With regard to visual perception, to take Dewey's example, there is a physiological basis for fundamental aspects of our sense, explaining why we are limited to the so-called visible spectrum and also why certain color categories seem to be universal.[19] However, biology is not the end of the story, as boundaries between color categories may differ as a function of learning, with different languages and cultures drawing them differently. We now know much more about this than Dewey did, but it strengthens aspects of his view, namely, that experience and training can measurably affect how one sees the world.

Dewey is an important figure for our discussion of habit because he drew not only from Aristotle but also from Darwin, as seen in his comment about the operation of bodily and moral habits being built upon instincts. He noted the original basis in instincts, in passing, in his discussion of learning to distinguish sensory input: "Our ideas truly depend upon experience, but so do our sensations. And the experience upon which they both depend is the operation of habits—originally of instincts. Thus, our purposes and commands regarding action (whether physical or moral) come to us through the refracting medium of bodily and moral habits."[20]

Dewey takes the evolutionary view for granted here, providing little in the way of analysis regarding how the process of developing habits works or how it serves to link biology, character, and action. To fill the gap, let us to return to Darwin.

Darwin on Instinct vs. Habit

It will be helpful to connect this discussion of habits to our earlier discussion about curiosity and our evolutionary account of its origin and development. We focused there on the idea that curiosity is instinctual, an evolved trait with adaptive value in particular kinds of environments, but that was only the first part of the story. Darwin distinguished instinct from habit, emphasizing how the latter built on the former. Instinct is genetic and shaped by evolution. Habits, on the other hand, are acquired within the lifetime of an individual. In later editions of the *Origin of Species*, Darwin began to stray from a strict selectionist view and allowed that some acquired habits might be inherited, but we will ignore that sort of Lamarckian view in favor of a more plausible one. What does make sense for our purposes is the notion of cultural transmission of habits; acquired traits may be passed on through imitation and training. Becoming a scientist starts with instinctual curiosity, but that must be developed through the cultivation of systematic habits.

We may get a sense of how this works using the scientific virtue that was the focus of the last chapter, namely, attentiveness. In the basic sense, we are born with a biological capacity to attend to the world. Even for the simplest of organisms, it is a selective advantage to be able to sense features of their environment, for many of these provide information that is useful, for example, for finding food or avoiding dangers. Evolution provides organisms with an instinctual capacity to attend to the world in at least a basic sense. For a microbe, this may be limited to minimal detection of a chemical gradient. In more complex environments, it can be advantageous to attend to one signal over another, making it useful to be able to discriminate. Moreover, if relevant aspects of the world may change at time scales that do not allow for hardwired, instinctual action to suffice, then the advantage turns to organisms that evolve the ability to learn. Learning involves being able to differentially change one's behavior on the basis of information acquired within one's lifetime. All of this is part of the account of the evolution of instinctual curiosity that we discussed previously. Here, however, we are focusing on how such instinctual capacities are trained by experience, beginning early in life, to function in the environments in which organisms find themselves.

It is important to recognize the complexity of this relationship, for habituation seems to reduce attention, at least conscious attention. As one is learning a new skill, like driving a car, one must attend consciously to every movement. After learning is complete and has become habitual, one drives as though unconscious—one can make a long commute and hardly remember it. To cultivate attentiveness itself as a habit, therefore, requires a special sort of training. A scientist must establish consistent procedures but cannot allow them to become completely routine. Researchers must remain attentive to possible anomalies or confounding factors. This is sufficiently difficult for our limited brains that scientists must develop other traits to increase the precision and reliability of their observations. This is part of the reason why Darwin was right to say that scientific habits need to be systematic.

This is especially so for the kind of habits that are relevant to scientific practice. We have now seen, in brief, how this works with regard to attentiveness; because science is based on observational evidence, being observant and attentive is a critical virtue for scientists. The point here is that to achieve excellence in the virtue as it is required for science, one must hone it so that it becomes habitual, systematic, and methodologically reliable.

The Well-Tempered Scientist

Meticulousness

There are many aspects to the development of a well-tempered scientist, but let us focus here on those that Darwin had in mind—the systematic habits that are conducive to answering scientific questions about the empirical world. Thinking of scientists as instruments to serve this purpose makes it easy to generate a list of characteristics that one ought to try to make habitual.

When Alfred Russel Wallace was asked why it was he and Darwin who discovered evolution, he speculated that it may have had something to do with the fact that they were both avid collectors. As boys, they both competed with friends to have the most complete beetle collection, and they continued their interest in collecting into adulthood, developing their hobby into a vocation. I was privileged to be shown some of the carefully labeled beetle specimens collected by both Wallace and Darwin that are among the treasures in the back rooms of the Natural History Museum in London. One needs a more highly trained eye than mine to be able to identify the fine differences among the hundreds of thousands of species of beetles that inhabit the earth, and both these scientists had clearly developed the requisite skills to a high degree. Obviously, beetle collecting was not a sufficient condition for discovering evolution, but one can readily see how the sorts of precise, systematic organizational habits one develops to be a top-notch collector would be useful for any scientific work. In his autobiography, Darwin wrote that "my habits are methodical"[21] and credited this trait as one of the reasons for his success. As one example, he described the systematic, meticulous way in which he organized all his research materials.

I keep from thirty to forty large portfolios, in cabinets with labeled shelves, into which I can at once put a detached reference or memorandum. I have bought many books, and at their ends I make an index of all the facts that concern my work; or, if the book is not my own, write out a separate abstract, and of such abstracts I have a large drawer full. Before beginning on any subject I look to all the short indexes and make a general and classified index, and by taking the one or more proper portfolios I have all the information collected during my life ready for use.[22]

Discovering and documenting patterns in the world, especially patterns no one has previously detected, requires that one be organized in a way that is both thorough and detailed. Darwin's systematic habits for data collection and organization reflected that mindset, and they served him well.

Being detail-oriented, organized, and systematically thorough are all ways of describing this kind of trait, but for our purposes it is reasonable to refer to these generally as meticulousness. This is not to say that the terms are synonymous; a meticulous researcher would certainly want to point out nuances among different forms of meticulousness. But perhaps that makes the point; the generic term will serve in that it carries the sense of being extremely careful, precise, and attentive with regard to technical details and thereby highlights the scientific aspects of the closely related concepts. Moreover, as we found in our survey, this is the term that scientists themselves most use when discussing this constellation of values.

Obsessive-compulsive? Being meticulous in this broad sense is a useful trait in all areas of science, but since we adumbrated it with regard to Wallace's comment about beetle collecting, it is worth dealing with an issue that someone might bring up regarding this trait. Collectors can be an odd bunch, it must be admitted, and collecting can sometimes edge into obsession. One might wonder whether the meticulousness of scientists is related to having obsessive tendencies. The root of the word comes from the Latin word *metus*, which means fear, and the term evolved to mean being overcautious about details. The word's current sense of being attentive to detail, careful, and precise is generally positive, but there are still echoes of the earlier meaning in synonyms like painstaking, fussy, and finicky. Habits taken to an extreme sometimes become compulsions, and scientists are indeed often thought of as exhibiting "oddball" behaviors and abilities. One need not think of idiot savants who can perform complex mental calculations in their heads, but there is some evidence that scientists tend toward Asperger's syndrome or high-functioning autism.[23] In a scientific field that demands that precise, and sometimes repetitive, procedures be followed, being more extreme in this kind of way may well be an advantage. For example, for many kinds of research, especially in a laboratory setting, cleanliness is critical. This is not to say that compulsive handwashing is in order, but habitual cleanliness is so basic that it is usually taught on the first day of every freshman science lab course. No food or drink! Tie back your hair! Wash your hands! More specialized habits are practiced in chemistry and biology research labs—neatly choreographed and meticulously executed routines for mixing chemicals or streaking Petri dishes—to avoid contamination and improve accuracy. One scientist colleague of mine explained that she had been diagnosed as somewhat obsessive-compulsive but that she did not mind as it helped her do her research better and get more done.

Instrumental reliability   There are many aspects of scientific methodology that are relevant to the issue of scientific reliability. Philosophers of science most often investigate abstract methodological issues like the nature of explanation and confirmation, but it will be useful to look first at an aspect of the material methodology of science. As a simple, initial example, let us focus on one feature as a useful point of reference, namely, scientific instrumentation.

Many critical advances in the history of science, including ones that launched a discipline or subdiscipline, depended on the development of some new instrument that extended researchers' powers of observation. The telescope that Galileo crafted opened the heavens to our eyes in ways that previously had been impossible; it became the basis of modern astronomy. Similarly, microscopes, especially through their development and use by Hooke and van Leeuwenhoek, revealed the previously invisible world of microscopic organisms, becoming the basis of microbiology. These instruments became iconic not just of their disciplines but also of science itself. Today, specialized technical instruments like particle accelerators, PCR (polymerase chain reaction) machines, and so on, are equally important in science. When instruments malfunction, scientific progress is retarded, so considerable expertise and expense goes into making them useful and keeping them working.

Specialized instruments require specialized knowledge to build and keep them reliable, but there are some general concepts that pertain to reliability for all kinds of scientific instruments. First of all, instruments must be calibrated—they must be set to give the same reading in the same conditions as other instruments, especially in comparison to some standard. Even an instrument as simple as a balance scale must be calibrated against a common standard weight if it is to be useful for taking measurements that scientists can rely on. Furthermore, any variability in the instrument's accuracy must stay within some acceptable degree of tolerance; it must be good enough for the purpose to which it is to be put. Moreover, it must be sufficiently robust to stay within the acceptable degree of tolerance for the requisite period of time that it will be used—the calibration interval—before it has to be recalibrated. Scientists also need instruments to be as free as possible of artifacts. Telescopes can suffer from optical distortions that observers must try to correct. Chromatic aberration, for example, is caused by the fact that a lens refracts different wavelengths of light slightly differently, so colors do not all focus at the same point. Creating lenses with elements made of different materials, with differing dispersions, is one of several techniques developed over the centuries to correct this kind of optical artifact. One could easily expand this list of factors that contribute to the reliability of instruments.

What is often forgotten in thinking about the reliability of science is that, as we noted earlier, scientists themselves are essential instruments in this practice, and we may properly think of this in an instrumentalist fashion.[24] The success of any experiment depends as much on the characteristics of the researchers as the equipment they are using. Scientists often require training that is as specialized as the instruments themselves to be able to use them. However, this is only the simplest way in which scientists themselves are critical components in experiment research. An experiment could as easily fail because the investigator was an inattentive observer as because of a computer glitch. Having top-of-the-line DNA sequencing equipment will be wasted if researchers fail to habitually wash their hands properly. These are simple examples, but the point is general: to have a better chance of achieving our scientific goals, we need to try to develop the traits that are conducive to excellent practice. Many of the relevant traits for excellent scientists will be quite analogous to the sorts of factors mentioned above. The "chromatic aberration" of a single scientist may be corrected by systematically combining multiple, diverse observers, so biases may be reduced. This, as we saw in chapter 3, is one function of the scientific community. Similarly, there ought to be factors that improve the reliability of individual researchers. What the virtue-theoretic account we have been developing does is provide a framework for articulating the sorts of characteristics that make for more reliable scientists. To fulfill the general scientific *telos*, scientists should develop the mental and physical habits that will increase their reliability, and these must be in balance. As an instrument must be properly tuned, so must the scientist be well tempered.

Meticulous Methods

My colleague's insightful observation about her own behavior highlights the feature that our virtue-theoretic analysis provides for what makes habits virtuous for a given practice, namely, the ways they help the practitioner achieve excellence so they may flourish in their vocation. Once one is sensitized to this perspective it becomes easy to see how meticulousness has critical connections to fundamental features of scientific methodology and practice. Being careful, precise, and attentive to details are all traits that improve not only the sorts of observations and activities that constitute beetle collecting and the like, but are equally important for the experimentalist. In both cases, they help researchers avoid the scientific equivalent of pilot error. Carelessness, sloppiness, or imprecision on the part of the scientist could ruin results of what otherwise was a well-designed study. In such cases, the instrument that failed was the scientist. This is not just about avoiding error; meticulous habits are part of what allows one to say that a scientist is getting reliable data. In this sense, meticulousness is conducive to reproducibility, which is another methodological touchstone for science. It is also relevant to responsible research practice, which requires not only scrupulous care in data collection and management but also systematic maintenance of records. In these and other ways, being meticulous makes one a better scientist for aspects of practice that are essential for scientific methodology. There is another critical sense in which meticulousness has a special role in science that is worth emphasizing, which involves science's preference for quantitative reasoning.

When you can measure   The special value given in science to quantitative data is well known. The nineteenth-century physicist William Thomson, better known as Lord Kelvin, famously expressed this value in an oft-quoted speech: "I often say that when you can measure what you are speaking about, and express it in numbers, you know something about it; but when you cannot measure it, when you cannot express it in numbers, your knowledge is of a meagre and unsatisfactory kind."[25] Robert Merton wrote a paper in which he and two other colleagues traced various misquotations of Kelvin's Dictum, as this is now often called, eventually finding the original source.[26] As Merton pointed out, simple phrases, slogans, and dicta can be very powerful and influential if they are able to encapsulate "programs of research or action." Misquoted forms of Kelvin's Dictum—e.g., "If you cannot measure, your knowledge is meager and unsatisfactory"—are probably more common because they are shorter and punchier, which hammers in the point more forcefully. Merton's phrase about programs of research or action was left unanalyzed, but it is apt, for it fits with our point that traits are virtuous for a vocation because of the roles they play in providing the structure (here for scientific research in general) that guides action.

Of course, pithy sayings can also be misleading in that they often oversimplify for effect. It would be wrong to claim that one cannot have scientific knowledge if it is not quantitative, as qualitative research may also be meticulous and systematic, and fully satisfy our curiosity. However, while we may agree that Kelvin's Dictum's ought not rule that out, we may still acknowledge that it expresses a characteristic scientific aspiration to ever higher standards of systematic precision, which is what quantitative reasoning makes plain. Scientific conclusions that may be drawn to the nth decimal place provide a measurable standard for meticulousness. Quantification makes it clearer when biases occur and helps scientists determine when they are getting better at reducing them. It makes it less likely that meaningful data will be lost in translation. And it reveals patterns and relationships in phenomena that may not be visible if observations are made at less precise degrees of accuracy. For all sorts of instruments, including scientists themselves, it improves the process of calibration and thereby improves reproducibility, and in this way supports the basic scientific requirement that conclusions be drawn on the basis of intersubjectively testable evidence.

There are any number of scientists one could cite who exemplify this trait, but there is no better exemplar than the eighteenth-century chemist Antoine-Laurent de Lavoisier. He is perhaps most remembered for discovering the role of oxygen in combustion, and for his scientific battle with Joseph Priestley over the latter's phlogiston theory, but even more significant than his specific findings was his pioneering role in making chemistry an exact science. It was Lavoisier who established the use of the chemical balance as a precise and essential instrument in chemistry, and his greatest accomplishment was turning chemistry from a qualitative to a quantitative science. It was his systematic work that established that matter is conserved through any chemical transformation—the law of the conservation of mass—which is emblematic of this kind of meticulous, systematic research. Lavoisier could spend months creating or adapting instruments that could isolate the reactions he wanted to study, and be sufficiently accurate to get the precision needed, and then months more conducting experiments to get sufficient data to draw a conclusion.

We might briefly reiterate that being well tempered as a scientist does not necessarily translate to other vocations. Some of the qualities and systematic habits that made Lavoisier exemplary as a scientist did not serve him well in his work as part of the *Ferme générale*, which collected taxes for the royal government. Lavoisier's innovation for the latter, not unlike his scientific instruments, was to commission a wall to surround Paris to control the transportation of goods into the city and thereby better collect customs taxes without loss. This did not endear him to radicals of the French Revolution and led to his execution by guillotine, proving the Darwinian point that fitness is always relative to an environment—traits that are highly adaptive in one setting may be maladaptive in another. However, some behavioral traits are transferable and useful in various circumstances, which brings us to another set of traits that Lavoisier had that are essential for the disciplined practice of science, namely, perseverance and patience.

It's Dogged as Does It

Darwin's Chief Virtue

Lavoisier, and any number of other scientists, could also serve as exemplars of the kind of perseverance one needs in science. Marie Curie is often cited as iconic in this regard; processing a ton of pitchblende over the course of a year to isolate one-tenth of a gram of radium chloride, required mental and physical determination of the highest order. However, for my money, Darwin is the epitome of the virtue, if only on the strength of the effort he put into his monumental work on the Cirripedia. The idea of examining barnacle after barnacle after barnacle under a microscope for eight years is exhausting even to think about. Darwin's house was filled with boxes of specimens, as his goal was to systematize both living and fossil species. The degree to which the project dominated his life and household during this period is reflected in the oft-told story about the time his young son visited a friend's house and asked, "But where does your father do his barnacles?" Darwin applied this kind of persistent attention and thought in all areas of science that he studied. Regarding geology, he wrote "a man must for years examine for himself great piles of superimposed strata, and watch the sea at work grinding down old rocks" in order to understand the "incomprehensibly vast" ages of time.[27]

Darwin reflected in his autobiography about what character traits made him a good scientist. After his love of science, he highlighted two related traits: "Unbounded patience in long reflecting over any subject" and "industry in observing and collecting facts."[28] Indeed, his desire to understand and explain what he observed gave him, he said, "the patience to reflect or ponder for any number of years over any unexplained problem."[29] Darwin was plagued by ill health through most of his life, so this kind of unflagging perseverance was especially critical for continuing his research in spite of his disability. He mentioned these traits again in a letter quoted years later by his son Francis, in which he apologetically explained why his poor health prevented his accepting an invitation to contribute an article to a journal: "My health is very weak: I never pass 24 hours without many hours of discomfort, when I can do nothing whatever. I have thus, also, lost two whole consecutive months this season.... At no time am I a quick thinker or writer: whatever I have done in science has solely been by long pondering, patience and industry."[30] Historian of science Janet Browne, one of Darwin's most insightful biographers, highlighted the importance of this aspect of his character:

> On another level, however, he may have sensed that the quality of perseverance was indeed the single most important feature of his personality, or at least the feature that had contributed most to his success in science.... He identified in himself the ability to persist, and retrospectively turned it into an all-consuming methodology and justification—that was *his* way of tackling questions. Whereas Huxley or Wallace might have relied on flashes of brilliance, or Hooker on his prodigious memory, it seems possible that Darwin may genuinely have regarded

his special talent for science to lie in mental determination. He had given his life to the ceaseless documentation of facts, pushing and pulling at species until everything fell into place. From time to time, he used the expression "It's dogged as does it," drawing on the phrase used by Anthony Trollope in *The Last Chronicle of Barset*.[31]

Persistent Terminological Variants

Browne's passage, with its suite of related terms, highlights a similar pattern as we saw in the previous section, where we noted a variety of traits related to meticulousness. Darwin most often spoke of patience and industry, but these are part of a constellation of related terms that name the kind of mental, and sometimes physical, determination that science demands. Persistence is a common term, but there is also tenacity and, more colloquially, stick-to-itiveness. Grit has recently become a popular term to express this sort of trait, and one also hears of the need for diligence and drive.[32] Philosophers, to whom one can always sell a distinction, would ordinarily tend to emphasize the differences between these concepts, but congruence is more significant for our current purposes, so we do not need to insist upon finding a single term.

In our survey of exemplary scientists, the terms perseverance and patience rose to the top. One could certainly quibble about their differences; perseverance seems to be a more active term, with the sense of carrying on in the face of difficulties or opposition, whereas patience is more often seen as passive, in the sense of waiting, perhaps while nothing is happening, until something finally does occur. Perhaps for this reason, people who are inclined to segregate virtues by gender suggest that perseverance is a more masculine virtue and patience more a feminine one. I see no good reason to make that sort of distinction. A few scientists in our survey did see perseverance and patience as very different concepts, but most seemed to regard them as simply different aspects of the same important trait, at least as they function in science.

Perseverance and Habits

There are many ways in which perseverance can be valuable in scientific practice, and many stories about different ways it is expressed in different fields, or even in different aspects of a research project. For example, it took Neil Shubin six years to secure funding and organize a research expedition that finally got him and his collaborators to the location in northern Canada that he had predicted would be a likely spot to find a Devonian-era transitional fossil linking fish and amphibians. It took almost another three years of searching before they found what turned out to be the first Tiktaalik fossil on Ellesmere Island. Then it took four months of careful work to start to expose the fossil in the lab and several more months of meticulous effort to fully prepare it for full analysis. Every phase of this investigation and discovery required patience and perseverance.

It is worth highlighting a connection of this to our discussion about habit. The ability to do something persistently is what habits are all about. Developing habits is a process that aims at persistent repetition. As Aristotle defined them, habits are repeating behaviors. Dewey improperly rejected this because he wanted habits to be seen simply as dispositions—meaning that under circumstance C a person is disposed to some behavior B, even if that disposition never has the chance to be repeated. However, while it is true that habits are dispositions, not all dispositions are habits. For instance, there may be physical or biological dispositions that were not formed by habit. Aristotle had it right, recognizing that habits must be developed by training. Habits are formed by imitation and repetition. It is useful to think of this as a cultural version of the Darwinian mechanism—as a form of copying. Put another way, habits are developed by replicating or reproducing some behavioral variant until it become a part of one's being. Through training and practice, a disposition forms that allows the behavior to become persistent and reliable. Thus is character developed.

As we previously discussed, everyone is born with instinctual curiosity. That predisposes one to at least a prototypical form of empirical investigation that is sufficient to learn patterns that help one survive in a variable world. However, this is not yet science in any robust sense. One must develop, as Darwin put it, the requisite systematic habits. This is what scientific discipline involves and what scientific training aims to cultivate. Anyone can get better through repetition and practice. This is not to say that everyone will excel equally. Some individuals may be born with stronger predispositions of one sort rather than others, and some may choose to work on developing certain habits rather than others. This point holds for any of the virtues important for scientific practice, but here we may stick with our current focus on patience and perseverance. It is easy to see these traits embodied in exemplary scientists in the service of their research.

The primatologist Jane Goodall is a wonderful example of someone with an exceptionally high degree of these qualities. She exhibited them at an early age, hiding quietly in the chicken coop for hours so that she could observe a hen lay an egg. Doing her initial observational work in Tanzania, she spent months of tracking chimps before they got used to her presence sufficiently to allow her to get good data. She spent decades in the field conducting her pioneering work, making discoveries that would not have been possible without this kind of perseverance.

Science is hard, and good science cannot be hurried. The idea of eureka moments is not entirely a myth, for scientific insights do occasionally occur like the turning on of the proverbial lightbulb. What is usually missing from tales of such cases, however, is that the insight typically came only after a long and arduous journey of discovery. We remember the discovery but often forget what it took to get there—an expedition filled with setbacks and failures through uncharted terrain that required unyielding determination and effort. This is true in any scientific field. Computer scientist Cynthia Dwork tells the story of meeting Jack Edmonds, a pioneer in the field and winner of the John Von Neumann Theory Prize, when he visited her department. She had finished her graduate school course requirements by that time and was beginning to wonder whether she might go on to do research. Perhaps he had some advice. "How did your greatest results happen? Did they just come to you?" she said she asked him. "He looked at me, and stared at me, and yelled, 'By the sweat of my brow!'"[33]

Objects of Perseverance

Having laid out the case for the importance of perseverance in science, we should acknowledge an obvious and reasonable objection, which is the converse of what was mentioned in the previous section. Patience and persistence are, of course, useful traits for the scientist but so are they useful for most any practice. Indeed, does not the idea of a practice imply that one needs to practice it to acquire the necessary knowledge and skills, and is not perseverance thus a virtue universally? Perseverance is no less important for the musician working through scales and etudes, for the quarterback struggling to bring a team together, or for the chef trying to master the subtle art of the omelet. Meticulousness is similarly useful for many practices, though perhaps not quite so broadly; its relationship with measurement and quantification does give it a closer connection to science. Still, we should admit that there does not seem to be anything about perseverance itself that is in itself particularly scientific. So, in what sense is it a scientific virtue? The answer is simply its object. Perseverance may be a broadly applicable character trait, but it may not be broadly transferable. Some individuals may have the virtue in so great a measure that they may persist in whatever task they set themselves, and we can agree that this is a good thing. Others, however, may be able to persevere in one kind of task or subject matter but grind to a halt out of indifference to another. To succeed in a difficult endeavor, one must at least have sufficient patience and persistence for that object. For scientists, that object is understanding the object of their curiosity and wonder.

Mental Discipline

Assiduity

The eighteenth-century Quaker physicist John Dalton is most known as the founder of modern atomic theory, but he seemed to be interested in everything, and his scientific work ranged broadly. Daltonism, a form of color-blindness affecting perception of green and red, was named after him, as he was the first person to characterize it scientifically in a 1798 publication. Dalton was a pioneer in chemistry, and the unit that is used to quantify mass at the atomic level—Da, or one dalton—is also named in his honor. He was fascinated in meteorology and for over half a century kept a meteorological diary that eventually topped 200,000 entries. Reflecting on his life's work, Dalton wrote: "If I have succeeded better than many who surround me, it has been chiefly, nay, I may say almost solely from unwearied assiduity."[34] The term assiduity nicely captures the qualities having to do with persistent, careful attention and effort that we have been considering in this chapter. We do still speak of being assiduous in some activity, but the noun form has so declined in use that it is no longer viable as a banner term to emulate. This is a pity, as it nicely captures both the notions of meticulousness and perseverance that we have focused on here as a way of examining the role of systematic habits in science.

These are not the only traits that a scientist should aim to make habitual, but they serve as key examples because they help make clear the notion of a disciplined practice. Developing the ability to precisely repeat procedures, making them habitual, is essential for the sort of evidence that science requires. Attentiveness, skepticism, and objectivity, as we saw previously, are basic virtues in science, and being attentive, skeptical, and objective in a meticulous and persistent manner is part of what makes them systematic habits.

William Whewell, who we encountered previously as the coiner of the term scientist, was another polymath; over the course of his career at Cambridge University he taught subjects ranging from mineralogy to moral philosophy and theology. Whewell is most known today for his two books on the history and philosophy of the inductive sciences. Though he mostly focused on mathematical and logical methods, Whewell did make some mention of the qualities that scientists ought to develop. He spoke of how scientific discoveries and improvements of "mental discipline" are the cause of every great intellectual advance in education and culture.[35] He explained how geometrical reasoning became a part of the discipline of the Platonic school in Athens, and how the two great notions of mathematics and jurisprudence drawn from the ancient Greeks allow us, through education, to raise our intellect "into a habitual condition, superior to the rudeness, dimness, confusion, laxity, insecurity, to which the undisciplined impulses of human thought in all ages and nations commonly lead."[36]

Whewell may have been overstating the case with regard to all the debilitating conditions that mental discipline can be credited with solving, but the general point is surely correct. Developing a discipline of meticulous thought and action, until it becomes habitual, provides one with dispositions that are necessary for the sort of observational and experimental practice that science requires. Both Whewell and Dalton had been students of John Gough, who was thought of as a living exemplar of these traits. Known as "the blind philosopher," in the sense of natural philosophy that we have discussed, Gough had lost his sight as a young child but subsequently developed a keen observational sense, which he applied to a variety of scientific topics. He learned to identify plants by touch and was able to differentiate closely related species that stumped sighted botanists. He published in mathematics, mechanics, chemistry, and physics. As an exemplary pioneer, Gough demonstrated the power of systematic habits.

Systematic habits are necessary because science is hard. Nature does not give up its secrets easily. Like any mining operation, mining data, and uncovering hidden truths, takes time and effort; one may have to sift through tons of rocks to find the nuggets of gold. To be quantitatively meticulous, one may think of this in terms of statistical significance; one must persist until one gathers enough data to get a significant test. It also relates, methodologically, to the patience a scientist requires to not jump to conclusions. Science demands that hypotheses not be accepted until they have earned their place with sufficient evidence.

Einstein's Paper Clip

The Nobel Prize–winning Italian physicist Carlo Rubbia attributed his success not to special talents but to intellectual persistence. Reflecting in an interview on why this trait was necessary for science, he said, "[T]he best quality of a true scientist is not sheer intelligence or Albert Einstein's approach, it is just persistence and having no fear of doing work which turns out to be useless."[37] As it turns out, Rubbia should not have listed Einstein as an exception to this rule. The mathematician Ernst Straus, who collaborated with Einstein, recounted a story that illustrates this quality in action in a quirky way:

> We had finished the preparation of a paper and were looking for a paper clip. After opening a lot of drawers we finally found one which turned out to be too badly bent for use. So we were looking for a tool to straighten it. Opening a lot more drawers we came upon a whole box of unused paper clips. Einstein immediately started to shape one of them into a tool to straighten the bent one. When asked what he was doing, he said, "Once I am set on a goal, it becomes difficult to deflect me."[38]

Einstein is reputed to have said that this was the most characteristic anecdote that could be told of him.[39]

What drives such focused perseverance in science? In his paean to Planck, Einstein said that it was the longing to behold the underlying harmonies of the world. As we saw previously, for the devoted disciple of truth, this is not a chore but a pleasure. Scientists happily cultivate those habits of mind that are needed for their chosen vocation.

This brings us to a point where we may summarize what have we learned about what it means to become a disciple of scientific truth and to enter into science as a vocation. We have seen that action arises from a combination of reason (which calculates cause-effect/means-ends relations) and emotion (passion, in Hume's sense, which motivates). Mental dispositions of both sorts are built on evolved capacities and instincts—our predispositions—that then must be developed by practice. Habitual dispositions are developed by taking instincts and adapting them to circumstances by intra-life learning. With regard to the scientific mindset, we all have an instinctual curiosity in at least some measure, but no one is born with the sort of mental discipline that is required to satisfy it with the degree of rigor that science requires. Science expects this and more because of its higher standards for evidence-based action. Furthermore, scientific habits must be systematic. By practicing systematic habits, we reduce errors and improve our reliability. Developing a scientific attitude is part of what makes this possible by tempering the scientist's mindset. One must develop it through repetition and practice until it becomes habitual. Mental discipline is essential because scientists may often work for years before they catch even a glimpse of some aspect of that harmony. Most of what one does in science fails, so one must be able to persevere. But something more is still required. We have seen one aspect of what it means for habits to be systematic—they must be appropriately disciplined. What remains is to examine how these habits relate to scientific methodology, for science is a methodological practice. We have looked a bit at the quantitative aspect of scientific reasoning, what Whewell called the mathematical part. We now need to look more closely at the logical part—science's inductive methodology—and the special kind of intellectual courage and intellectual humility that should attend it.

6 Courageous Humility

I abhor your pretentious "insight." I respect conscious guessing, because it comes from the best human qualities: courage and modesty.

—Imre Lakatos[1]

Science: A History of Failure

Unfulfilled Dreams

The history of science is littered with failed hypotheses and premature pronouncements of finality. In his book *Dreams of a Final Theory*, physicist Steven Weinberg relates a story of physicist William Thomson, Lord Kelvin's notorious recommendation at the turn of the century for physics students to redirect their attention to other sciences, since nothing of significance remained to be discovered in that field beyond more precise measurements.[2] The story was meant as a cautionary tale about scientific hubris; we may not only be ignorant, but even ignorant of our own ignorance. (Metaknowledge is knowledge about knowledge, so perhaps this kind of ignorance about ignorance ought to be recognized as metaignorance.) Weinberg shared a Nobel Prize for work showing how electromagnetism and the weak nuclear force can be seen as following from the same underlying theory—what became known as the Standard Model. The advance was significant, but it was also seriously incomplete. Weinberg knew this very well. His book reviewed the state of the field, showing how much had been learned but also how much more remained to be resolved. The idea of a "final" theory that would provide the next level of unification in physics remains an unfulfilled dream. It is not the only one; there are many ways that one can experience failure in science. Weinberg wrote his book as part of a lobbying effort to fund a superconducting supercollider to help test the next-generation hypotheses about subatomic physics. The hope was that Congress would provide funds for it to be built in Texas. This effort failed as well.

One may state the matter simply: the history of science is a history of repeated failures. One can almost guarantee that in investigating any scientific puzzle one's first hypothesized explanation, and often one's second and third as well, will be wrong. Reflecting on his own scientific career, Charles Darwin called attention to this important feature of his experience, writing that "with the exception of the Coral Reefs, I cannot remember a single first-formed hypothesis which had not after a time to be given up or greatly modified."[3] A scientist has to be ready to fail and to keep failing until at last, with perseverance and perhaps a stroke of luck, one line of research stops failing. As we shall see, this takes a special sort of intellectual humility. Darwin said of himself, "With such moderate abilities as I possess, it is truly surprising that thus I should have influenced to a considerable extent the beliefs of scientific men on some important points."[4]

It also takes a special sort of courage, though that case will take a bit more work to establish. I want to call particular attention to the courage needed to follow the evidence where it leads, even when it leads away from one's preferred hypothesis, though science regularly requires this kind of courage. As we shall see, this is a fundamental scientific virtue and a pervasive feature of scientific methodology; one must be ready to have one's pet hypotheses slain again and again. There is also, I submit, a base level of courage that is required to live with the constant uncertainty and failure that is part and parcel of being a scientist.

Uncertainty and Failure as a Method

The philosopher Charles Sanders Peirce argued that epistemic uncertainty is an essential feature of scientific inquiry. Fallibilism, as he defined the term, is the view that "one cannot attain absolute certainty concerning questions of fact" in that any of one's current beliefs could turn out to be mistaken.[5] This is a point about knowledge generally, but it is especially salient when speaking about scientific knowledge. The nature of scientific evidence always leaves open the possibility that even apparently well-confirmed hypotheses could be overturned. Peirce saw this as a good thing, because it is uncertainty that drives continued inquiry.

This connects to an important aspect of the scientific mindset. Weinberg's and other stories of scientific failure are not meant to be taken as discouraging, and scientists do not see them as such. Quite the opposite. Scientists love such stories, seeing them as uplifting and inspiring. They remind us that even when we think that we have all the answers, there is still more to be learned.

Our interest now is to examine how this functions as another aspect of the mindset of scientists and of the character of science. Darwin is once again a pioneer in recognizing the connection. Partly out of necessity he made this an intentional mental habit: "I have steadily endeavoured to keep my mind free so as to give up any hypothesis, however much beloved (and I cannot resist forming one on every subject), as soon as facts are shown to be opposed to it."[6] It is this attitude and its relationship to scientific methodology that we will investigate in this chapter. My thesis here is that science requires a particular kind of intellectual humility and intellectual courage that embraces uncertainty and failure and that these are intimately connected to essential features of scientific methodology.

Intellectual Humility

Ignorance and Confusion

We have said that science begins in wonder and curiosity, but we might equally say that it begins in ignorance and confusion. Curiosity arises regarding something about which I have a question. It is about something that puzzles me. I do not know the answer, and I want to find it out. In science, this is the first step in a process that, with attentiveness, perseverance, and perhaps a bit of luck, may lead to a discovery. It was Socrates who first taught that recognizing one's own ignorance about some question was the first step to wisdom, and it is no less true in science.

Evolutionary biologist Richard Dawkins credited one of his science teachers in school, Ioan Thomas, with teaching this important lesson in a memorable way. Dawkins told the story of how Thomas came into class one day and asked, "What animal eats hydra?" One by one they tried to guess, and Thomas became more and more excited as each failed. Dawkins said that by the time he reached the last student in the class they all "were agog for the true answer" and finally asked him, "Sir, what animal does eat hydra?" He let the question hang in the air and then answered, "I don't know…I don't know…And I don't think Mr. Coulson does either." Thomas then interrupted his colleague's lesson, dragged him in from his room next door, and posed the same question: "What animal eats hydra?" Sure enough, even Coulson did not know. Dawkins said that he never forgot that inspiring teacher who, in this wonderful lesson, made every student understand that it is OK to not know the answer.[7]

Thomas's lesson conveyed even more than Dawkins directly identified. The fact that there is no shame in not knowing something in science is an important lesson, but it is only the first of what was a many-layered master class in scientific values. Thomas conveyed something deeper about not knowing, which is evidenced by the way in which Dawkins told the story of that classroom lesson. Thomas conveyed the excitement that the recognition of an unanswered question can bring to a scientist. He expressed increasing excitement as he went around the room, and it became clear that no one knew the answer. Moreover, that excitement translated into curiosity; the students became "agog for the true answer," just as scientists do when they find an interesting open question. What would a parent not give to have their child taught by such a science teacher, who knew how to stimulate curiosity and the desire to know?

I would argue that there were additional layers of meaning in Thomas's lesson that shows just how gifted a science teacher he was. Dragging in the other science teacher who, likely being in on the plan, also admitted that he did not know the answer, made the subtle point that one cannot simply look to authorities for answers. With the realization that authorities do not automatically have the answer comes the scientific point that empirical matters must in the end be found out not by asking an authority but by asking nature itself. One imagines the next step in the process in which students and teacher together begin a series of observations to find out the answer. (Dawkins told just such a story of Thomas's predecessor at the school, who cut short a faculty committee meeting and grabbed his binoculars when a young student timidly knocked on the door to say that there were black terns down by the river.) Equally significant, this informal lesson undermined traditional classroom power dynamics. In the traditional model, the teacher is an authority figure who dispenses knowledge. Many teachers rely on that power differential, so it takes a degree of confidence and courage to admit that one does not know something and to step out of the role of being the authority. As scientists, we all are equal before nature and all beholden to the evidence. Admitting that one does not know something, and being willing to follow the evidence where it leads, is the kind of intellectual humility that stems from the nature of scientific curiosity. Thomas certainly knew far more than he was letting on, and one imagines that he found a different way to make the point in other years, but in this deep lesson he was modeling the values that make science what it is. These are far more important things for the science student to learn, I submit, than it would have been to simply be told what eats hydra.

The stereotype of the scientist is the brainy know-it-all. Amateurs think of scientists as the individuals who can recite the digits of pi to the hundredth decimal place, who spout the Latin genus-species names of plants, who recount obscure facts about the life cycles of insects. Scientists do often have this kind of knowledge, but that is not what makes them scientists. Such knowledge is what some previous scientist discovered, which now provides the background to their real work: to discover something new. It is what they do not know, and what they do to try to find out, that makes them scientists. Most of a research scientist's life is spent in a state of confusion, searching for an explanation of some puzzling phenomenon, grasping for possible solutions that are just out of reach, or trying to figure out even how to formulate a question in a way that gives one some small chance of answering it.

William Thompson, Lord Kelvin, identified this state of confusion as a scientific opportunity: "When you are face to face with a difficulty, you are up against a discovery."[8] The difficulty is an empirical problem, and science is all about problem solving. It may begin with the recognition that something is odd. Hmm, that's curious, one thinks. This is unexpected. This is an anomaly. What is going on here? What's wrong? It is the initial failure to be able to explain something that provides the opportunity. In solving what began as a puzzle or problem, one has made a discovery. Indeed, the trickier and deeper the problem, the more significant the discovery would be.

Actually, science demands even more. It is not just that scientists must be comfortable admitting when they do not know something, they must also be ready to acknowledge that something they thought that they did know was wrong.

I Was Wrong

As a case in point, I recall the presidential address given by the evolutionary geneticist H. Allen Orr for the Society for the Study of Evolution. Much of his talk was about being shown wrong by the data in some of his work on the genetics of speciation. "I wrote four papers against role of genetic conflict for post-zygotic reproductive isolation," he said, "but I eventually recognized that I was wrong." The audience of fellow biologists nodded empathetically and with appreciation. Orr's presentation exemplified an attitude that is part of what it means to be a scientist. Every scientist has experienced setbacks and failures of one sort or another and understands that they must always be ready to bow before the evidence. We already noted Darwin's statement about his own sense of this and how important it is in science to be ready to abandon even beloved hypotheses if the evidence goes against them. Jane Goodall began her study of chimps thinking of them as gentle and peaceful but had to reject that preconception after she observed troop war and infanticide.

This readiness to defer to the evidence is related to our previous discussion about scientific method as a form of bias control. As scientists, we know that we can be biased, so we put in place methodological constraints to reduce the chance that we will be tempted by hypotheses we wish for rather than for ones that are true. Our goal is to be led by the evidence. We therefore simply must accept that being wrong, and being shown to be wrong, is part of the scientific bargain. It goes without saying that this requires a certain measure of intellectual humility, but it is a kind of humility that is distinctive to science.

Humility to Evidence

In the pilot study for our survey of exemplary scientists, we included humility by itself in a long list of possible virtues we asked our subjects to respond to, but it did not score very highly. Scientists can sometimes be jerks like anyone else, we heard, and in certain ways may even be arrogant, just as Einstein had been. This was not just a descriptive claim, as our respondents did not necessarily think that arrogance hurt someone's ability to do good science, again like Einstein. But when we asked subjects to explain their reasoning, we discovered a subtler view. While arrogance as a personality trait could be tolerated, it was viewed as negative if it meant that scientists asserted their own preferred views in the face of counterevidence. When we narrowed the notion of humility to the sense of humility to evidence, to try to capture this sentiment, the ratings immediately shot up.

This special notion of humility to evidence, rather than humility in general, relates to our previous discussion about authority. The flip side of humility to evidence is the dismissal of authority. Recall Einstein's statement about Galileo's passionate fight against any dogma based on authority.[9] With regard to this issue, Galileo himself made an explicit connection between rejection of authority and scientific humility to evidence, writing, "In questions of science the authority of a thousand is not worth the humble reasoning of a single individual."[10]

Modesty in science For many scientists, this manifests itself in an admirable modesty in how they comport themselves generally with their colleagues. Isaac Newton prefaced his *Principia*, as revolutionary a treatise as there ever was in science, with a self-deprecating request that readers might forgive and correct his errors: "I heartily beg that what I have here done may be read with forbearance; and that my labors in a subject so difficult may be examined, not so much with the view to censure, as to remedy their defects."[11] However, not all scientists exhibit this kind of modesty in their personalities. Einstein, at least in his younger days, was arrogant and self-centered, with little regard for others who he did not see as his intellectual equals. This is another example of a basic distinction we have pointed out before. There is a fundamental conceptual difference between description and prescription, between personality and character. As we discussed previously, our interest in this book is on the latter, normative, aspect of science.

To be intellectually humble is to recognize that one's knowledge is incomplete and fallible, and it therefore requires that one be open-minded. What does this mean in science? Being open-minded has the positive effect of allowing one to be ready to add something new to one's store of knowledge. Again, this stems from acknowledging that one could be wrong. With such an attitude, a scientist ought to be able to consider countervailing data, revise or reject ideas that others may simply take for granted, and entertain novel hypotheses that may open up new lines of investigation. There are, of course, limits to this; as has often been said, it is good to be open-minded, but not so open-minded that your brains fall out. This is usually meant as a humorous way to indicate that one should not waste one's time on pseudoscience and conspiracy theories. But the general point behind it involves the notion of humility to evidence that we have been discussing, namely, that scientists must always aim to constrain their thoughts by the data.

Scientists, of course, do not typically use this terminology, but once one is sensitized to the concept, one finds the notion of humility to the evidence expressed everywhere in science. Richard Feynman, for example, referred to this when talking about the need for a scientist to learn nature's own language: "If you want to learn about nature, to appreciate nature, it is necessary to understand the language that she speaks in," he said. "She offers her information only in one form; we are not so unhumble as to demand that she change before we pay any attention."[12]

Shoulders of Giants

A related aspect of intellectual humility that is worth mentioning has to do with the fact that scientists do not stand alone in their research. It is rare that a scientist can investigate any aspect of nature without relying on the work of previous investigators. Acknowledging this is not a matter of self-deprecation; in science, the debts we owe to our colleagues is so obvious that even the most revolutionary scientists sincerely acknowledged their reliance on the findings of their scientific predecessors. Isaac Newton himself was not expressing false modesty when he wrote, "If I have seen further than others, it is by standing upon the shoulders of giants." Robert Merton took special note of Newton's expression of this norm in science, devoting an entire book to an exploration of its history and significance, including its relationship to the important issue of reputation and attribution of credit to discoverers in science.[13] Even Newton was no stranger to scientific priority disputes, but one's desire to be properly recognized as a discoverer does not negate the real feeling of indebtedness to one's scientific predecessors. The kind of intellectual humility represented by Newton's tribute is a proper part of the scientific mindset in that it recognizes what one owes to one's scientific colleagues, past and present.

Humility vs. Timidity

We should briefly deal with a possible concern about intellectual humility. One might worry that there is a danger in calling humility a virtue in science in the sense that it may make the scientist too hesitant to draw a conclusion. Some scientists have indeed pointed out a weakness of this sort that they have observed in some colleagues who become timid in their research, never feeling that they were ready to publish. It may be the case that some scientists are overly cautious in this way, but that is more likely the result of a perfectionist personality. In any case, the criticism arises from a misunderstanding of the meaning of intellectual humility. In science, meekness is not the appropriate way to understand this virtue. The idea is that one should be humble to the evidence, which means that one should follow it where it leads. Humility to evidence does not imply that one should invariably hold back from assenting to a conclusion. Quite the contrary. If the evidence is insufficient, then science insists that one not jump to a conclusion, but as evidence accrues in favor of some hypothesis, then science demands that one stand ready to accept that as well, acknowledging a finding as such. Scientific methodology is all about this process of gathering and weighing evidence so that one may reduce possible biases when assessing its value, so it is worth taking a short excursion into what philosophers of science call confirmation theory with an eye to seeing how it relates to this basic scientific virtue.

Question. Hypothesize. Test. Repeat.

Lakatos's Conscious Guessing

Imre Lakatos, the philosopher of science and mathematics who taught at the London School of Economics (LSE) in the 1960s until his death in 1974, used the term "conscious guessing" to refer to what he took to be a key element of the method of science. The term itself obscures aspects of the procedure, but the general idea is easy to explain. The notion of guessing here does not mean that something is put forward without any basis whatsoever. He was not referring to "mere" guesses. Rather, he was referring to hypotheses—conjectures that are put forward for test. Hypotheses may be thought of as possible answers to some scientific question. As we saw, there are many open questions in science stemming from the many things we do not know and the many empirical puzzles that remain unsolved. Furthermore, for every question, there may be many possible hypotheses to sort through before finding any that adequately answers the question or solves the puzzle. We do not need to delve into the process by which scientists come up with hypotheses; in some cases they may truly be guesses but even these are necessarily educated guesses in that they must be informed by what is already known in the field.

Lakatos's claim that these hypotheses must be "conscious" was also an odd choice of words, but the idea is that they should be offered with a particular kind of intent. Putting it in these terms, what he was saying is that scientists must *hypothesize with intent*, which is to say, with the intent of boldly testing the hypotheses' worth. This was his point of contrast with accepting something just by "insight." Tests will more often than not show that one's hypothesis was wrong, in which case one must go through the process again and again. Perseverance certainly comes into play here, but Lakatos alerts us to two other virtues that are particularly relevant in this process—courage and modesty.[14] We have already seen, in our discussion of intellectual humility, how modesty comes into play in this cycle of testing; scientists must be humble in the face of failed tests and be ready to give up even what they may have thought was a brilliant idea if the evidence goes against it. It is the high likelihood of failure that makes courage important as well, and we will consider that in more detail shortly.

Lakatos highlighted these virtues because he took the idea of testing seriously. A hypothesis was not subjected to a real test if there was not a real possibility that it could have been found to be false. Lakatos's emphasis on the importance of the possibility of falsification comes from the influence of another philosopher of science, his LSE colleague Karl Popper. Popper's particular view of scientific testing—Falsificationism, with a capital F—was very influential, but it was flawed in a serious way, and we need to dive briefly into some deeper philosophical waters before we can appreciate how a virtue-based philosophy of science can help us understand, evaluate, and perhaps even improve scientific methodology.[15]

Popper and Falsificationism

Popper was searching for a formulation of scientific reasoning that could be expressed in deductive logic. In *The Logic of Scientific Discovery*,[16] he argued that science was characterized by what he called Falsificationism, which was a testing procedure based on a deductively valid argument form that he thought provided science with a clear and logical justification for its conclusions. The argument runs in the following manner. Starting with the hypothesis (H) you want to test, deduce some observable prediction (OP) that should follow if it is true. This gives a conditional sentence "If H, then OP" as the first premise of the argument. Then do the test, and see what happens. If OP turns out to be false, then H must be false as well. Why?

Because if OP is a necessary consequence of H, which was what deducing the prediction ensured, then it would have to be true, so long as H is. Put another way, in a situation where H is true, you cannot not get OP, so if one's prediction fails, then so does one's hypothesis. In this way, H is disproven. A simple example: Say Abe claims to be able to lose three pounds a week by adding a special supplement to his diet. Sue is doubtful that the supplement works as advertised, so Abe agrees to let her test his claim over the coming month. With an easy calculation they predict that at the end of the month they should observe at least a twelve-pound drop when Abe steps on the bathroom scale, compared to his weight today. If at that time the scale shows no change (or a gain, or any drop less than twelve pounds), then Abe's claim has been proven false and he has to pay up. This argument form (If H then OP. Not OP. Therefore not H.) is called *modus tollens*, getting its name from the fact that the form involves denial of the consequent. It is deductively valid, which means that if the premises of the argument are true, then the conclusion must be true as well. Popper thought that he had found the kind of argument that science needed. Using an evolutionary analogy, Popper argued that Falsificationism provides a rational framework for understanding science as a process by which false hypotheses would be eliminated in a relentless competition for survival.

On Popper's view, a real scientist makes bold conjectures and then seeks out challenging tests that would conclusively refute them if they are wrong. Einstein's hypothesis about relativity may have sounded strange, but it made risky and precise predictions that, if they had been shown to be false, would have disproven his theory. On the other hand, it would be relatively simple to offer a vaguely stated or slippery hypothesis and to propose easy "tests" that it could be expected to slide through. Popper criticized Freudian psychology and Marxism on this basis, saying that they were stated so loosely that they could never be falsified. That is not real science, he thought, and he went on to argue that falsification was a sufficient criterion to demarcate science from pseudoscience.

One important consequence of this kind of approach, however, was a somewhat counterintuitive result that hypotheses could be disproven but never proven. How so? If the observable prediction turned out to be true (rather than false, as in the case above), that does not guarantee that the hypothesis is true as well, because there may be other ways for OP to be true, even if H is false. Abe's weight at the end of the month may turn out to be twelve pounds lower, but that could be the result of some other factors that had come into play that month—perhaps his heightened focus that month on his food intake—and not because of the diet supplement. This argument form has a name as well—*modus ponens*, which gets its name from the fact that the form affirms the antecedent. It is deductively invalid, which means that even if the premises are true, the conclusion may not be. Because he insisted on deductive validity, Popper rejected this form of argument, proposing that science can never conclusively say that a hypothesis has been verified but only that it has been corroborated, by which he meant that it has stood up to tests and not yet been falsified.

Like the view of Logical Positivism that he criticized, Popperian Falsificationism was clear and precise, so like Positivism it became easy to identify places where it failed. One must hand it to Popper for living by his own lights and putting forward a bold view that was amenable to being shown faulty. For example, modus tollens turns out to be a faulty model for scientific reasoning even on its own terms, because it leaves out auxiliary hypotheses. Auxiliary hypotheses are required for the initial derivation, but they are usually simply tacitly assumed. This means that if a prediction is not borne out, it may not be the fault of the original hypothesis at all but rather the fault of one or another auxiliary. This is an example of a general issue that was identified by philosophers Willard Van Orman Quine and Pierre Duhem, that hypotheses can never be tested in isolation because empirical tests always involve auxiliary hypotheses.[17] For example, a tacit auxiliary hypothesis in the dietary supplement example above was that the scale was working; if it had failed or was not properly calibrated each time, then the negative reading was not the fault of the supplement hypothesis, and it should not have been rejected. Thus, in a straightforward sense, falsification is no more deductively certain than confirmation, and a hypothesis that has been "disproven" is really just mostly dead; it could, in principle, be revived depending on subsequent evidence.

Flaw in Popper's Model   The flaw in Popper's Falsificationist model is all too easily carried over into the evolutionary model, especially when one's knowledge of evolution stops with a simplistic understanding of the notion of survival of the fittest. The phrase is catchy but often misleading, as it makes one think of an individual's fitness only relative to its death. One sees this mistaken view exemplified in the gruesomely unsympathetic "Darwin Awards" that each year "commemorate those who improve our gene pool by removing themselves from it," mostly by dying from acts of unusual stupidity.

From a theoretical point of view, however, death is a red herring. After all, in the long run we are all dead. To see the problem, imagine a science fiction scenario in which future scientists use some combination of genetic engineering and temporal stasis technology to create a "perfect" living cell that is effectively immortal but is unable to reproduce itself or effect the reproduction of any other living organism. That it survives for eternity may be existentially significant, but in the Darwinian game of life its evolutionary fitness is nil. On the other hand, the humble mayfly that lays thousands of eggs in its few hours of life that spawn the next generation has high fitness despite its brief existence. Nor does evolutionary success even require that one reproduce oneself, for indirect effects (for example, on others with whom one shares genes) can also have profound effects. Thus, it is not death per se that makes an evolutionary difference, except as the limiting case, when it typically shuts off chances to effect reproduction. What matters for fitness is not whether one survives, but rather what one has contributed to life before, or even after, one dies.

Strict falsifiability is, similarly, a red herring for scientific methodology. While the general idea that the competitive scientific process selects out false hypotheses is on the right track (as we saw earlier in our discussion of how norms of the scientific community help make the process self-correcting), it is a mistake to think of this as survival of the fittest in the sense of killing off the false ones until only the true ones survive. Because science's evidential logic is inductive, hypotheses may never be falsified with certainty just as they may never be confirmed with certainty. The guarantee that one gets with scientific evidence is never 100 percent certainty but rather credibility that approaches that limit to varying degrees. As we saw before, part of what it means to be meticulous in science is to be as careful as it is reasonable to be for the strength of conclusions that is required.

What do we mean by "reasonable" here? This involves a humble assessment of our rational capacities. There are theoretical and practical limits to our mental abilities (we have what Herbert Simon called "bounded rationality"), so we must assess evidence pragmatically. As in evolution, the evidence must be good enough. What evidential tests do is reduce the likelihood that false hypotheses will propagate very widely or for very long, and increase the likelihood that true, or at least truer, ones will.

Inductive Reasoning—Method of Hypothesis

Given that the scientist's goal is to find out things about reality, it follows that one should adopt methods that allow reality to make as big a difference as possible in the outcomes. That is, different methods allow reality to more or less constrain observations, and scientists should pick those methods that maximize the constraints from reality and minimize the effects of other factors. As noted previously, randomized controlled experimentation is the gold standard for this purpose. Popper was wrong to think that this was captured by his deductive Falsificationist framework. The general idea that one should look for observations that can function as real tests was on the right track, but we need to think about this in terms of inductive reasoning.

Because he wanted to limit science to deductively valid arguments, Popper rejected what is arguably the most common form of evidential reasoning in science—the Method of Hypothesis. In practice, this method looks very much like what Popper was recommending but with an important difference. Having come up with a hypothesis that could explain whatever puzzling question one is investigating, one tests it by checking for things that one should observe if it is correct. In a straightforward case, one looks for observations that ought to follow in a predictable way under some set of repeatable conditions if the hypothesized model is true and then judges the hypothesis depending upon which way the observation test turns out.

One way that this differs from Falsificationism is that the Method of Hypothesis holds that positive outcomes of such a test may be taken as *confirmation* of the hypothesis. Confirmation in this sense does not mean deductive proof but rather inductive support. Again, inductive evidence never gives absolute certainty; depending on the strength of the test, the evidence may be weakly or strongly positive. If the outcome of the test is negative, the hypothesis is disconfirmed, with the same caveats. And even if the evidence is strong, it always remains possible, at least in principle, that new evidence will overturn what we thought we knew. A basic dose of humility is thus a prerequisite for any form of inductive reasoning.

Caveats: H-D Method and Paradoxes of Confirmation   We should pause briefly here to head off a possible confusion about the Method of Hypothesis. The Method of Hypothesis is sometimes equated with and explained in terms of the Hypothetico-Deductive (H-D) Method, but this is a misleading conflation. Hypothetico-Deductivism should be thought of not as equivalent to, but as rather one possible form of, the Method of Hypothesis. The H-D Method is what the Method of Hypothesis would look like for someone who thought it should be put into the same kind of logical framework that Popper used. In particular, it requires that the observational prediction that one checks in a test of one's hypothesis should be something that is logically derived from the hypothesis. Hypothetico-Deductivism posits that it is this logical relationship that makes the observational prediction evidentially relevant to the hypothesis. This would have been very useful if it had worked, but the H-D Method fails for the same sorts of reasons as Falsificationism. Without additional constraints, it becomes subject to a variety of technical problems that are known as the paradoxes of confirmation, including the Paradox of the Ravens and Goodman's Grue Problem.

The Ravens' Paradox arose within what seemed to be a straightforwardly simple idea of how to confirm a universal generalization of the form "All X's are Y," such as "All ravens are black." If that hypothesis is correct, it seems like one ought to be able to confirm it by going out to observe a bunch of ravens and finding that they are black. The idea is that one may support such a hypothesis by confirming instances of it, where an instance is defined logically as a statement that fulfills both elements of the generalization, such as, "This thing, A, is a raven and is black." Keep doing that with B, C, and so on, and the generalization gathers further support. It is the logical relationships among the sentences that make those observations evidentially relevant to the hypothesized generalization. So far, so good. Or so it seems. Then it was pointed out that from the point of view of deductive logic, "All ravens are black" is equivalent to "All nonblack things are not ravens." That seemed OK at one level, until it was noticed that one should now be able to test that hypothesis by checking its instances, that is, "This thing, O, is not black and is not a raven." What is an example? How about a white shoe or a green vase? But one can observe such instances without going outside at all. Indeed, one could rack up "evidence" for the raven hypothesis without ever seeing a single raven and never leaving one's room. That seems clearly wrong; a scientist ought not be able to do indoor ornithology in this way. Such observations are irrelevant to the hypothesis, which suggests that something was wrong with the logic we used to determine that this should count as evidence.[18]

The paradox of Goodman's Grue Problem goes deeper down the rabbit hole and is broader in its scope but shares similar problems. Think again of a universal generalization, this time, "All emeralds are green." Observing a green emerald would seem to support it. But now let us consider a different language that includes the predicate "grue," which is defined logically as something that is green if observed before time T or is blue afterward. If my observation of emerald E is before time T, then it seems that both the green and the grue hypotheses about emeralds are equally confirmed. But then what color should we say that emeralds will be after time T? Again, it seems as though logic and language have led us astray with regard to what evidence is relevant to what.[19]

This is not the place to get into the complex details of the paradoxes of confirmation, but it is worth making an additional point about how they are tied to similar paradoxes in logical accounts of scientific explanation. To relate just one classic example, if we approach the question of scientific explanation in terms of logical relationships, it seems reasonable to think that explanations in science involve the ability to logically derive statements of the thing to be explained (the explanandum) from statements that comprise the explanation (the explanans), where the latter include statements of the relevant natural laws and the initial conditions at play in the case to be explained. For instance, why not say that one had a good scientific explanation of why the shadow of a flagpole was a certain length at a certain time of day by virtue of being able to logically derive it from the laws of optics together with the height of the flagpole, given the position of the sun in the sky. The paradox that was pointed out in the flagpole example took a similar form as the cases we briefly described above—the logical derivation is symmetrical in that one may calculate the length of the shadow from the height of the pole, given the other factors, but also the height of the pole given the length of the shadow. But while it is reasonable to say that the height of the pole explains the length of the shadow, the reverse does not generally hold in the same way. Explanations are asymmetrical. Thinking that one should look to logical relationships to determine explanatory relevance leads to the same sort of problems that we saw before.

A solution to the explanatory symmetry paradox in the flagpole and other counterexamples to the proposed logic-based evidence relation involved switching to one that focused on causal relationships. Philosopher of science Wesley Salmon's ontic account of scientific explanation holds that scientific explanations involved not logical relations within language but rather relationships within the causal structure of the world.[20] In the flagpole case, the explanation goes one way and not the other because the flagpole causes the shadow, and not the other way round. The asymmetry of explanation follows the asymmetry of cause-effect relationships. My own view is that we can solve the paradoxes of confirmation in the same way and that we should approach scientific testing naturalistically, starting from ontic, causal relationships.[21] Evidential relevance, like explanatory relevance, depends on cause-effect relationships as well. This form of the Method of Hypothesis is thus very different from Hypothetico-Deductivism; we may not simply look at any logical prediction from a statement about some model we hypothesize, but rather that we need to consider what would be a predicted effect if the model is correct.

A bit of caution is needed here: we must recognize that cause-effect relationships are complicated and that the way we speak about them is typically simplified. In reality, there is more than a two-place relationship between cause and effect, for there are invariably multiple causal factors that are potentially involved in producing any given effect. To sort these out requires isolating some factor of potential interest from others in a given situation. This is what controlled experiments are all about. A set of background variables is fixed in an experiment so that one can vary just the independent variable (the possible cause) and observe what happens to the dependent variable (the effect of interest). If the model is correct, one expects to see the effect produced in the presence of the cause and not otherwise. Returning to our original point about the scientists' goal of finding out about reality, controlled experimentation is the best method for having reality make as clear a difference as possible in the outcomes of a test.

Lessons from Kuhn

There are a couple of additional possible confusions about science's inductive methodology that are worth heading off.

The first is to think that following scientific testing procedures will inevitably lead to a single hypothesis. Theory is underdetermined by evidence, as philosophers put this point, which is to say that we should not expect a test by itself to fix exactly what we should accept. Scientific progress is not necessarily a linear accumulation of knowledge. It sometimes goes through periods of revolution where scientists must rethink their assumptions and strike off on a radically new path. The most influential scholar to articulate these issues was the historian and philosopher of science Thomas Kuhn. In his book *The Structure of Scientific Revolutions*, Kuhn argued that scientists mostly operated within what he called a paradigm, which assumed a set of basic theoretical concepts that structured scientific investigation in the field. In these periods of what Kuhn called "normal science,"[22] scientists worked to answer questions and solve puzzles within that theoretical framework. However, very occasionally in the history of science there came points when anomalies—mismatches between the theoretical expectations and empirical observations—became too numerous and puzzles became too hard to solve in a reasonable way, so the paradigm had to shift before progress could resume. In such a scientific revolution, previously accepted findings and concepts may be eliminated or revised to such a degree that the old and new theories were purportedly incommensurable. Kuhn described such revolutions as scientific gestalt switches that caused one to see the world in a profoundly new way.

Kuhn had originally trained as a physicist, so his first exploration of this involved the Copernican Revolution and the shift from a geocentric to a heliocentric paradigm. This was not just a matter of whether the earth or the sun was at the center of our planetary system but involved other fundamental issues about the physical makeup of the universe and the forces that governed orbital and other motion. This revolution required overturning the scientific understanding of physics and astronomy that went back to Aristotle, whose physical account had remained serviceable for over a millennium, during which period we thought that we knew that the earth stood motionless at the center of the universe. We were wrong about this, and we were wrong for a very long time. This is another sense in which intellectual humility is warranted in science—it is not just that we must bow to the evidence, in specific ways we must also be willing to reassess what we take to be the rules of evidence.

The second confusion that we need to head off involves what lesson should be drawn from Kuhn's account of scientific revolutions. Kuhn argued that conflicts that arise during a paradigm shift often "cannot be resolved by proof" and that "in the absence of binding criteria" the decision about how to proceed ought to be left to the "collective judgment" of the community of specialists.[23] Keying in on such passages as well as his talk of paradigm shifts, gestalt switches, and incommensurability, many scholars concluded that the lesson to be drawn was that science was subjective, relativist, and perhaps in some basic sense even irrational. This interpretation of Kuhn became an important element of radical postmodernism, which held that science had no special privilege in determining matters of empirical truth and questioned whether truth was worth pursuing at all.

This interpretation of Kuhn served a useful corrective purpose. It was important to recognize that progress in science is not linear, that scientific testing is not reducible to deductive logic, and that it is foolish to expect it to provide a master narrative that conveys the one true picture of the world. However, in the end, the radical interpretation went too far and overstayed its welcome. We have already discussed why truth, understood in a more nuanced way, is a useful and important concept, and why it is reasonable to claim that scientists value objectivity in the search for truth. In the next chapter, we will discuss some of the claims of postmodernists and see some ways that they went wrong. Here I only want to suggest that this strand of Kuhnian interpretation was misguided, having drawn the wrong lesson from Kuhn's work. Kuhn thought so too. He was startled by such interpretations of his views and said that they manifested "total misunderstanding" of his position.[24] This is not the place for a full discussion of this fatal error or the reasons for taking a different path, but I do want to lay out at least a rough outline to show how the approach we have been taking fits with and extends Kuhn's own views.

The first point is that Kuhn held that scientists do have a shared basis for choosing among competing theories. In an important paper that responded to his critics, Kuhn identified five characteristics of theories that he said serve as bases for assessment: accuracy; consistency (both internal and with other related and currently accepted theories); scope (a theory's consequences should extend beyond the data it is required to explain); simplicity (organizing otherwise confused and isolated phenomena); and fruitfulness (by which he meant fecundity for further research). Kuhn wrote that these five criteria and others of the same sort "provide the shared basis for theory choice."[25] However, although Kuhn saw these as constitutive of science,[26] he did not think that they were sufficient to determine scientific choice.[27] Again, the point is that theory choice cannot be reduced to a simple algorithmic procedure that leads to a single answer.

Kuhn suggested that philosophers of science originally expected that there would be an "algorithm able to dictate rational, unanimous choice" among theories that would completely eliminate individual differences such that every scientist would be compelled to accept.[28] While he did not rule out this possibility entirely, he explained why philosophers no longer expected to find such an algorithm—it seems unlikely that all individual choice criteria could be stated unambiguously and, more importantly, that if more than one proved to be relevant, an appropriate weight function for each would be needed for their joint application.

Today we can say a bit more about this, given advances in computer science. Evolutionary computation implements an algorithmic procedure that enables choice among possible models. In nature, this happens automatically. It can also be set up to happen automatically in a computer. This is not the place to get into the details, but one may optimize on a single criterion or set up multicriteria optimization. The main problem is not whether there can be an algorithm. There can be. The problem is thinking that we should expect an algorithm (or any scientific procedure) to produce a "unanimous"—that is to say, single—choice. We should embrace pluralism. Thinking evolutionarily, we know that different models (adaptations) will be useful given different goals (e.g., in different environments). In certain situations, scientists may reasonably care more about accuracy than scope and so fit to a tighter curve for some specific purpose. These are pragmatic issues. This is where values come in. Kuhn saw that, writing: "I am suggesting, of course, that the criteria of choice with which I began function not as rules, which determine choice, but as values, which influence it."[29]

This is the view that we have been articulating in this book, though I have argued that the set of values extends into areas that go beyond the values involved in theory choice that Kuhn listed. Resolving which hypothesis to go with may be the most important step in the scientific process, so it is not surprising that this is where philosophers have focused their attention. However, there are many other aspects of scientific practice that are critical for getting to that step and these, too, are structured by values. We have been focusing on the character virtues that are important for exemplary science in part because they have been unduly neglected but also because they provide the most natural way to introduce virtue theory as a framework for seeing how values work generally in science and in other vocations.

Kuhn recognized that different "disciplines are characterized, among other things, by different sets of shared values."[30] Philosophy and engineering are different from science, though still relatively close to it compared to literature. As he rightly put it, "Milton's failure to set Paradise Lost in a Copernican universe does not indicate that he agreed with Ptolemy but that he had things other than science to do."[31] Even scientists can legitimately differ in their particular scientific interests. For one scientific investigation, accuracy may be paramount, and for another that may be sacrificed in favor of simplicity, perhaps for the sake of speed of calculation. Again, this does not mean that the one disagrees with the other in any dangerously relativist sense—they all still care about discovering truths about the world—but rather that they had different things they wanted to do and needed their work to be true enough for each purpose.

The approach we have been developing, which takes confirmation and explanation to be philosophically justified by ontic rather than linguistic relations, also helps address Kuhn's problem of incommensurability and partial communication.[32] Kuhn almost saw this, but did not quite get its import, saying that

> despite the incompleteness of their communication, proponents of different theories can exhibit to each other, not always easily, the concrete technical results achievable by those who practice within each theory. Little or no translation is needed to apply at least some value criteria to those results.[33]

This relates to our previous argument for the primacy of demonstrations. Kuhn said that "the exhibit of impressive concrete result will persuade at least a few of them that they must discover how such results are achieved. For that purpose, they must learn to translate, perhaps by treating already published papers as a Rosetta stone or, often more effective, by visiting the innovator, talking with him and watching him and his students at work."[34] Kuhn, again without seeming to recognize the evolutionary opening, said that "[some of these scientists], if the new theory is to survive, will find that at some point in the language-learning process they have ceased to translate and begun to speak like a native. No process quite like choice has occurred, but they are practicing the new theory nonetheless."[35] Kuhn did not explore what this process might be, but I contend that the approach we have been developing fits the bill. On my view, what has happened is a replication event—they have copied (perhaps with some errors) the model by practicing the associated behavior and modifying their way of thinking to accord with it. This is not choice per se, but rather reproduction. Language comes later.

Order of Justification—Methods Follow Values

The order of justification is important, and I want to be clear which direction I think the justification goes. The issue is complex because there are two notions of justification that are relevant but that are easily confused. The first is the sense that applies prior to language where a course of action may be said to be warranted or justified for some purpose because it is causally effective. This is the sense of justification that applies in the ontic notion of scientific explanation and confirmation—causal relationships determine the relevance relations. This is the sense in which one would say that a test has justified a model, for the correctness of the model explains why it works. However, there is a second important sense of justification that occurs in the realm of the asking and giving of reasons. This is a kind of justification that occurs in language and involves providing a theoretical account. The difference between these two senses tracks, to a first approximation, the difference between knowing how and knowing that. The relationship between these deserves separate treatment that I must defer. My claim here is that values are fundamental to both and that values provide the normative standards for judgment and justification.

Scientific methods should be followed because they follow from scientific values. More than that, scientific methods should be developed, judged, and refined in relationship to the core scientific values, which is always what they are meant to serve. Regarding the justification of these values, Kuhn wrote, "Though the experience of scientists provides no philosophical justification for the values they deploy (such justification would solve the problem of induction) those values are in part learned from that experience, and they evolve with it."[36] The philosophical justification is not itself a scientific task, but it cannot be divorced from the practice of science.

Returning to the question at hand about the order of justification, it is not that one should be intellectually humble and courageous because these traits follow from scientific methods. Lakatos had it right that the opposite is the case, though the influence from Falsificationism led him astray about this and also caused him to miss Kuhn's real insight. He thought that Kuhn's view reduced science to "mob psychology."[37] In that matter, he was wrong; Kuhn was arguing for the primacy of values, not psychology. The approach that we have been articulating in this book is Kuhnian in this sense. Kuhn admitted that he had no answer to how a value-based enterprise of the sort he recommended could develop as science does, repeatedly producing powerful new techniques for prediction and control.[38] However, he thought that this put the approach in no worse a position than traditional rule-based views, as none was in any better position to solve Hume's general problem of induction. We have not taken on that philosophical chestnut here either, but the value-based approach has provided plenty of new conceptual ground to explore.

The fundamental scientific motivation is to discover truths about the empirical world. To satisfy our curiosity requires that we eschew authority and resolve to heed the evidence alone—to courageously follow it where it leads and to humbly accept what it shows. This means that we must follow practices and methods that provide real tests. Furthermore, it means that scientists should behave in ways that exemplify these values.

We clapped our hands red   What does this look like in practice? There are many ways it is expressed in different circumstances, but I will give just one example from Richard Dawkins, who immediately recalled this story when I told him about my thesis about the virtues of the scientist over lunch at New College, Oxford, many years ago. The story involved a zoology professor during Dawkins's undergraduate days, and it had made such a lasting impression on him that he had recounted it at least twice in his writings.

The professor, a senior lecturer in the Oxford zoology department, had long held that one feature of the cell was an artifact and did not actually exist. However, a visiting American lecturer presented evidence powerful enough to convince even this skeptic. Dawkins recalled how after the lecture this professor, the "elder statesman" of the department, "strode to the front of the lecture hall, shook the American warmly by the hand, and declared in ringing, emotional tones: 'My dear fellow, I wish to thank you. I have been wrong these fifteen years.'" It was not just his admission of having been mistaken that was noteworthy, but also the spontaneous response of the audience of scientists and science students who burst into applause. "We clapped our hands red," Dawkins related, "Can you imagine a government minister's being cheered in the House of Commons for a similar admission?" [39] Dawkins cited this episode as evidence that "scientists, more than others, impress their peers by admitting their mistakes."[40]

Someone might object that the attitude expressed by Dawkins's professor is not distinctive and is really just like good sportsmanship when the other team wins. However, although both share a praiseworthy, positive feeling of respect and good will, there is a significant difference between these two practices. Sportsmanship is expressed when the loser acknowledges a deserved victory by their opponent. The good loser is thought of as honorable, whereas a sore loser is thought of as unworthy of the sport. However, in both cases the loser remains the loser because the goal for each player in the game was to win over their opponent. That is the central point of the game. Not so in science. Although there can be a fierce rivalry to be first in science, and to be recognized as the discoverer, as we have noted previously that is only a secondary goal and an external good. The primary goal in science—its central, guiding purpose—is to find out what is true. In that sense, there are no losers in science when a discovery is made, for everyone has their curiosity satisfied. This does not mean that one might not feel disappointment at not having made the discovery oneself, or even envy for the other's priority. However, with regard to one's primary goal, this is no loss, but a win for every scientist. The audience's response that Dawkins described is an authentic expression of this deeply shared community value.

Intellectual Courage

Intellectual Courage in Science
This takes us back to Lakatos's point, which was an insight worthy of respect. Courage and modesty do indeed provide at least part of a rationale for something that is essential for scientific methodology.

Exemplary scientists rated courage somewhat lower than other virtues in our survey. This was somewhat unexpected, and I attribute it to authentic modesty on the part of most scientists. Of course, courage is important for science, many said, but more so in the past when scientists still had to struggle against established powers. Many mentioned the courage of Galileo, who was tried by the Inquisition and threatened with torture for arguing that the earth moved around the sun, which challenged the authority of the Church's teachings that the earth was fixed at the center of the created universe. Others mentioned the courage of scientist whistleblowers, like the biochemist Jeffrey Wigand who made public the scientific evidence of health dangers of smoking that the tobacco industry had kept hidden,[41] who risked their jobs in bringing the truth to light. Others mentioned the courage that is required to lodge complaints against scientific colleagues if one discovers that they have been fabricating or falsifying data, or engaging in some other form of scientific misconduct. But most thought that courage was not typically an important feature of science.

Courage is so strongly associated as the core character virtue of knights and soldiers that it is sometimes hard to see its broader application. And yet, even setting aside extraordinary cases like Galileo, whistleblowers, and the like, there is a clear sense in which courage is important in scientific contexts and may even be seen as overlapping the classic knightly virtue. Central to every knightly story and legend is some series of trials that would-be knights must pass to prove their worthiness. Can one pass the test of the sword? How keen are one's senses, and how dexterous are one's fingers? Indeed, this is the basic notion of proof—it means to undertake and to pass a test. The more severe the test, the worthier the knight. Courage is not the only virtue that a knight requires, but it is essential for every test, because failure and defeat are always possible. Considering courage in this way, the parallels to the vocation and practice of the scientist become clearer.

Intellectual courage is required in the character of scientists, who may not be passive or complacent but who should daily test their mettle. As we have seen, this is what is expected by the idea of humility to evidence—scientists must submit their ideas to severe tests and stand ready to give up even their favored hypothesis if the evidence goes against it. Scientists require courage to confront real tests of their hypotheses, especially knowing that their precious ideas may be incorrect. Scientists require courage in their relation to scientific competitors, including as we saw, readiness to honorably accept the verdict of the evidence presented by someone who may scoop them or prove their ideas wrong. Scientists require courage to proceed with their research, knowing that most tests will end in failure. Even the gathering of data may require courage, including physical courage if that task involves dangers to one's body, as it sometimes does. Scientists regularly put their health at risk, and it is not uncommon for them to show their scars or trade stories of close calls in the lab or in the field. One could easily extend this list.

Aristotle's account of courage in general   Courage is essential not only because of the nature of scientific practice but also because of the nature of virtue. Straightforward examples involve courage in warfare, but Aristotle speaks of its application for all emergencies that involve death—one knows that performing the right action puts one at personal risk and it is natural to be frightened, but it is courage that allows one to overcome that fear. Facing death is, of course, the extreme case, but courage in at least some degree is needed to do the virtuous thing in any situation that involves confronting risk and overcoming fear.

A coward is someone who is never able to overcome fear to do what should be done when confronted by a challenging situation. However, courage does not mean that one should rush into battle or plow forward no matter what the circumstances; only the foolhardy rush in where angels fear to tread. Such persons are rash, and that is hardly a virtue. Courage stands balanced between cowardice and rashness. The truly virtuous person has the practical wisdom to judge when action is required, and courage allows one to follow through, whether the challenges are large or small. On Aristotle's view, being fearless, in the strict sense of not experiencing fear at all, is not a good thing, for fear is an appropriate feeling in fearful circumstances. The person of courage is not fearless but rather is able to act appropriately in spite of fear, in the service of the final good of excellence and human flourishing. To be courageous is to fear the right things, in the right manner, and at the right time, from the right motive.

Curiosity Conquers Fear

How does this apply for scientists? In science, the goal is to discover truths about the empirical world. And, as we have seen previously, curiosity has an emotional and motivational element, overcoming fear of the unknown by its interest in the unknown. This is an emotional intelligence that drives scientists. They are explorers and frontier seekers. As we saw, in many cases the instinct of curiosity increases the chance of finding new resources and allies. Through some mechanism or other, curiosity conquers fear in nature.

Is it that curiosity substitutes for courage in such cases? Various writers have thought so. The novelist James Stephens seemed to hold that kind of view, saying, "Curiosity will conquer fear even more than bravery will."[42] Victor Hugo offered a similar view, saying, "Curiosity is more powerful than fear."[43] Hugo took curiosity to be "a form of feminine courage,"[44] but I see no good reason to think that it is gendered as he suggests; particular cultural and historical circumstances may be more or less conducive to the training or expression of one or another virtue. Our interest is not that question but rather whether curiosity is a form of courage. There are some reasons to think that it is not. For example, one may be curious to know what is in the cave but not have the courage to enter to find out. This suggests that they are separate virtues.

On the other hand, it also seems that someone whose curiosity about finding the answer is higher would, accordingly, be more likely to proceed. So, perhaps what curiosity does is not bolster courage but reduce fear. Marie Curie seemed to hold that view, famously opining that "Nothing in life is to be feared, it is only to be understood."[45] This seems to be a characteristically scientific way of putting the issue, making it sound like scientific courage is a sort of intellectual fearlessness.

One finds a similar sentiment expressed by Richard Feynman as he reflected on the pervasive doubt and uncertainty in science that we have been considering in this chapter. "I can live with doubt and uncertainty," Feynman said, "I think it's much more interesting to live not knowing than to have answers which might be wrong."[46] Again we hear the connection to curiosity—not knowing is more interesting than having false answers. He goes on to explain the emotional basis of this quintessentially scientific attitude: "I don't feel frightened by not knowing things, by being lost in a mysterious universe without any purpose, which is the way it really is, so far as I can tell. It doesn't frighten me."[47] Here Feynman is touching on matters that edge into religion, but it is not clear exactly what he is claiming about his state of mind. The literal meaning suggests that curiosity simply reduces the feeling of fear. On the other hand, what Feynman is expressing here is commonly recognized as a kind of existential courage in the face of a possibly purposeless universe. So, perhaps this is not fearlessness at all, at least not in Aristotle's more precise sense of the term. It may be that higher curiosity simply enables one to screw up one's courage enough to be able to overcome one's fears or existential dread.

It is useful to see how virtue theory helps set up this question and provides a framework for proposing and judging alternative possible answers, but we do not need to resolve this issue now. For our purposes here, it is not important to figure out which of these possibilities Curie or Feynman had in mind, or to make the case for a particular way of expressing their point. Whatever turns out to be the best way to express the mechanism, the key point is that these virtues are given a structure in science by their relationship to curiosity. Curiosity requires intellectual courage, and sometimes physical courage as well.

The First Step to Wisdom

Socrates and Science

As Socrates taught, to recognize and accept that one does not know is the first step to wisdom. In the *Apology*, Plato described the trial of Socrates, in which he was tried by the court of Athens for not believing in the Athenian gods, for corrupting the youth of the city with his relentless questioning of the political and cultural norms of Athens, and for being "a curious person, who searches into things under the earth and in heaven."[48] He faced his death, to be administered by drinking poison hemlock, with courage and equanimity. No evil can happen to a good person, either in life or after death, he argued. The courage he displayed is the same that he embodied throughout his life. It was based on a humble recognition of his own ignorance and a relentless search for answers to pressing questions, even in the face of adversity. As for death, perhaps it is a state of nothingness and complete unconsciousness, in which case it is to be like a perfect night's sleep. Or perhaps in death the soul moves on to another world where other departed souls have gone before, in which case he may continue his conversations and his search into true and false knowledge. While alive, we are in no position to know which is the case, but Socrates stood ready to face the test, willing to see what discovery it might bring.

Although science does not usually require courage or intellectual humility to this degree, we have seen that the mindset and methodology it aspires to is very much in line with Socratic virtue. Science is a practice where ignorance and failure are essential to the enterprise. Continuing one's exploration in the face of constant failure and doubt does require courage. I have argued that, in science, curiosity is the root of courageous humility. This should not be surprising in that there is a close connection between curiosity and confusion; because curiosity involves new things that we do not yet understand, most of a scientist's research career is spent in a state of confusion. Intellectual humility involves a willingness to question oneself and to change one's mind, acknowledging when one is wrong. The methodology of science is based on trying to do that systematically. The secret to being an exemplary scientist is being comfortable with being confused. The Socratic first step is recognizing the things that one does not know and then being eagerly willing and unafraid to let the evidence take you to where it will. This, I suggest, is the special sort of courageous humility that

Lakatos spoke of as the best human qualities that were embodied in the methodology of science.

Follow Humbly Where Nature Leads

Thomas Huxley, Darwin's bulldog advocate whose larger-than-life personality was hardly an example of personal humility and who had little positive to say about religion, was nevertheless moved to speak of the importance of humility to evidence in religious terms. Reflecting on the proper scientific mindset in a letter explaining why he could neither affirm nor deny the religious doctrine of immortality, he wrote that science teaches an important truth about what is needed to make discoveries about the world: "Science seems to me to teach in the highest and strongest manner the great truth which is embodied in the Christian conception of entire surrender to the will of God. Sit down before fact as a little child, be prepared to give up every preconceived notion, follow humbly wherever and to whatever abysses nature leads, or you shall learn nothing."[49]

7 Creative Conflict

The world of poetry, mythology, and religion represent the world as a man would like to have it, while science represents the world as he gradually comes to discover it.

—Joseph Wood Krutch[1]

Culture Wars

Darwin's Tomb: Two Views

It is not only British royalty and churchmen who are buried in London's Westminster Abbey. Literary giants including Geoffrey Chaucer, Charles Dickens, Rudyard Kipling, and Jane Austen are interred or memorialized in Poets' Corner. Under the same soaring roof are buried some of the greatest scientists, including Joseph Lister, the father of antiseptic surgery, and Charles Lyell, whose research helped establish that the age of the earth should be measured in millions of years, and Sir Isaac Newton himself. Charles Darwin is buried here as well, next to Newton, beneath a white marble stone with a simple inscription giving his name and dates of birth and death, which belies the significance of his discovery but was in keeping with his modest character. The Rev. George Prothero, canon of Westminster, who preached the funeral sermon to the many friends and dignitaries who attended, spoke of Darwin as "The greatest man of science of his day" but also as someone who "possessed a sweet and gentle disposition" and who was "entirely a stranger to intellectual pride and arrogance." We have already seen how intellectual humility is a scientific virtue, but Prothero was thinking in religious terms and he saw fit to link Darwin's character traits to virtues that he took to be essentially Christian. Noting his "pure and earnest love of truth" and how Darwin put forward his discoveries "with the utmost modesty," knowing that others would find them challenging, Prothero concluded that "[s]urely in such a man lived that charity which is the very essence of the true Spirit of Christ."[2]

This is a remarkable statement, especially in light of the rancor that Darwin's findings engendered in some Christians who felt that evolution threatened their faith. I observed one example of this hostility during a visit to Westminster Abbey that I have described elsewhere.[3] A young woman, probably in her mid-teens, had noticed Darwin's tomb and, in a loud voice with a noticeable accent that placed her as coming from the American South, called her mother over to look. "Why would *he* be buried in a church!" she said with clear disdain. To my astonishment, she then gave Darwin's gravestone three sharp kicks. Her father appeared a few moments later and they pointed out Darwin's tomb to him. Loudly uttering his own opinion—"This is not where he belongs"—he too stomped his heel on Darwin's grave. Having studied creationism, I knew that fundamentalists view evolution as evil and see Darwin as someone who questioned the authority of Scripture and what they take as the absolute Word of God, but it was still shocking to witness this desecration of Darwin's tomb.

Will science, religion, and poetry ever be able to peacefully coexist under one roof, or is conflict inevitable? As we noted earlier, C. P. Snow identified the gap between the cultures of science and literature half a century ago, and there has been rather little progress in the intervening decades to bring them together. Conflict between the cultures of science and religion has continued for centuries. There are many possible explanations offered for such conflicts, the most common being that they involve struggles for power and dominance not unlike those among rival states or parties. Following his leadership in the overthrow of Charles I and the royalists, Oliver Cromwell, who credited his victories in battle to God's providence, became Lord Protector of England, Scotland, and Ireland, and upon his death was buried in Westminster Abbey. A few years later the royalists regained power, exhumed his body, and then hung and beheaded it before burying him in a pit. The poet John Milton had lauded Cromwell in verse while he was alive, and some two centuries after his death the essayist Thomas Carlyle touted him as an exemplar of the heroic Great Man. Is cultural conflict of a kind, with each side caring only to keep or gain power, and the winners, however temporary, writing and rewriting history with the help of its favored poets?

Some would call this view of cultural conflict a form of cynicism, while defenders would call it realism. Although there is certainly some element of power that is at play in these and other culture wars, in general this is too simple a view. At a deeper level, such conflicts are not primarily about power but rather about values.

Clash of Cultures

Cultures are structured by norms that are so taken for granted that they are often not recognized as such until they are seen in contrast with different ones. Norms simply seem, well, normal. Cultural norms are expressed in practices that include characteristic behaviors and expectations. The rituals of food and drink, the conventions of politeness and deference, the customs of daily life that fix the borders between familiar and foreign ... these are learned at our parents' knees and then invisibly rule our lives as adults. Sociologists focus on these kinds of practices—social controls that provide stability, predictability, and familiarity. Philosophers, however, look at cultural norms at the level of values. What do members of the culture care about? What do and should they aspire to? Who are their heroes? What is their central, guiding purpose? Practices express these values. Because the values themselves are often tacit, it is not always easy to identify them, but they have a coherent logical structure of their own.

For the culture of science, as we have seen, the core values involve the search for empirical truths. The methods and practices of science are aimed at the honest satisfaction of curiosity. Testable hypothesizing, measured observations, crucial experiments, and other distinctive procedures are the warp and woof of evidence-based reasoning. The practice of such methods and behaviors, with their distinctive movements and materials, is the visible culture of science. Less visible, but equally important, are the traits of the scientists who perform these—the mindset they seek to embody to help them excel in their practice. The virtues that stand at the intersection of values and practices constitute the character of science.

However, cultures can come into conflict. Indeed, tacit values most often rise to the surface and are articulated when one culture comes in contact with another that has different basic values. It is in such circumstances (such as in interdisciplinary research) that one first notices such differences, when conflicts arise over norms that previously were simply assumed within each group. It is significant that Robert Merton's paper on the norms of science was written in response to antiscientific, nationalistic challenges in the years before World War II. The scientific norm of universalism, for example, which holds that scientific claims should be judged on impersonal criteria without reference to a scientist's race or religion, was seen in high contrast with Nazi views about rejecting the supposedly distinctive "Jewish science."

This chapter will examine a few ways in which scientific culture has come into conflict with other practices. We will first look historically at the conflict between science and religion, and then at the tension between science and the humanities that Snow identified. As part of the latter, we will look in particular at the clash between science and postmodernism and try to discern its causes. A virtue-theoretic approach provides a fresh perspective on these long-standing conflicts. It may offer a useful way to think about points for negotiation and potential terms of peace. There is much that can be said about this, but I will focus on one difference in basic values that seems significant for explaining some of these conflicts, namely, the different roles of, and the tension between, curiosity and creativity as they relate to the place of truth in their practices.

Presume Not God to Scan

In the Beginning

For many Christians, the conflict with science begins right at the beginning, in Genesis, where God created the heavens and the earth.[4] For those who read it "literally," the Bible is thought to provide a divinely given account of the creation, structure, and history of the world, so if science contradicted that story, then it was science that had to be rejected. The individuals who stomped on Darwin's tomb in Westminster Abbey likely held this sort of creationist view. Of course, those who privilege Scriptural texts over scientific evidence do not stop with evolution; they reject a host of scientific findings from astronomy, biology, geology, linguistics, and beyond.

This is not to say that those who reject science in this way agree with each other on all points. Even a "robust" view of Scripture leaves room for interpretation. Take the issues of the age of the earth and Noah's Flood, for instance. Young Earth Creationism (YEC), which holds that the Bible reveals that Creation occurred only 8,000 to 10,000 years ago, typically also believes in a global flood and appeals to it to explain various geographical features of the world. Old Earth Creationism (OEC), which interprets Genesis in ways that allow the days of Creation to encompass eons of geological history, is more likely to view the flood as local or "tranquil," so that it did not leave global traces. Young Earth and Old Earth creationists oppose each other as vehemently as they oppose evolution. Intelligent Design Creationism (IDC), which includes both Young-Earthers and Old-Earthers, tried, with only limited success, to forge a strategic alliance between the groups by agreeing to temporarily set aside such differences in order to fight against evolution as their common enemy.

So, what unites creationists besides their opposition to evolution? In addition to their agreement in taking the authority of Scripture over the evidence of science, the key points of agreement have to do with what they take to be at stake. Creationists fear that accepting evolution and a scientific picture of the world leads to a loss of values. They take morality as emanating from God's authority. God is the divine legislator. In bringing forth the world and everything in it through His creative power, God's word defines not only what is, but also what ought to be—what is right and what is wrong. To question biblical dogma is, on their view, to undermine morality itself.

With such disagreements with science regarding both facts and values, religious antagonism to science seems inevitable.

Science vs. religion? The stereotypical view of the relationship of scientific and religious worldviews is that they are opposites. Even foes. Darwin's own religious faith faded over the years, though he was never antagonistic toward religion. The same could not be said about Thomas Huxley, the pugnacious advocate of evolution, who spent much effort countering religious opposition to science. "Theology and Parsondom," he complained in a letter to a correspondent, "are in my mind the natural and irreconcilable enemies of Science." He believed that "a new Reformation" was nigh that would accommodate the scientific view and said that he wished to live a few more decades so he might "see the God of Science on the necks of her enemies."[5]

In the United States, the view that science and religious dogma were at war was promoted forcefully by the historian Andrew Dickson White, one of the founders and first president of Cornell University, who had been a politician and served as a U.S. diplomat before turning to education. White put forward his conflict thesis in his two-volume *A History of the Warfare of Science with Theology in Christendom,*[6] which argued that dogmatic Christian theology had hindered the progress of every field of science, from geology to chemistry, physics and medicine. One chapter dealt with the shift from interpreting the heavens in terms of supernatural signs to seeing it as governed by natural law. As for Darwinian evolution, he described it as affecting the theological world "like a plough into an ant-hill"[7] and went on to quote several dozen examples of the attacks from eminent clerics from a wide variety of Christian denominations in England, France, Germany, and America, who denounced Darwin and his findings in the harshest of terms. Even more than John William Draper's earlier book on the topic,[8] which articulated a similar thesis, White's book is responsible for the view that science and religion are inevitable adversaries. White was also influential in making science a standard part of the university curriculum, setting up Cornell as a model; it was to be "an asylum for Science—where truth shall be sought for truth's sake, not stretched or cut exactly to fit Revealed Religion."[9]

Later historians have argued that White's warfare thesis depended on a skewed history of the relationship between science and religion that overemphasized the instances of conflict and ignored confluences. Fair enough, but some critics equally oversimplify their case against him. White's own view was not antireligious in the way in which it is now commonly made out. For instance, White was explicit that his argument was against dogmatic theology, not religion in general or even any particular denomination of Christianity. He called one American Anglican cleric's view that if evolution is true, then Christian beliefs must be a lie, "an exceedingly dangerous line of argument."[10] Even the founding of Cornell was hardly an example of the warfare thesis, though it was the vehement religious objection to its proposed charter (which explicitly prohibited control by any single political party or religious sect) that had galvanized White's original writings on the topic. In fact, the vision of "a place where the most highly prized instruction may be afforded to all—regardless of sex or color," that might "turn the current of mercantile morality" and "temper & restrain the current of military passion" and, yes, where truth could be sought for truth's sake, were explicitly in line with the Quaker beliefs of his cofounder, and the university's namesake, Ezra Cornell. White mentioned this in the introduction to his book and pointed out that it "certainly never entered into the mind of either of us that in all this we were doing anything irreligious or unchristian."[11] As for the book itself, White went out of his way to say that his thesis was not about religion per se but rather against the kind of religious dogma that saw science as a threat. With such dogma excised from religion, there was ample room for cooperation:

> My conviction is that Science, though it has evidently conquered Dogmatic Theology based on biblical texts and ancient modes of thought, will go hand in hand with Religion; and that, although theological control will continue to diminish, Religion, as seen in the recognition of "a Power in the universe, not ourselves, which makes for righteousness," and in the love of God and of our neighbor, will steadily grow stronger and stronger ... Thus may the declaration of Micah as to the requirements of Jehovah, the definition by St. James of "pure religion and undefiled," and, above all, the precepts and ideals of the blessed Founder of Christianity himself, be brought to bear more and more effectively on mankind.[12]

This is hardly the view of someone who thought that warfare was inevitable. White's goal was to identify a dangerous strand of dogmatic religious opposition to science, and it can hardly be claimed that he was wrong about that. Indeed, there are several basic reasons why religion of this ilk would prosecute a cultural war against science.

The Galileo affair The most straightforward reason involves cases where there is a conflict over matters of fact. If a religion holds as dogma that some aspect of the world was created in a certain way, then scientific findings that show something different can be threatening. When Martin Luther (whose insistence that Christians are "bound by the Scriptures" over even the Pope ushered in the Protestant Reformation five hundred years ago) learned of the Copernican view that the earth moved around the sun rather than being fixed at the center of Creation, he judged it in relation to what he took to be a biblical truth. He is reported to have said of Copernicus: "The fool wants to overturn the whole science of astronomy. But as Holy Writ informs us, Joshua bade the sun stand still, and not the earth."[13] Copernicus's view was indeed revolutionary in every sense of the term, and in the early years someone with faith in the traditional biblical view could dismiss Copernicus as a fool. However, that became hard to do as the evidence for it grew stronger and the revolution took hold.

The continuation of this episode—the Galileo affair—is a classic example of the "warfare" between science and religion because the Church saw Galileo's research as a direct challenge not only of religious authority but also of the basis of Christian belief. Robert Bellarmine, a cardinal of the Catholic Church and a central figure in the Catholic response to the Reformation, considering how the Church should respond to Galileo's work, wrote that there was no danger in saying that it was better mathematically to *suppose* that the earth moved around the sun, but that it became a serious threat to say that this picture was factually *true*:

> To say that on the supposition that the earth moves and the sun stands still all the appearances are saved better than on the assumption of eccentrics and epicycles, is to say very well—there is no danger in that, and it is sufficient for the mathematician: but to wish to affirm that in reality the sun stands still in the center of the world, and only revolves upon itself without traveling from east to west, and that the earth is located in the third heaven and revolves with great velocity about the sun, is a thing in which there is much danger not only of irritating all the scholastic philosophers and theologians, but also of injuring the Holy Faith by rendering false the Sacred Scriptures.[14]

Bellarmine was unwilling to consider that Scripture could be wrong, but he did concede that theologians would need to rethink their interpretations if they were to see "a real demonstration" that the sun stands at the center of the world and that the earth revolves around it. However, he doubted the possibility of such a demonstration and concluded that "in case of doubt one should not abandon the Sacred Scriptures as they have been expounded by the Holy Fathers."[15]

Galileo's reply to Bellarmine's argument is an exemplar of the application of science's epistemic values. While showing that the heliocentric model fits the observations is insufficient to demonstrate its truth conclusively, he said, in the end "no other greater truth can or ought to be sought in a hypothesis than its correspondence with the particular appearances," which the alternative models fail to do.[16] In his *Dialogue on the Two Chief World Systems*, Galileo had presented evidence that went well beyond what Copernicus had offered, and his pioneering telescopic observations overcame the major objections that had been made against the Copernican model. Galileo still did not have a way to show directly that the earth moves, which he freely acknowledged, but in his new account of physics he was able to explain why we would not detect its motion in our ordinary experience. As for science's relationship to religion, he offered a view that reconciled the two; in his *Letter to the Grand Duchess Christina of Tuscany*, he wrote that Scripture teaches "that the glory and greatness of Almighty God are marvelously discerned in all his works and divinely read in the open book of heaven."[17] One should use science to consult the Book of Nature directly to discover the workings of Creation.

However, although the elements of a peace accord were already in place, this conflict between science and religion was not resolved in time to help him. The Church was not yet ready to reinterpret its understanding of Scripture. Brought to trial by the Inquisition for suspicion of heresy, Galileo was found guilty. Though he lost the battle, the episode came to be seen as a turning point in the culture war—the scientific worldview was in the ascendancy and its evidence-based approach became seen as the trustworthy methodology for answering questions about the natural world.

Methodological naturalism   A second major factor in this conflict involved the most important element of science, which is that it requires that scientific explanations of natural phenomena be understood in terms of natural laws. Recall that science was originally called natural philosophy because of this conceptual innovation; science, as a point of method, does not allow appeal to the supernatural in its explanations. Science explains lightning by the laws of electricity rather than as thunderbolts thrown by Zeus. It explains pestilence in terms of the germ theory of disease rather than as punishment for sin. This may not have seemed problematic at first, but it set up a potential conflict. Once this pattern of explanation gained traction, it did not take long before scientists began to look for similar explanations for features of the world that the Bible had explained as God's direct creative acts. Such a methodology seemed to leave God out, replaced by impersonal natural laws. Strictly speaking, methodological naturalism says nothing about God as a metaphysical being, but it may still be seen as threatening, if only because appeal to supernatural explanations is common in religion, especially in religions that retain a premodern view. An insistence upon supernatural explanations for specific features and phenomena of the world in terms of God's intelligent design is a central element of the creationist agenda and creationists' ongoing attacks upon science.[18] However, though such views remain a significant cause of conflict today, especially in the United States, it reflects a particular theological viewpoint that hardly represents mainstream theological views that see no necessary conflict with science.

There are any number of ways one could illustrate this, but to address that worry about natural laws one need only look to the pioneers of the scientific revolution itself. The seventeenth-century natural philosopher Robert Boyle was one of the founders of modern chemistry and a key figure in the scientific revolution. He was also a devout Anglican who saw the scientific picture of a world run by natural laws not as threatening to his Christian faith but as supportive of it. Boyle and other natural philosophers viewed the world as composed of corpuscles (elementary particles similar to what the ancient Greeks had called atoms but allowing of division into smaller units) whose motion and interactions determined material phenomena. Thinking of how this model of the world might fit with religious beliefs, Boyle wrote:

> It is intelligible to me, that God should at the beginning impress determinate motions upon the parts of matter, and guide them, as he thought requisite, for the primordial constitution of things; and that ever since he should, by his ordinary and general concourse, maintain those powers which he gave the parts of matter, to transmit their motion thus and thus to one another.[19]

Boyle was responsible for the well-known clock analogy for understanding the workings of nature: the world runs by natural laws like the mechanism of a clock, with God as the divine clockmaker. Think of the world as a machine that can run on its own. Boyle was not alone in thinking that such a view fit better with the idea of a wise and glorious God who could create an intricate, lawful mechanism that worked as intended from the beginning, without the need for subsequent managerial oversight or fussy assistance to control the motions of the parts.[20] Boyle did allow that God could, and perhaps sometimes did, intervene miraculously in nature but thought that no part of science.

This theological notion of God as creator of natural laws provided a framework within which science and religion could be seen as working together. God created the laws, and scientists worked to discover them. This is not to say that the reconciliation occurred easily in all areas, for there was theological inertia that had to be overcome with regard to a variety of specific beliefs about creation. The creation of organic life remained a stumbling block for many, but after Darwin's discovery of the law of evolution by natural selection, even on this topic the religious came to share Boyle's view that a theological picture of God as the creator of natural law was ennobling. To give just one example, in the second edition of the *Origin*, Darwin quoted a letter he received from the Anglican clergyman and naturalist Charles Kingsley expressing just this point:

I have gradually learnt to see that it is just as noble a conception of Deity, to believe that He created primal forms capable of self-development into all forms needful *pro tempore* and *pro loci*, as to believe that He required a fresh act of intervention to supply the *lacunas* which he himself had made. I question whether the former be not the loftier thought.[21]

Different Mindsets

This brief review covers the historical culture wars in familiar ways. Battles over specific empirical facts and battles over modes of explanation were fought again and again. For the most part, the terms of a peace settlement have now been accepted by both sides. So why do fights continue to break out? I suspect it is in part because differences have not been fully worked out with regard to values. A virtue-theoretic analysis may help identify where points of conflict remain and perhaps suggest ways to resolve them.

In laying out his theory of virtue, David Hume dismissed the "monkish" virtues in a biting aside:

Celibacy, fasting, penance, mortification, self-denial, humility, silence, solitude, and the whole train of monkish virtues; for what reason are they everywhere rejected by men of sense, but because they serve to no manner of purpose; neither advance a man's fortune in the world, nor render him a more valuable member of society; neither qualify him for the entertainment of company, nor increase his power of self-enjoyment? We observe, on the contrary, that they cross all these desirable ends; stupefy the understanding and harden the heart, obscure the fancy and sour the temper. We justly, therefore, transfer them to the opposite column, and place them in the catalogue of vices...[22]

Hume was opining about the utility of certain religious virtues and practices for general human flourishing, which is not our topic here. However, it is worth considering how far the pattern of conflict extends. At least at first glance, it does seem that there is little overlap between religious and scientific values.

None of the three theological virtues—faith, hope, and charity—clearly figure in the set of scientific virtues. As we noted previously, Merton did identify communism—the view that scientific knowledge was not owned even by the discoverer but was to be freely shared with researchers—as one of the institutional norms of science, but that is quite different from Christian charity. Epistemic charity is to be eschewed, for a scientist may not presume or donate warrant to the cause of a theory—it must earn its own way. Similarly, science rejects the idea that one's hope (i.e., what hypothesis one might wish to be true) should make any justifiable difference to what is in fact true. Hope, in a more general sense, might be argued to fit with what is the generally positive, optimistic attitude of science, with its focus on progress, but that would be a charitable interpretation. And faith, which is arguably the most important of the theological virtues, would be seen as a vice for a scientist. Scripture makes the blindness of faith a virtue, "For we walk by faith, not by sight."[23] Put another way, the claim is that we should live not by the appearances of things seen but by faith in the unseen, namely God; we are to see the world through the eyes of faith. Faith is understood here, by definition, to be quite the opposite of what science holds. It is "the substance of things hoped for, the evidence of things not seen."[24] Of course, rather than thinking of faith in relation to particular propositional beliefs, one may think of it more in terms of where one puts one's heart and energy.[25] A few scientists occasionally speak of having faith in scientific methods or that their perseverance will eventually sort out the data, but mostly this seems to be a way of expressing confidence in the long-term efficacy of evidential methods. In any case, as we have discussed, when it comes to determining matters of empirical fact, science finds dogmatic faith to be anathema. The scientific mindset eschews appeal to authority, so it is not surprising that an institution like the Church, which is organized on obedience to authority, might find science to be threatening.

There is slightly more room for agreement if one looks at the so-called seven heavenly virtues, which include chastity, temperance, charity, diligence, patience, kindness, and humility. The stereotypical view of scientists takes them as being too focused on the intellect to be interested in sex, but that is not the same thing as chastity. However, scientists do affirm patience as a virtue in research. Diligence may be taken as overlapping both with perseverance and meticulousness. As we have seen, scientists do not necessarily find humility in a general sense to be important for science, but they do highly value the intellectual virtue of humility to evidence. Kindness and temperance are not special virtues in science, but scientists could certainly affirm them as general virtues.

Mystery and mindset However, there are other differences between the scientific and religious mindsets regarding the values they take to be central that can lead to mutual antagonism. In particular, some religious believers see scientific curiosity as undermining wonder at the glory of creation by reducing it to mundane explanations. Consider the rainbow, which for the Christian is, as a sign from God, a reminder of a promise after the Flood to never again destroy the world. While most Christians are happy to read this symbolically, creationists take the story of Noah's flood as factual. An explanation of the rainbow solely in terms of the natural laws of optics, some would say, removes its mystery and undermines its miraculous nature. It is true that scientists have no truck with miraculous "explanations," but that does not mean that they do not appreciate the mysterious. Einstein, for example, wrote:

> The most beautiful experience we can have is the mysterious. It is the fundamental emotion that stands at the cradle of true art and true science. Whoever does not know it and can no longer wonder, no longer marvel, is as good as dead, and his eyes are dimmed. It was the experience of mystery—even if mixed with fear—that engendered religion. A knowledge of the existence of something we cannot penetrate, our perceptions of the profoundest reason and the most radiant beauty, which only in their most primitive forms are accessible to our minds: it is this knowledge and this emotion that constitute true religiosity. In this sense, and only this sense, I am a deeply religious man. ... I am satisfied with the mystery of life's eternity and with a knowledge, a sense, of the marvelous structure of existence—as well as the humble attempt to understand even a tiny portion of the Reason that manifests itself in nature.[26]

However, scientists think about this very differently than those who object to the disenchantment of the world. When scientists are shown the discovered mechanism behind some natural mystery, their response is typically an expression of delight: "Very neat," "Wonderful," or even, "That's so cool!" Moreover, such expressions are not about the discoverers, though there is appreciation for their achievement as well, but about the magician of nature itself, and how it pulled off the trick.

Scientists appreciate mysteries as opportunities for discovery. Methodologically, investigations are built around unanswered questions, unexpected patterns, and anomalies within theories or experiments. This helps explain what to outsiders is an odd attitude that scientists have even with regard to their own findings. For example, scientists would be very interested to find that a problem they thought had been resolved is once again in play, as it reopens the search. A recent case in point was the experimental work at CERN's Large Hadron Collider that provided evidence for the predicted Higgs boson. Theorists were happy to have been shown to be right. But they would still have been happy if the experiment had gone against expectations, which would have indicated that there was a deeper puzzle to solve.

So, scientists love a mystery as much as the next person. What they do not abide is mystery for its own sake. Wonder does not require mystery. What they most object to is mystery-mongering, as this opposes the basic goal of their practice. The scientist sees a mystery as something that might be solved. One pauses to wonder and then to investigate. That is what makes the scientist feel alive. Curiosity killed the cat; satisfaction brought him back.

Curiosity—a Christian sin?   At first glance, at least, we have to admit that science does not seem to place a very high value on many characteristically religious virtues. Even where there is agreement about some particular virtues, these are viewed differently. It is equally easy to find cases where religious believers have taken as dim a view of science's virtues as Hume had of the monkish virtues. Indeed, the core scientific virtue of curiosity has been seen not as a virtue at all but as a vice—a sin—by important Christian theologians throughout history. Consider Tertullian, the second-century church apologist, who insisted that "No curiosity is required of us after Christ Jesus, no investigation after the Gospel."[27] This did not mean that Tertullian thought that a Christian needed no knowledge beyond what the Bible provided, but it was wrong and dangerous to deny what it did provide. Do not question; just believe. Even Augustine, who also took faith to be epistemically primary but who moved toward the idea of "faith seeking understanding," was nevertheless dismissive of curiosity, comparing it to "that lust of the flesh which lies in the gratification of all senses and pleasures." He warned against "a certain vain and curious longing, cloaked under the name of knowledge and learning."[28]

Even after the Reformation, such attitudes persisted. Martin Luther wrote that if one were to rely on and believe the judgement of reason, then God's word and the articles of Christian faith would be seen as "foolish, weak, absurd, abominable, heretical, and devilish things."[29] The reference here is to a major theme of 1 Corinthians, which suggests that skeptics will take believers to be fools, but that it is virtuous to be "fools for Christ." Why? Because human reason cannot hope to understand God's ways or see the world as it really is. "Because the foolishness of God is wiser than men."[30] On such a view, it is science that is the real fool for doubting when it ought only to believe. This sort of religion sees science is a threat.

As with its rejection of certain empirical facts about the world, we might trace this antipathy to what will become a core scientific virtue back to Genesis. Eve's sin was all about this conflict. Eve is characterized as disobedient, but she might rather be seen as curious. Scientists would certainly identify with Eve's inquisitive nature and her desire to seek knowledge even against the command of authority. Genesis depicts curiosity as a temptation to be resisted. Science not only says "Don't resist," but it goes further, viewing curiosity not as a vice but as a basic virtue. For science, Eve's curiosity does not make her a sinner but a role model. If anything, it makes her a candidate for science's matron saint.

The difference of mindsets between science and religion should not be surprising from a virtue-theoretic point of view, as religion and science have distinctive aims and methods. As Galileo put it, quoting an eminent ecclesiastic, "the intention of the Holy Ghost is to teach us how one goes to heaven, not how heaven goes."[31] It would be quite inappropriate to try to artificially force the ideals of either one into a Procrustean bed of the other.

Dialogue Model

Of course, the stereotypical view of the inevitable conflict between science and religion is by no means universal. Most Christians have long accepted Galileo's contention that the earth moves around the sun, and the Catholic Church formally acknowledged its error in 1992, in time for the 350th anniversary of Galileo's death.[32] Except for creationists, few Christians think that their faith or morality is threatened by accepting the scientific findings supporting the 4.5-billion-year age of the earth. Although a sizable portion of Americans still reject evolution, nearly a third by recent count,[33] mainstream denominations, including the Catholic Church, no longer view it as problematic. In fact, in some sense, such accommodation to science is hardly a recent occurrence; many of the early scientists who supported Darwin's discovery were clergymen naturalists.

There is even reason to think that religion has sometimes been a help to the advancement of science. Before his work on scientific norms, Robert Merton was already famous for his thesis that the rise of early experimental science had roots in Protestant Pietism. Puritan religion in the seventeenth century, especially in England, provided fertile soil for the development of science.[34] Merton did not suggest that Puritan doctrines were the causal factor at play so much as some of the sentiments that it emphasized.[35] Many scientists found ways to integrate their religious faith and their scientific research, such as those who held some variation of the view that science helped one read the Book of Nature, or simply that studying nature was a way to glorify God's Creation even if one could never hope to understand His ultimate purposes.

A more common model is that science and religion are in some sense independent of one another because they deal with different subject matter. A variety of complementarity views have been promoted from both the side of science and religion. Paleontologist Stephen Jay Gould's notion of what he called "nonoverlapping magisteria"—often abbreviated with the acronym NOMA—is perhaps the most well-known and well-discussed recent example. According to NOMA, the magisterium of science is the realm of facts while that of religion is the realm of values. (Significantly, this is the same division that we saw as one possible explanation for the gap between science and the humanities.) So long as each magisterium stayed within its proper realm, they could complement one another without coming into conflict. Albert Einstein's assessment may be the pithiest statement of this kind of view: "Science without religion is lame; religion without science is blind."[36] To take just one example from the side of religion, Martin Luther King, Jr. identified a variety of differences that he saw as complementary: "Science investigates; religion interprets. Science gives man knowledge, which is power; religion gives man wisdom, which is control. Science deals mainly with facts; religion deals mainly with values. The two are not rivals."[37]

Conflict. Integration. Independence. These are three of the models of Ian Barbour's classic typology of the relationship of science and religion.[38] Each model, he argued, is problematic for different reasons. The fourth model he offered—dialogue—is different from the others in that it focuses on a process of mutual discussion without settling on a particular conclusion. It involves an exploration of differences and similarities among the presuppositions, the methods, and the concepts of each. A virtue-theoretic analysis provides a new perspective for such a dialogue. It helps explain issues where science and religion continue to have a hard time seeing eye to eye. It also helps identify places where the two may come together. This deserves a fuller treatment than is possible here, but it is worth briefly considering a few further points of comparison.

Virtues as bases for dialogue? Religions more often are based in following revelations or charismatic leaders, so obedience to authority is far more prominent a virtue in their eyes than scientific humility to evidence. But there are exceptions. The fourteenth Dalai Lama, spiritual head of Tibetan Buddhism, regularly emphasized the value of scientific evidence. "[T]o defy the authority of empirical evidence," he once said, "is to disqualify oneself as someone worthy of critical engagement in a dialogue."[39] Richard Rohr, a Franciscan priest and ecumenical writer, commenting on the need for personal transformation that takes one beyond one's identity as a member of some particular denomination, said, "I think my great disappointment as a priest has been to see how little actual spiritual curiosity there is in so many people."[40] It is exceedingly rare to find the terms spiritual and curiosity linked in this way, but this new interest is a promising sign.

There are, of course, various other cases where some religious and scientific views overlapped with regard to one or another important virtue. The Religious Society of Friends (Quakers) hold truth as a core value, and the historical name of the sect was Friends of Truth. In his classic book *The Varieties of Religious Experience*, William James called Quakerism "a religion of veracity."[41] Arising in seventeenth-century England in tandem with the Scientific Revolution, Quakers rejected the special authority of the Bible and regularly spoke in terms of what one could "test" and come to know "experimentally." George Fox, its founder, cautioned those who quoted Scripture without speaking from their own experience, "You will say Christ saith this, and the apostles say this, but what canst thou say?" William Penn saw the Pennsylvania Colony as a "holy experiment" in something like a modern scientific sense, where Quaker principles would be put to the test.[42] In line with Merton's thesis, it comes as no surprise that Friends were well represented among the pioneers of science.[43]

We have already discussed skepticism and its expression in Descartes's method of doubt, which is part of his long argument for the existence of God, and one may find elements of other scientific virtues in unlikely religious places. The great variety of religious views means that it is very likely that, even with regard to scientific values, one can find numerous such exceptions to the stereotypical view of inevitable conflict. As evolutionary ethologist and primatologist Frans de Waal put it, "The enemy of science is not religion. Religion comes in endless shapes and forms." He noted that many people of faith have an open mind and have no issue with science whatsoever, concluding that, "The true enemy is the substitution of thought, reflection, and curiosity with dogma."[44]

The evolutionary biologist David Sloan Wilson, writing about what it would take to find a belief system that combined the best of religion and science, recognized this and, indeed, also spoke of it in terms that fit well with our virtue-theoretic framework. "The first step," he wrote, "is to decide that factual knowledge is a virtue—sacred, if you like—and that value systems must treat statements of fact more respectfully in the future than the past."[45]

No God's-eye view   This is wise counsel, but we should be under no illusion that achieving it in a mutually satisfactory manner will be easy. While truth-seeking is the most obvious common value and it would be better for us all if truth were valued more highly, we must acknowledge that this will not end all dispute, if only because science and religion often differ with regard to their methods for validating truth claims. The curiosity of scientists focuses on what we can learn evidentially, using scientific methods. This leads to a lack of patience with, and even lack of interest in, questions that are not amenable to empirical test. But surely curiosity extends even to depths we know we may never plumb.

On the other hand, some theists hold that only God, as omniscient creator, can see the truth, and science can never pretend to that. Human intelligence can have no God's-eye view of the world. ID creationist Nancy Pearcey, to give just one example, insisted that religion govern matters of fact as well as values. She demanded, as the title of her book suggested, that Christianity be seen as providing "Total Truth."[46] Her title drew from the work of the Evangelical Christian theologian Francis Schaeffer, who proclaimed that "Christianity is not a series of truths in the plural, but rather truth spelled with a capital 'T.' Truth about total reality, not just about religious things. Biblical Christianity is Truth concerning total reality—and the intellectual holding of that total Truth."[47] Such believers conclude that we must give up reliance upon our own limited view and have faith in the view offered to us by God. Not that this resolves the epistemological problem. One might wonder, if our intellect is so limited as to not make out even natural truths, how much more difficult it would be to properly judge among the possible supernatural truths offered by the many possible gods. Appeals to grace and revelation are not satisfactory answers for the scientist, as these return to an unscientific basis in authority over evidence.

Some theists have proposed that mathematics provides the God's-eye view. Does not the fact that the laws of nature obey mathematics suggest that God was a geometer, laying out the circumference and vectors of the world? Perhaps it should not be a surprise that we owe analytic geometry and the Cartesian coordinate system to René Descartes, who compared his ontological proof for the existence of God to a geometrical demonstration. The formal certainty that mathematics and deductive logic provide seems to take one closer to an absolute notion of truth, but even that falls short of a God's-eye view. The twentieth century's preeminent logician Kurt Gödel's proof of the incompleteness of any formal mathematical language undermined the hope that math will provide a complete picture. Gödel did offer his own version of Descartes's ontological argument, but even if one thought that it was sound, the best it would do would be to show that a God's-eye view exists, but not what it is. Perhaps this is sufficient for some religiously minded scientists, but that is a theological conclusion, not a scientific one.

Science does not pretend to provide a God's-eye view. Scientific humility to evidence, tied to the notions of scientific attention, observation, and empirical test, stands in stark contrast to extreme religious views with impossible standards. Again, science is not after Truth in any absolute, metaphysical sense. The scientists' view is situated squarely within the world itself, with ourselves as evolved creatures within it. Mathematics may offer us a glimpse of eternal truths, but it takes an extra step to connect such formal truths to the empirical world. The logic of science, as we have previously discussed, is not deductive but inductive, and inductive conclusions are never absolute.

The atheist philosopher Bertrand Russell saw these different stances regarding truth as fundamentally incompatible ways of thinking about the world. In his classic book *Religion and Science*, he wrote about how this was but one reason for what he took to be a basic conflict between their worldviews: "A religious creed differs from a scientific theory in claiming to embody eternal and absolutely certain truth, whereas science is always tentative, expecting that modification in its present theories will sooner or later be found necessary, and aware that its method is one which is logically incapable of arriving at a complete and final demonstration."[48] However, even with regard to this issue, the great variety of religious views allows the possibility of contact or at least mutual accommodation.

The pioneering American astronomer and educator Maria Mitchell was just one example of a scientist who saw no problem reconciling the tentative nature of scientific findings with a religious spirit, writing: "It is the highest joy of the true scientists...that he can reap no lasting harvest—that whatever he may bring into the storehouse today will be surpassed by the gleaners tomorrow—he studies Nature because he loves her and rejoices to 'look through Nature up to Nature's God.'"[49] Mitchell's view encompassed much of what we have already seen about the scientific mindset—the passionate drive to discover, the intellectual humility in recognizing that scientific knowledge is tentative rather than absolute, and the emotional satisfaction of adding to the store of knowledge about the world—but adds to it a religious lens. Scientists do not need that lens and many study nature for its own sake, but for some it provides an additional motivational element. She expressed the latter sentiment by quoting Alexander Pope's classic eighteenth-century poem "An Essay on Man,"[50] which offered a view of humanity's place in the universe that appealed both to religious believers and to scientists.

Go, Wondrous Creature!

In his poem, Pope wryly expressed the epistemic situation of creatures who hang between knowledge and ignorance, mind and body—"Thought and Passion, all confus'd"[51]—unsure whether to deem ourselves more God or beast. Some might give up in such a circumstance, but the picture he painted was the very opposite of defeatist. Rather, it was a call to joyfully study Creation, unapologetically self-aware that there will be limits to our reason. Pope recognized that we can have no knowledge of a supernatural God's purposes and that the power of science is to study the natural world and our place in it. Today we would put this in terms of methodological naturalism and bounded rationality, but Pope's poetic essay expresses the ideas with a more literary flair. He referenced the ancient Greek injunction "Know thyself," which Plato highlighted in various dialogues. In the *Phaedrus*, a dialogue between Socrates and Phaedrus about the pull between passion and reason, Socrates appealed to this Delphic injunction when he said that he did not disbelieve stories about the afterlife or the actions of the gods but explained that there was no point wasting time trying to explain them (or explain them away) when it was difficult enough to try to understand ourselves.[52] Pope affirmed this attitude. "Know then thyself, presume not God to scan," he wrote, "The proper study of Mankind is Man." Pope saw science as a noble calling, and his admonition to follow where it leads was inspiring to believers and unbelievers alike.

> Go, wondrous creature! mount where science guides,
> Go, measure earth, weigh air, and state the tides;
> Instruct the planets in what orbs to run,
> Correct old time, and regulate the sun;
> Go, soar with Plato to th' empyreal sphere,
> To the first good, first perfect, and first fair;
> ...
> Go, teach Eternal Wisdom how to rule—
> Then drop into thyself, and be a fool![53]

The last heroic couplet points again to Socrates and other wise fools and likely also to the mystic notion of God's fool, nicely taking the scientist's lofty aspirations with a healthy, virtuous dose of self-skepticism. Charles Darwin's grandfather, the physician, natural philosopher, and poet Erasmus Darwin, whose book *Zoonomia* speculated about the laws that governed organic life, defined a fool as "a man who never tried an experiment in his life."[54] Fair enough, but I think that Charles Darwin himself expressed the scientific attitude with the proper combination of self-deprecation and delight: "I love fools' experiments; I am always making them."[55]

The Science Wars

A Bridge Too Far?

Snow identified the two-cultures gap as a divide between scientists and literary intellectuals. Today, we more often speak of this as a divide between the sciences and the humanities. The latter includes classics, history, languages, philosophy, and so on. One could examine Snow's thesis from any of these vantage points, but the gap does seem greatest when one focuses on art, drama, and literature because of their special status as creative disciplines. The artist and the author share a privileged place in the cultural landscape in large part because they are recognized as being so critical to creating it. Theirs is quite a different form of cultural transformation than that of scientists, as the latter includes not only ideas but also technology. Though art and literature have clearly benefited from science and technology, which has provided them with new media for expression, Snow was certainly correct in pointing out that their relationship was often strained. It has even, on occasion, broken down into outright hostilities.

This takes us to what became known in the 1990s as "the science wars," a period in which science found itself under attack from the humanities by an extreme form of social constructivism and "postmodernism" that, as the name implies, asserted as one of its central themes that the modern era of science was over. Although this may at first seem to be a different battle, it follows a similar pattern to the one we saw in the war between science and religion. As in that case, we do not need to get into the well-known details of the conflict but will just give a broad-brush review of the issues. Both involved a dispute about their respective cultural standing. Both involved fundamental epistemological questions about the basis of knowledge. Both involved disagreements about authority and power. And both, I suggest, involved a deep difference of view about values, especially the relative significance of curiosity and creativity.

Postmodern critical theory Both conflicts also had significant political overtones, which naturally added to the rancor. As extreme religious conservatism is associated with the political far right, radical postmodernism is associated with the political far left. It has its roots in an influential group of German social theorists including Max Horkheimer, Theodor Adorno, Herbert Marcuse, and others that became known as the Frankfurt School, who began with a Marxist political view, to which they added elements from Hegelian political philosophy, Freudian psychoanalysis, and antipositivist sociology. Horkheimer called the approach *critical theory*, which he explicitly contrasted with "traditional" science-centered theory, which was based in observation and that attempted to draw lawful generalizations about the world. More directly relevant to the science wars were French philosophers like Roland Barthes, Gilles Deleuze, Jean-François Lyotard, and especially Jacques Derrida, who initially were reacting to problems with a very particular sociological view—Structuralism—but who then, making an appeal to Thomas Kuhn's and Paul Feyerabend's work, went on to offer a broad, radical view of culture and knowledge that rejected science and its picture of the world, as least with regard to scientific claims about objective truths about nature. They saw what they called the "crisis of representation" as a liberation, for it allowed language and expression to dominate.

Derrida's notion of "hypertextualism" is an example of this idea. Derrida subsumed everything within a system of textual and linguistic differences.

Forget about the real world; the "outside" is always to be put in quotation marks and relegated to the category of "otherness." This approach, in Derrida's terminology, "deconstructs" the ordinary way that one thinks of the relationship between the world and texts, claiming that knowledge is no more than a linguistic construct that is fundamentally unstable. Barthes went further, denying even the primacy of the author. The author is dead once a text is produced, he argued, and the text then belongs entirely to the reader. It is the reader who constructs its meaning. We need not get into the details of these and related views here; the basic message is that texts are primary, and the world is constructed in the play of language and power with no external anchor. Thusly is science demoted; creativity trumps curiosity, for there is nothing of the "other" that one can hope to discover. These theorists repudiated the modern world view, and they proposed that we had entered, in Lyotard's term, a postmodern era.[56] Scientific "truth" is thus no more than a story told by those who currently have the power to imperialistically enforce their preferred narrative.

A similar theme played out in the field of science and technology studies (STS) through the influence of social constructivists like David Bloor, Barry Barnes, Trevor Pinch, Steve Woolgar, Harry Collins, and others. Here, too, the critique of science often returned to Kuhn and Feyerabend for support of their contention that science was socially constructed. Bruno Latour, a French philosopher and sociologist, was an especially influential representative of the constructivist critique. With Woolgar, he gave what amounted to a radical reinterpretation of research in a scientific lab. What was going on in science, they wrote after an ethnographic study of a research lab at the Salk Institute, was not discovery of truths about the world but quite the opposite—the creation or "construction" of objects. Explicitly drawing upon literary criticism, Latour interpreted science not as an objective search for facts but as the subjective social production of scientific "texts." Scientific activity, on this account, is composed of "the construction and sustenance of fictional accounts."[57] Some of these may become "reified" or "materialized" through the process of scientific dispute and negotiation, but the so-called reality of scientific entities and of "nature" itself was a consequence of the settlement of scientific dispute, not its cause. Latour had been inspired by Bloor's "Strong Program," and even as he came to reject other aspects of it, he retained its view that there was no reason to treat "true" or "false" scientific theories any differently—both were simply the product of social forces.[58] It is ironic that this field was called the sociology of scientific knowledge (SSK), for at least in its dogmatic form, its constructivism left science with no knowledge that could be called "true" without scare quotes.

The Sokal affair I have glossed over many differences among these various positions in this brief review. Indeed, Bloor saw Latour as a traitor to science studies though the latter saw himself as taking it to its next level. But these inside-baseball disputes are minor compared to the science wars as a whole that their various critiques of science gave rise to. Scientists had mostly ignored critical theory, social constructionism, and postmodernism generally until a few felt compelled to respond to their direct, general attacks. The counterattack was led by biologist Paul Gross and mathematician Norman Levitt in their book *Higher Superstition: The Academic Left and Its Quarrels with Science*.[59] Philosophers of science responded as well.[60] The STS program that I was a member of changed its name back to HPS to distance itself from a field that, by the influence of its most extreme version, had become seen as absurdly divorced from actual scientific practice and even from reality.

The conflict played out through the 1990s not only in books and academic journals but even to some extent in the popular press. The science wars are now mostly ancient history, though postmodernism lives on in some humanities departments and flare-ups occasionally continue to erupt in local battles. The science wars as an episode, however, is viewed in retrospect as not having ended with a surrender, a settlement, or even a bang, but with a belly laugh. In 1996, the postmodern critical theory academic journal *Social Text* published a special issue on the science wars. Included was an article by physicist Alan Sokal titled "Transgressing the Boundaries: Towards a Transformative Hermeneutics of Quantum Gravity" that argued that physical reality was merely a social and linguistic construct and "dogma imposed by the long post-Enlightenment hegemony over the Western intellectual outlook."[61] The article was a purposeful hoax—an experiment to test whether leading postmodernists would detect the nonsense that he deliberately laced it with, or lazily accept it on the basis of its fawning support of the postmodern agenda. Sokal revealed the hoax after the science wars issue appeared and scientists had a good last laugh. They then returned to their labs, with many concluding that the humanities really was irrelevant to their research. Creating a world through the social construction of texts would be way easier than the slogging but invigorating work of figuring out the real one, but that is no way to satisfy one's curiosity.

We could end this overview of postmodernism here, but I want to briefly describe one further case that illustrates the problem with the postmodern view through a real example of its application that links this to our discussion of religion. I have in mind a specific attack on science that arrived in the late 1980s in the form of a postmodern form of creationism.

Creationist postmodernism Intelligent Design creationism (IDC) is a strange hybrid of Christian fundamentalism and postmodernism. As such, it embodies the problems we discussed about religious dogma in previous sections and also those we have been considering here about radical literary critical theory. I saw IDC as a reduction to absurdity for postmodernism. My work on this new creationist movement began as a critique of the dogmatic religious right, but because of the way IDC incorporated postmodern theory, it became part of one of the final skirmishes of the science wars.[62]

Philip Johnson, a UC Berkeley law professor, was the godfather of the IDC movement, and it was postmodernism, which he had first encountered through critical legal studies, that formed the theoretical framework for his rejection of and attack on evolutionary science.

> The newly fashionable postmodernist model of knowledge is not based on scientific empiricism, but on literary criticism, and its philosopher kings are Richard Rorty, Stanley Fish, and Jacques Derrida. Knowledge in this model is relative to culture, and no single picture of reality has absolute authority over rival cultural understandings.[63]

Johnson's original title for the book where he laid out his attack on evolution was "Darwinism Deconstructed." There and elsewhere, Johnson repeated the standard postmodern misunderstanding of Kuhn. He cited the epistemological anarchism of Feyerabend, whose "anything goes" view we discussed in chapter 3, as another postmodern precursor who revealed that science was ideology.[64] He spoke of Rorty as showing that "Darwinism" means that there is no objective reality and that truth is no more than a kind of hegemony of one group over another. We have already seen, at least in outline, why this is a fundamental misunderstanding of evolution—far from undermining reality and truth, evolution provides a way to understand how we can have inductive knowledge. But evidence makes little difference to a radical postmodern view that says science has no special "privilege" over other cultural narratives in determining what is true about the world. Using standard postmodern language, ID creationists claimed that evolution was merely a "creation story" that the "scientistic priesthood" used to enforce their "imperialistic," atheistic worldview through a "program of indoctrination" to maintain their power and cultural dominance.[65]

Though they opposed evolution, creationists were eager to appropriate scientific language to give a scientific veneer to their enterprise. Evolution supposedly violated the second law of thermodynamics or the no-free-lunch theorem. ID creationists went even further, holding that only a supernatural intelligent designer could create complex living things, rejecting the common theological notion we discussed previously that God could have created natural laws that served that purpose—Darwin's law, for example—without the need for direct divine intervention. Never mind that scientists could point to observations of evolution in action working just as predicted; ID creationists ridiculed the notion that evolution was supported by overwhelming evidence. Why think that there is any real difference that can demarcate scientific knowledge from religious beliefs; both are just narratives with no privileged view about matters of fact. On the postmodern view, we are also post-truth.

Of course, in the final analysis, Johnson thought that he could escape the postmodern loss of objective truth, as he believed that the Bible provided the otherwise-impossible God's-eye view. God is the all-knowing author, creating the world by fiat. Only supernatural intelligence has authorial power to create. This is a theological-literary interpretation of Creation. Johnson's favorite Scripture trumpets that viewpoint: "In the beginning was the Word, and the Word was with God, and the Word was God."[66] This passage is more poetic and mysterious than Genesis 1:1, but creationists read both in much the same way, as indicating the primacy of God as Creator. And because this is a creator with a capital C, one gets not just truth, but truth with a capital T. In the end, it is the authority of the Book, with a capital B, or at least their interpretation of it, that should take up arms to reclaim the power that science had usurped.

Except for this divine escape clause, one cannot say that IDC took postmodernism to an extreme, for Johnson simply appropriated and applied it. As he explained it to some political science department colleagues, he was a postmodernist and deconstructionist just like them but simply aiming at a slightly different target. Using this approach, IDC attracted a few academic supporters from social constructivists who still haunted Science and Technology Studies, but radical postmodernism was already on the wane after the Sokal Affair, so IDC never reached critical mass as a cross-over movement. Fortunately, postmodern critical legal studies was not a factor when a policy that opened the door to including IDC in public school science classes went to trial and was soundly defeated.[67]

Different Points of View

This quick review of the science wars should have been as familiar as our overview of the conflict between science and religion. Twenty years after Sokal highlighted publicly that the radical postmodern emperor had no clothes, the episode is seen as marking the end of an extreme, ridiculous academic movement. And yet, just as fights between religion and science still break out, the gap between the sciences and the humanities has yet to be fully closed. My hope is that a virtue-theoretic approach can help sort out at least some of the points of conflict and perhaps suggest possibilities for a peaceful way forward.

Snow sought a way to a third culture because he thought that the wicked social problems we faced demanded the unified efforts of both the sciences and the humanities, but after decades of failure, one might wonder whether his idea of a bridge between the two cultures was a pipe dream. Maybe the reason for the gap between the sciences and the humanities is that these just are essentially different sorts of vocations. Could it be that, in a very basic sense, the cultures of science and the humanities are unbridgeable? For the most part, they have different goals, so it should not be surprising that they have different methods and take different virtues to be central.

A few individuals, like Snow himself, felt comfortable crossing between the two cultures. Jacob Bronowski, for instance, a mathematician, scientist, and historian, wrote that he "grew up to be indifferent to the distinction between literature and science, which in my teens were simply two languages for experience that I learned together."[68] But others continued to see these as very different points of view. In his banquet speech at the Nobel Prize ceremony, Swedish biochemist Sune K. Bergström pointed out what he took to be one essential difference. Making reference to an earlier speech given by the winner of the Nobel Prize for literature, he homed in on a key difference between authors and scientists: "The literature laureate of this year has said that an author can do anything as long as his readers believe him. A scientist cannot do anything that is not checked and rechecked by scientists of this network before it is accepted."[69] This is the difference we have been investigating between creativity, which seems free from every constraint but that of authors' relationship to their readers (or perhaps not even that, if one follows Barthes), and curiosity, which aims at discovering reproducible truths about the world.

Another way to describe this difference is that science is primarily interested in discovering general facts, while the humanities is primarily interested in expressing particular points of view. Scientists mostly seek points of commonality; artists mostly highlight individual differences. This accounts for much of their miscommunication, their talking past each other. It may also help explain the differences in how even shared virtues are expressed.

Meticulousness, for example, may be shared at some level, but in very different ways. Authors aim to eliminate every typographical error, but they do not necessarily care about quantification. For scientists, on the other hand, the ideal mode of expression is mathematical. The mathematician G. H. Hardy thought that this gave math a solidity that surpassed great literature. "Greek mathematics is the real thing," he wrote. "The Greeks first spoke a language which modern mathematicians can understand…So Greek mathematics is 'permanent,' more permanent even than Greek literature."[70]

Attentiveness is another shared virtue, because both scientists and writers must be observers. However, while literature values the expression of a writer's viewpoint for its own sake, that is of little interest to science. Scientists examine individual observations for patterns from which they may draw generalizations. They seek commonalities that indicate a deeper structure—the causal processes that explain how the world works not just here, but elsewhere. Consilience is an indication of a true discovery. This is not to say that scientists do not appreciate particulars—as we noted previously, they typically have a special fascination with their study organism—but even if they care only about beetles, or maize, or streptococcus, they aim to discover something true about the species generally. This is the nature of scientific curiosity.

Similarly, in science, objectivity—the attempt to eliminate personal biases in assessing the evidence—is a virtue because science aims to base its conclusions on evidence that anyone could check and confirm interchangeably.

Literature, on the other hand, sees subjectivity as a virtue. It strives to express the individuality of characters and their distinctive points of view. Authors aim to create memorable characters, which involves the expression of subjectivity.

This difference goes beyond the content itself; writers also seek their own individual voice. This is true for nonfiction as well as fiction writers. The Pulitzer Prize–winning writer John McPhee taught a factual writing course at Princeton in which he emphasized the importance of structure. However, he also highlighted the uniqueness of each writer, and one of the pleasures of reading a piece by McPhee is the pristine clarity of his language and understated elegance of his craftsmanship. McPhee is best known for his writings about geology and the environment, but it remains the case that writing about science is not the same thing as writing *as* a scientist. It is significant for us that McPhee's course was offered in the journalism program as the vocational virtues of journalism are mostly congruent with those of science—that is, they both share the central goal of getting the facts right. Investigative journalists, especially, are like scientists in their dogged pursuit of the truth, digging for information, then checking and double-checking their facts to be sure these are reported accurately. They rightly are offended when their reporting is dismissed as "fake news," for that is as much a contradiction of the basic values of their vocation as the idea of "fake data" is for science. The writing in straight news reporting is not unlike that of scientific research papers, where the emphasis is on conveying the information clearly and accurately; but in other venues, nonfiction writers have room for individual stylistic expression. Developing a creative voice, however, is not a value for scientists.

Indeed, scientists try to remove their individuality from their professional writing. Scientific reports are formulaic. The goal is to be not subjective but intersubjective so that any other individual could replicate the observation. This observational interchangeability is an important part of what it means to say that science is universal. This is not to say that scientists aim for a view from nowhere, which is a red herring, but rather for views that in principle could be shared by anyone. Replicability, as noted previously, is a critical value in science but not so much in literature. The majority of novels may be predictable plot-driven stories filled with oft-repeated cardboard-stock characters, but such tales are rightly dismissed as hack writing; exemplary literature aims higher.

Interestingly, although McPhee's course may be thought of as a "science-writing-as-literature" class, he disavowed the label and viewed it as "a plain writing course."[71] It was originally titled "The Literature of Fact," though he later changed it to "Creative Nonfiction," perhaps to highlight the special artistic challenge of presenting facts in a creative, compelling way without crossing the line to fiction. This different stance regarding the value of creativity is critical; it is a cardinal virtue for authors and artists but not for scientists.

Creativity and discovery   It behooves us to pause here to consider whether creativity plays a more central role in science than we have so far considered. Einstein is often quoted to the effect that "The intuitive mind is a sacred gift and the rational mind is a faithful servant. We have created a society that honors the servant and has forgotten the gift." Elsewhere he seemed to make the point even more forcefully: "Imagination is more important than knowledge." Is this so? Have we forgotten the gift?

Certainly, there have been scholars who have considered creativity to be essential in science. To give a couple of examples, historian Evelyn Fox Keller spoke of the role of creativity in science in her biography of Nobel Prize–winning biologist Barbara McClintock: "Scientific knowledge as a whole grows out of the interaction—sometimes complex, always subtle—between individual creativity and communal validation."[72] There has also been some research about the role of creativity in science. Dean Simonton, a professor of psychology, argued that creativity in science involves the integration of chance, logic, genius, and zeitgeist.[73]

However, Simonton and Keller see this, properly, as part of a larger process that keeps creativity in check, which is to say, within the bounds of what can be checked. For Simonton, the combinatorial process of chance serves as the primary cause with the other elements acting as constraints. And Keller emphasizes that individual creativity requires the validation of the scientific community for it to become scientific knowledge. We have previously discussed the distinction between the context of discovery and the context of justification that underlies this idea—creativity may be involved in coming up with an idea in the first place, which must then be subject to test.

Creativity might also be argued to be important in the latter step, specifically in the design of experiments. For example, a researcher may think of a novel, original way to test a hypothesis that previously had seemed too difficult to test. One might plausibly call these creative experiments, though it is much more common to speak of clever experiments, and that is probably a more apt term.

Thus, it cannot be said that creativity is never considered, but its significance remains questionable. In our survey of scientists, creativity did not rank very highly. In our pilot study, some scientists did mention it as important for science, but the number was not large enough to make the top ten list. In the full study, it came up relatively infrequently. What might account for this?

One possibility was mentioned in passing above; in many cases creativity is not really the best term to describe what is going on in the context of discovery. People often cite eureka moments in science—Kekule's dream of the structure of benzene, for instance—as examples of scientists' creativity. However, such moments, even brilliant ones, are not a form of creation but rather a form of insightful discovery. Thus, in most such cases, even Einstein's, insightfulness is the more relevant term in science than creativity. But, as we have previously discussed, by itself, insight counts for naught; to qualify as science it must be testable insight.

Creativity—a scientific vice? This initial review shows that science regards creativity with ambivalence. Even those scientists who speak positively of it seem to be using the term to refer to concepts, such as insightfulness, that are arguably different. Though creativity can be useful in special circumstances for science (as most any virtue could be for any vocation), creativity is not a central value for science. This may seem surprising at first, but a virtue-theoretic analysis may shed some light on what may be going on. As in the comparison with religion, the aims of science are different from those of literature and art.

Because discovering truths about the natural world is the central, guiding purpose of science, curiosity is a core virtue in a way that creativity is not. An author or artist may be inspired and create a vision of the world, but a scientist can never be satisfied with that. A scientist can also be inspired and have some insight, but it may be regarded as knowledge only after it has been put to the test, and the evidence has been weighed and found to support it. Mere insight, like purported religious revelation, should not be valued on its own by science because it remains at the level of authority.

True knowledge is power, not the other way around. Dogmatic postmodernism, like dogmatic religion, is an authoritarian view, and science does see that as vicious.

However, we should not let extreme views block the hope of a third culture. Perhaps we cannot hope to build a bridge without first finding some common ground in the midst of the stream. My main interest in this book has been to seek that from the side of science by digging down to what I hold is a bedrock of values, but to complete the task one must do the concomitant work from the side of literature. As in the case of resolving the conflict between science and religion, that will require that facts be treated more respectfully than they have in the past.

Model Dialogue

In a straightforward sense, the postmodern attack on science failed and science won the war. The sorry effect of postmodern science studies was to burn bridges that were being built between the sciences and the humanities. Scientists saw this "critical" perspective, at least in its common radical form, as naive about the actual practice of science and not worthy of serious attention. Sokal's hoax revealed how ridiculous it had become. However, as in the conflict between science and religion, some combatants refused to consider possible peace terms, so it cannot be said that this is entirely behind us. One sees the unfortunate effect of this in academia, with the continuing isolation of some humanities departments from the sciences, to the detriment of both. How are we to get them working together again?

Snow described the two-cultures gap as involving a failure to communicate. It was, he observed, as though both sides spoke different languages. To be able to work together, they need to learn how to talk to one another instead of talking past one another. This requires a model for communication that builds on at least some point in common. If there is to be a dialogue, it must be grounded. Let me briefly suggest a model worth developing for how that might be done in a way that respects what each brings to the task.

Social constructivists hold that knowledge is social. In an ordinary sense, in most cases this is true and nonproblematic. The problem arises in the dogmatic claim (sometimes sly, sometimes overt) that it is just social. To turn the view around, it is equally true to say that all narrative is natural but equally problematic to say that it is just natural.[74] However, if a more moderate view is taken, one side need not annihilate the other. The episte-mology of curiosity I outlined in the opening chapter begins with the know-how of beings in the world. We are social beings, so our knowledge is social, but this is not separate from the world. Rather, our knowledge evolved within it. This suggests a possible model for how to structure a dialogue between science and literature—it must be a form of communication that recognizes that authors and scientists alike are natural beings who are in and of a real world. We have mostly examined this from the point of view of scientists, but it should hold for authors as well.

The essayist's goal of expressing truth in nonfiction is clear. That is why it is called factual writing, after all. When one speaks of using "creative license" in telling a story, this usually means that one is free to depart from the facts, but McPhee's example shows that creativity does not require such a departure, as creativity may come into play in other ways, such as in how authors structure their tales. One may "play with the facts" in other ways, just as one may play with language, but even in fiction an author cannot succeed who is too fast and loose with them.

I stand with those who hold that truth plays a role even in fiction. It is an expression of ourselves, in the world. Poets as well as scientists look for and aim to convey truth, though perhaps of a subjective rather than objec-tive sort. This is so even in the most fantastic stories involving magical worlds, mythical creatures, and characters who purportedly bear "no iden-tification with actual persons, living or dead." Does anyone besides lawyers take such disclaimers seriously? Even the alien worlds of science fiction and fantasy are not and could never be entirely "other," in the sense of being unconnected to reality. Might "science fiction" in this sense be an oxymo-ron? Classical science fiction keeps close to scientific reality, breaking no physical laws. Speculative fiction may go further, but readers still expect a lawful consistency to a created world, even if the laws are bent or changed from the actual world. Even science fantasy must retain a connection to reality if it is to be of any interest.

Putting this another way, naturalism in literature is optional, but meth-odological naturalism may not be. Literature need not be seen as represent-ing the actual world, but the possible worlds it creates must be sufficiently nearby. On this view, literature may be seen as a different mode of pre-senting aspects of truth or possible truth, sometimes real, sometimes hypo-thetical. To complete the scientific analogy, we may think of the latter as hypotheses about possible worlds that are more or less similar to our own. Considered in this light, using creative license to intentionally depart in particular ways from what is observed "in the wild" is not unlike what the laboratory scientist does. In both cases, such simplification can serve as a way to isolate factors that one aims to investigate by highlighting them in contrast to other factors. When done well, such a process can be at least as illuminating in fiction as in science. Within such a framework, scientists and writers may be able to agree with Huxley's contention that "Science and literature are not two things, but two sides of one thing."[75]

Rachel Carson, whose transformative book *Silent Spring* launched the environmental movement, was both a scientist and an author. In her address at the ceremony where she received the National Book Award, she spoke to the possibility of this overlapping concern: "The aim of science is to discover and illuminate truth. And that, I take it, is the aim of literature, whether biography or history or fiction. It seems to me, then, that there can be no separate literature of science."[76] Carson's view here unites with McPhee's— they saw themselves not as science writers, but as writers simpliciter, aiming to shine a light for their readers.

Fact and Fiction

The dogmatic social constructivist view held that there is nothing to the world but texts and nothing to texts but interpretation, leaving nothing but the interpretation of one's group or, in the final analysis, my own interpretation as the reader. This is no better than solipsism. Never mind what writers might have meant or hoped to convey or the relationship they aimed to establish. Are there really no facts of the matter about texts? I think most writers, even while granting readers their freedom of interpretation, would say that there are and would protest that the postmodern pronouncement of the death of the author was negligently premature.

As a writer, I do have an aim in mind when setting words down on a page. I aim to communicate ideas to readers. I hope to replicate some thoughts in their heads. The relationship between writer and reader is a form of cultural reproduction, the inheritance of memes (in Richard Dawkins's original sense of the term, on analogy to genes). Furthermore, because I am (at least in part) my thoughts, it is in such a process of reproduction that a writer may be said to live on by virtue of their writings, as those thoughts are copied into the brains of readers, excite their synapses, and become embodied once again. Naturally, this is not to say that I am always fully aware of everything that is conveyed in my writing. The subconscious will no doubt contribute to my expression in ways that I may not even recognize until someone points it out to me. Also, there may be implications to my words that I failed to notice. Thus, the need for good editors and for insightful literary critics. Either way, there is something for readers to find, and most writers are more than a little offended by someone who says that there is no fact of the matter for their readers to discover.

Of course, there are many circumstances that may undermine this possible relationship between author and reader. For instance, I may fail in my goal of communicating because I am a poor writer. Maybe a reader cannot find the truth I hoped to convey because I was obscure or rambling or insufficiently conversant in this language. In that case, I need to work harder to improve my craft. Or perhaps you are a poor reader. In that case, you may need to work harder to learn how to give a text a close reading. You may need to acquire prerequisite knowledge that the writer takes for granted, such as vocabulary to understand their terms, background information to understand the context, or a broad familiarity with other writings to get their allusions.

Often there is simply a mismatch between writer and reader. An author cannot write for everyone simultaneously. Authors write differently, or at least they ought to, depending upon the audience they wish to reach. The dominant form of writing in science is the technical report of scientific findings, which are published in professional journals. Scientific papers are written for other specialists, and they typically assume that their readers share a sufficiently high level of expertise. Because of specialization within scientific disciplines, even scientists can have a difficult time understanding published research outside their own areas of research. It takes a different sort of writing to communicate a finding to nonspecialists. Thus, the need for science journalists and explainers who can translate technical jargon into language that is accessible and meaningful for a broader readership.

In other cases, there is a different kind of mismatch, such as when the writer and the reader simply have different interests. My goal may be to convey ideas about my scientific discovery, whereas you as a reader may not care about that and instead want to read and interpret my book to reveal something about my psychological makeup, or something about cultural assumptions in the period in which I worked. Someone might even read them with an eye to highlighting aspects of literary analysis and criticism, such as the case of an anthology of essays with different interpretations of Stephen Jay Gould and Richard Lewontin's "The Spandrels of San Marco and the Panglossian Paradigm: A Critique of the Adaptationist Programme" paper on exaptation in evolutionary biology.[77] If some of these essays seem to quite miss the point of that article, it may be that they simply were not interested in that question.

To say that there is a fact of the matter about my intentions as a writer in no way requires that you read to satisfy my intentions. Differences of interests are perfectly reasonable. Feminists may read Jane Austen to examine gendered cultural assumptions about relationships between women and men. Philosophers may read her as an expression of a certain approach to ethics. There is even a school of literary analysis that reads to reveal Darwinian patterns in reproductive strategies. None of these may claim to be "the" correct interpretation, for all may be correct given the question they pose. We should be pragmatically pluralist about this. The point here is that it does not even make sense to pose these or other interpretive questions if there were not some fact of the matter to be found, at least in principle, regarding the author's intentions or other aspects of a piece of writing—there may be truths to be found in a text even if you do not care to discover them.

I confess to finding it vaguely embarrassing to rehearse this defense of what seems like a self-evident view. But vestiges of dogmatic postmodernism remain, and there will be some who will object even to this, so due diligence requires that we briefly reply to some possible remaining objections.

Reply to objections For instance, what about artists and writers who claim to embrace the extreme postmodern view, saying that they leave interpretation entirely up to the viewer? In some cases, a writer may aim to make a (philosophical) point about perspective, interpretation, and the role of audience belief in the creation of art. To this end, they may write with intentional ambiguity, aiming to create in the reader/viewer a state of suspension between possible interpretations (e.g., was the character dreaming or not?). More power to them. But that just pushes truth back to the meta level, as the author's intention was to make (for instance) a point about interpretation. What about cases where the author/artist claims to have no intention to communicate a view to readers at all? I see this as disingenuous or superficial or lazy. Could there really be true writers who simply do not care about reaching any audience? If sincere, such individuals should stick to writing personal diaries. Moreover, even in such cases, there may be meaning in a personal text. Even a stream of consciousness journal can carry meaning from the subconscious or from links of association.

And what about the other person in the equation: what about readers who do not care about an author's intention? Perhaps they simply want to react to the experience. Again, more power to them. We may reasonably have different goals as we engage with art or nature for that matter. Not everyone needs to investigate the world like a scientist. It is perfectly legitimate to enjoy nature's surfaces. It is fine to appreciate the sunset and the magic hour for their colors.

New patterns capture our interest and stimulate our curiosity until their complexities are understood. At that point, they are regarded with a different kind of appreciation … not as a puzzle to be solved, but, as Einstein put it, as a harmony to behold. Our interest is held by phenomena that stand on the edge of chaos (think of fire and waves) that clearly follow some sort of law but one so complex that one cannot completely grasp it. On the other hand, a truly random text with no selective force would have little chance of sustaining our interest.

Some go further and claim that the scientist's view is impoverished because scientists are interested in figuring out why things are as they are rather than just appreciating them. Richard Feynman tells a story about an artist friend who had such a view. Feynman had worked out an exchange whereby his friend taught him drawing in return for learning about physics, so they talked regularly about art and science. Feynman said he found his friend's view about science to be "a little goofy." For the scientist, figuring out the explanation for some phenomenon—the reason why it is what it is—adds to appreciation, not subtracts. This is right, but it may be that we can say more about both views in a way that connects this issue back to our previous discussion.

Mystery-Mongering

What might lead someone to think that something is lost when science discovers the explanation for some beautiful puzzle? It cannot be that there is a loss of beauty or some other aesthetic factor, for all the physical patterns and relations that constitute an aesthetic quality remain for the scientist.

Discovering the laws of optics and the physical explanation for the rainbow do not dim its colors. Some scientists may not care about the aesthetic qualities of rainbows just as an artist may not care about the causal explanation, but for the scientist with an aesthetic sense there is no reason to think it is diminished.

However, some have a more religious sort of objection to the scientific point of view. Is not the magic lost when the rainbow is reduced to optics and angles? Admittedly, this is one kind of relation that actually is lost when a scientist discovers the answer to a mystery. Perhaps the claim is that the feeling of loss is akin to the sense of disappointment one may experience when a magician's trick is revealed and you learn how the magic is done. "Is that all there is to it?" one may think. Maybe what is behind the objection to a scientific explanation is that a wish for true magic has been dashed.

Some felt this kind of disappointment when it was discovered how the seventeenth-century Dutch painter Johannes Vermeer had used clever technology in creating naturalistic perspectives in his paintings. Vermeer's work was revered for its "detached precision" and otherworldly quality, the "almost magical sense" that he was "painting not objects but light itself."[78] It turns out that that was probably exactly what he was doing. In *Vermeer's Camera: Uncovering the Truth behind the Masterpieces*,[79] architecture professor Philip Steadman argued that Vermeer achieved his photo-realistic effects with the help of a camera obscura box, an optical system that was the precursor of the photographic camera. (The painter David Hockney and physicist Charles Falco extended the thesis and argued that many other of the old masters, as far back as the fifteenth century, secretly used lens and mirrors as technological assists.)[80] With such a system, the image to be painted would be projected onto the back wall of the box, allowing Vermeer to trace the image directly onto his canvas. Steadman's meticulous research, which included reconstructing Vermeer's studio and a cubicle-type camera obscura, plus calculations showing how the setup would result in images the same sizes as the final paintings, led to a battle among art historians between those who were convinced by his work and others who defended the view that Vermeer's technique was the result of his unassisted, preternatural genius.

Walter Liedtke, curator of European paintings at the Metropolitan Museum of Art (including its five Vermeers), said that he did not oppose the notion that Vermeer "in some way responded to the camera obscura" but that he did oppose "drastic devaluations of the role of art."[81] Here one sees another similarity with extreme religious views: the worry that science undermines values. In this case, it seems to involve a notion that technology takes away from the genius of the artist. Such masters are pictured, not unlike religious sages, as inhabiting rarified airs beyond the world of mere mortals. As creators, they acquire a godlike state. Indeed, one often speaks of great artists and poets as being *possessed* by their muse. When taken to the extreme, the cult of the artist and attribution to them of special superhuman powers is little different from veneration of saints who are supposed to perform literal miracles.

But this takes us back to fantasy, and for the scientist it substitutes mystery-mongering for real wonder. Vermeer likely learned about the camera obscura from the Dutch optics pioneer Antony van Leeuwenhoek, developer of the microscope and now credited as the founder of the field of microbiology. Why should that collaboration with science diminish his accomplishment? Galileo learned about lenses, built a telescope, and turned a party novelty into a scientific instrument that revolutionized astronomy and physics. Vermeer learned about lenses, built a camera obscura, and turned a scientific instrument into an artist's palette that did the same thing for painting.

The American magicians Penn and Teller produced and directed a film about the Vermeer controversy that featured Steadman and inventor Tim Jenison, who figured out that adding a couple of small mirrors to the camera obscura setup made for a more practical method of copying the color and tone of the scene. Speaking about the experience, Teller said that he learned that in the world of art "there are people who really want to believe in magic, that artists are supernatural beings"—that Vermeer and great masters like him could just walk up and do what they did. Teller found the attitude absurd; to see Vermeer as "a god" makes him "a discouraging bore," he opined. "[A] great work of art should seem to have magically sprung like a miracle on the wall," he said, but what gets the artist to that miracle is concentration, aggravation, and pain. Seeing Vermeer as a genius artist and an inventor, combining art and science, is heroic in a way that others can see themselves as able to emulate: "Now he can inspire."[82]

As we saw, postmodernists often tried to make their point against scientific objectivity by saying that there is no God's-eye view of the world.

(Perhaps not, but truth-with-a-capital-T is too high a standard, nor is it a view that science claims to provide.) They rejected science by lumping it with religion (which they also rejected). But in rejecting the possibility of objective methods for distinguishing reality from fantasy, they simply substituted their own authority for God's. In this way, among others, radical postmodernism is equivalent to the extreme religious views we discussed above. Little wonder that scientists dismissed it as higher superstition. By rejecting the possibility of an evidence-based approach to truth and reducing knowledge to power relations, it is a basic form of mystery-mongering. Little different from the religious view they reject, they remain in the created, artificial world of the matrix of language.

We have never been postmodern   It is ironic that postmodernists, whose defining mission was the rejection and ridicule of Positivism, not only retained but expanded its reliance upon language as the source of justification of knowledge. Part of where Logical Positivism went wrong was thinking that logical relationships among sentences could be the locus of justification (rather than ontic relationships). As we saw in the previous chapter, paradoxes of confirmation for the Hypothetico-Deductive method arose from looseness even of logic, which could not capture evidential relevance relationships. Much effort was expended trying to find ways to tighten logical constraints. I confess to judging this a hopeless task. Language is a modeling system, and many languages are possible. Why should we ever have thought it plausible that the logical relationships within one language, even a particularly beautiful formal language, would fix justification? This would be like insisting that justification in geometry had to be Euclidean. Such a view might have seemed reasonable when no alternative was available (though mathematicians always did have nagging concerns about Euclid's postulate set), but it lost plausibility once the consistency of hyperbolic and elliptic geometries was established. Radical postmodernists drew the wrong lesson from the failure of Logical Positivism in the same way they drew the same wrong lesson from Kuhn. Rejecting the possibility that any interpretation was privileged over any other except by virtue of the authority of those in power, they simultaneously rejected the possibility of any objective narrative. And because they took language, narratives, and interpretation as basic, they went all in with a form of relativism that leaves us all talking past one another. Such a Tower of Babel is a recipe for anarchism and for conflict that is of a kind with religious wars, with one privileged revelation pitted against another.[83] Science properly rejects this perverse perspective.

Post-textualism: The primacy of experimental demonstration   Like the knight who seeks to be invested by deeds rather than words, for science, the experiment (e.g., the demonstration) is primary. Knowledge comes from experimenting with nature, first in an evolutionary sense and later by scientific methods. The evidential relevance relationship is ontic. Show me so I can try (test) it myself. Even Einstein's scientific theorizing was not based primarily in language but in thought experiments, the mental elements of which were visual and occasionally "muscular."[84] Again, language comes later.

Scientists must struggle to put their findings into words. Thus, the need for technical vocabulary. Scientific publications do not themselves provide the evidence for a finding; rather they *report* on the evidence, which is to say they are an attempt to echo the demonstration so that someone might be able to reproduce it themselves. To reproduce is to cause to happen again. Scientific papers are signs of and pointers to the evidence, not the evidence itself. This is why Latour's analysis of laboratory life in terms of the production of texts and narratives was superficial. Analysis that focuses only on the exchange of words, as though science were only a literary activity, is an example of the postmodern fallacy. Words are but shadows on the wall of the cave. They are never sufficient to someone who has not been trained (through experiences) to understand the meanings of the technical terms.

The Revenge of Truth

Great literature is as universal as science. Paradoxically, it may be recognized as such more easily, because of the difficulty of science. In his Nobel Prize acceptance speech, Max Delbrück (who, with his colleague Salvador Luria had proven in an elegant experiment that genetic mutations are random) reflected on the different ways in which artists and scientists view the world and are viewed by others. He recounted a story of a special meeting of the Connecticut Academy of the Arts and Sciences for which a poet, a composer, and two scientists were invited "to 'create' and to 'perform.'" While everyone attended and enjoyed the music and poems, only other scientists attended the performances of the two scientists. This asymmetry was fitting, Delbrück opined, and properly humbling to scientists. Scientists may think of their experiments as compositions and expect that their results will be eternal, but they are only fooling themselves.

> The books of the great scientists are gathering dust on the shelves of learned libraries. And rightly so. The scientist addresses an infinitesimal audience of fellow composers. His message is not devoid of universality but its universality is disembodied and anonymous. While the artist's communication is linked forever with its original form, that of the scientist is modified, amplified, fused with the ideas and results of others and melts into the stream of knowledge and ideas which forms our culture. The scientist has in common with the artist only this: that he can find no better retreat from the world than his work and also no stronger link with the world than his work.[85]

He closed with a quote from a comic play by the Danish writer Karen Blixen (better known by her pseudonym Isak Dinesen), in which a witch casts a spell over the characters in an inn such that every lie that is told becomes true before morning. The name of the play? "The Revenge of Truth." Here literature makes the relevant point for us: the truth will out.

Speak Truth to Power

Alternatives to Violence

Reflecting on one of the differences between the two cultures, the nature writer Joseph Wood Krutch wrote that poetry, mythology, and religion represent the world as people "would like to have it," while science represents it as we gradually come to discover it.[86] Krutch's oft-quoted formulation suggests by its contrast that the former set involves just fantasy. Moreover, his nihilistic view of the modern temper and generally pessimistic view of scientific progress led him to view the human condition as essentially tragic. If we are limited to wishful thinking, then there would be little to gain from any of these viewpoints.

But there is an alternative notion that may help avoid this bleak assessment. The idea of the world as we "would like to have it" may be thought of not as a fantasy world but rather as a world of ideals and aspirations. Science focuses on discovering the contours and constraints of the world, but we also need a picture of the future as we would hope to build it. Without a vision of such a future, we will not know how to direct our efforts. In this sense, we should add literature, art, philosophy, and the rest of the humanities to the list to complete the fields that Snow saw as standing across the gap from the sciences. Their job is to use imagination and reason to create and evaluate the world of future possibilities so that we may choose wisely where to head.

One additional way that literature does this that is especially relevant to our topic is in its depiction of scientists. Often this comes with a warning, as in Mary Shelley's classic depiction of Victor Frankenstein. Occasionally, it sets out an image that captures the scientific spirit of skepticism and utter honesty, as in Carl Sagan's Ellie Arroway.[87] That said, outside of Sinclair Lewis's *Arrowsmith*,[88] there are relatively few nuanced depictions of scientists in literature. Perhaps this is because novelists are taught to speak from their own experience, and few have any real experience with science. Sagan was himself a scientist, and Lewis had the assistance of a scientist (Paul de Kruif, a microbiologist, received 25 percent of the royalties for his work on the book), which may explain why these go beyond the stereotype of the scientist. However, even stock characters of scientists in literature can be revealing. Whether negative or positive, scientific stock characters reflect a virtue-theoretic view, largely Aristotelian, with negative portrayals focusing on scientific vices. The way that one may look to literature as a way to understand and to critique the scientific character is worth a more detailed discussion, but that must wait for another occasion. Here, the point is that for such accounts to serve this purpose, they must be recognized as containing at least a germ of truth with regard to their subject.

Language does create. But it does not build out of nothing. It does not create ex nihilo. Like evolution, it builds on existing things, adding, subtracting, rearranging, and tinkering with them. Unlike evolution, it can do so with foresight. In this way, it really can create new things. The point here is that doing so requires building materials and a connection to reality. To create a world as we would like to have it will require cooperative effort.

Snow was right that the solutions to the wicked problems that face society cannot be solved by science alone or by the humanities alone. The gap between facts and values and between discovery and creativity must be closed so that the cultures can work successfully together. Interdisciplinarity is essential if we are to survive. In this book, I have mostly focused on building the bridge from the side of the sciences by showing how values are not foreign to science, but rather are integral elements of its nature that help structure its methods and practices. But one needs to build from the other side as well, and this discussion has outlined some ways that facts do matter to the literature and humanities. And, yes, truth as well.

Finding the Words

I began this chapter by suggesting that it might be helpful to examine the problem of culture conflict in terms of the concepts of creativity and curiosity. Along these lines, I would like to make one further point to help find common ground between the cultures, which is to suggest that writers not be cast exclusively in the role of creators. We ought to take seriously the notion that writers are engaged in a process that involves *finding* the right words. Granted, writers do occasionally coin a new term, but mostly they are searching for the right word from among ones that already exist. Of course, finding the perfect word is only a part of the process that really is about arranging words into completely original sequences, so in this sense writing does involve creation of something that did not exist before. However, if this form of novelty were all there is to creativity, then being creative would be way too easy; stringing words together in a novel way does not a novel make. Mostly it would lead to nonsense, infinite monkeys notwithstanding. If the goal is to communicate something, and that is as true for fiction as for nonfiction, then something more is required.

Think of sentences as chords that will cause some resonance (or not) in receptive minds. Or as possible programs that may (or may not) run to produce a desired outcome. Whether writers are conveying setting, plot, and character, or information and argument, they aim to produce an effect in the minds of their readers, and different sequences of words will have very different effects. This means they will need to struggle to discover sequences that will have the intended effect. And, yes, I think that "discover" is the right word here. There are an infinite number of potential combinations in the vast search space of possible sentences; only an infinitesimal fraction of these will work at all, and fewer still will work well. To bag an especially worthy sentence or even an apt phrase, one must patiently stalk them in the thickets of language and hope to catch them in the wild. Serious writers are as ruthless in their selection of words and phrases as evolution itself, knowing that most may need to be killed off before one will be found that is robust and fertile enough to survive and reproduce in some other brain. Editing is a process red in tooth and claw and ink.

Finding the right words to convey an idea is powerful. This is as true in science as in literature. Scientific terminology is essential to express the novel phenomena scientists discover and the specialized techniques they use. *Adaptation. Fitness. Niche construction. Pleiotropy. Zygote.* Such lovely technical vocabulary will always be needed in science, but it will never be sufficient, if only because science needs translators and explainers who can find words that will reach a broader audience. As Carl Sagan recognized, sometimes you need to send a poet. The right words—words that can reproduce themselves and change the mental ecology of other minds—can have far-reaching effects. It is in this sense that the pen is mightier than the sword; a sword works by eliminating bodies one at a time, but powerful words reproduce in mind after mind, forever changing how their bodies thereafter will behave. This is the way of the peacemaker.

Conflict Analysis

We have examined Snow's culture gap in terms of a difference between those who discover and those who create. This difference comes down to whether curiosity or creativity is taken to be the more central value in one's practice. The one practice seeks reality, the other creates reality. Truth is essential to the first project but seems optional in the second.

For this latter group—the literary intellectuals in Snow's formulation—language is primary. Everything begins with the word. With their words, authors create worlds. A legislator is able to create laws through words (i.e., just by speaking) because they have the authority to do so. This is the sense in which both may dictate. This is the power of the author.

But for the former group—the scientists—language is secondary. It is too unconstrained. It can formulate expressions that refer to no real thing, like square circles. I may or may not be able to believe six impossible things before breakfast, but I can easily say them. However, I surely cannot do them.

This is a key reason that scientists, who are law finders, and politicians, who are lawmakers, often come into conflict, especially if the power to dictate is taken to be absolute and turns one into a dictator. A legislator may assert that the law of gravity no longer shall hold, but the law of gravity will pay no heed. And neither should you, if you know what is good for you.

Feel free to replace the gravity example with evolution or climate change or vaccination.

Might makes right vs. right makes might   It may not be possible to avoid all wars, but it hardly makes sense to assume in the beginning that truth is nothing more than the say-so of those in power. This does not mean that the powerful do not often try to use their power to dictate facts or deny reality. We can thank deconstructionists for highlighting some of the ways in which such deceits are carried out. But we cannot accept their analysis that one may do no more than fight to replace one narrative with a different one, for that merely substitutes one assertion of power with another. This is the surrender of ideals not only to the cynical view that war is inevitable, but also that it is right—that might makes right.

It is ironic that those who defend a might-makes-right view when analyzing cultural conflicts often do so in Darwinian terms. It is a serious misunderstanding of evolutionary theory to think that it endorses a might-makes-right view. That is a dangerous warmongering fallacy. The mistake most likely arises from a misunderstanding of the phrase "survival of the fittest," as the colloquial notion of being fit involves strength. But that is not the biological notion of fitness, which rather has to do with how well an organism's traits fit with particular environments. Being fit in the biological sense means getting something approximately right about the world. Put another way, its model of the world is truer. The world provides resistance that selects out models that are not as true—that do not fit as well. That is what gives an organism whatever fitness advantages it has. In some environments, strength is not as adaptive as speed, or endurance, or flexibility, or curiosity. Thus, the real Darwinian lesson is not that might makes right, but that right makes might.

So, what is one to do when those in power try to dictate truth? Just as nature resists the false, so should we in such cases. While it is not possible to disobey true laws of nature, one can disobey authoritarians who try to dictate something unjustly. This was Thoreau's wise recommendation. Civil disobedience involves a principled rejection of authority in cases where authorities have abused their power. This is also what is behind the Quaker notion of speaking truth to power.[89] The idea of the efficacy of nonviolent resistance is that truth can be stronger than mere power. And that notion requires taking truth seriously.

Truth Shall Set You Free

I have suggested that radical postmodernism is problematic in much the same way as religious dogmatism. Both reject the possibility that human beings can have justified methods for ascertaining truths about the world, and so both view truth claims in terms of power and authority. Though wrong in their extreme, their motivations were laudable. Horkheimer took a theory to be "critical" in the sense that it aimed "to liberate human beings from the circumstances that enslave them."[90] This is, of course, the same goal that drives the narrative of the Bible, and it is a goal that we all should share, whether believers or not. However, if criticism is to have value for emancipation, it must have a basis in truth. To combat oppression, we must be able to call it out when we see it. The truth is needed to set you free. But what hope is there of freedom if one denies that human beings can work together to reject what is false and approach what is true without resorting to war?

War is contrary to the scientific mindset in that it is the final, bald assertion of authority. No wonder that Einstein, who despised all forms of authority, was a pacifist. Pacifism is often misunderstood as being passive or understood only in a negative sense (e.g., as a principled refusal to fight), but this truly misunderstands the general principle, which is an active attempt to find peaceful resolution of conflict through alternatives to violence. The scientific appeal to evidence should be thought of as one example of how such an idea can be implemented in practice. A duel will not determine the boiling point of water. We do not fix the atomic weight of gold by battle.

We will need to discover creative solutions to conflict. We might begin by applying this philosophy to the culture clashes that we have been investigating in this chapter. By keeping our eyes on the prize, we may have a better hope of working through Snow's battle between the two cultures and White's warfare between science and religion. This may require a revolution in thinking on a par with the scientific revolution itself. If we learn to see this task in moral terms, even in the scientific case, then the imperative to speak truth to power takes on a new significance. "It is," as the authors of *Speak Truth to Power* put it, "the practical effort to overcome evil with good."[91]

As Thoreau wrote: "Action from principle—the perception and the performance of right—changes things and relations; it is essentially revolutionary."[92] But revolutions can reject the divine right of kings and yet not fall into anarchism. This is true politically, and it should be true intellectually as well. We human beings can never have a perfect God's-eye view of absolute Truth, but that should not dismay us, for truth in lowercase is sufficient for this natural world. As evolved rational animals we have reason enough for scientific researches and hopefully virtue enough as well.

8   Scientific Vices

When it is actually natural science that speaks, we listen gladly and as disciples. But it is not always natural science that speaks when natural scientists are speaking.

—Edmund Husserl[1]

Bad Apples

By Their Fruits?

Eve's curiosity reflects a core scientific value. Scientists see little attraction to being in a blissful Eden if it is based in ignorance, so expulsion from that kind of walled garden is no punishment. The world provides them with endless room for exploration, and the apple of discovery is sweet. The idea that upon eating from the tree of knowledge "your eyes shall be opened" is a tempting enticement. Yet, we should acknowledge that there is potential for danger should scientists take too seriously the claimed second effect, namely, that "ye shall be as gods."[2]

Anyone with even a passing acquaintance with the history of science knows that its fruits are not always divine. Science is not perfect, and scientists are not gods. Scientists are fallible human beings, who are subject to the same temptations and weaknesses as the rest of us mere mortals. The physicist Murray Gell-Mann, who won a Nobel Prize in 1969 for his research on elementary particles (we owe the quark model to him and George Zweig), noted what should be an obvious point in this regard. "[T]he practitioners of science are, after all, human beings," he wrote, "They are not immune to the normal influences of egotism, economic self-interest, fashion, wishful thinking and laziness."[3] Nor, we may add, are scientists immune from prejudice. Science has seen its full share of sexism, racism, jingoism, and so on. However, in this chapter I want to focus not on such general human flaws, important as they are, but rather on vices that, rightly or wrongly, are more specifically associated with science. Gell-Mann mentions some examples of these as well: "A scientist may try to steal credit, knowingly initiate a worthless project for gain, or take a conventional idea for granted instead of looking for a better explanation. From time to time scientists even fudge their results, breaking one of the most serious taboos of their profession."[4]

The usual view is that such misconduct is the exception rather than norm and not a reflection on science itself. It is difficult to get good data on this kind of issue, of course. Studies of scientific misconduct show that the number of cases of misconduct is small but not negligible.[5] Whatever the actual numbers, it is incumbent upon the scientific community to take the issue seriously, as every case diminishes the trustworthiness of science. What are we to make of such transgressions? Is it just that there are bad apples in every bushel? Is this just part of human nature? Or might there be something about science itself that can spoil some fruit?

Scientific virtue theory may provide a useful perspective from which to analyze such questions and perhaps to offer some new solutions. It has already shown us that science has a moral character structured by values that set standards of excellence—the ideals we aspire to. As we shall see, it also provides a way to understand not only virtue but also vice. So far, we have looked mostly at virtues that make for exemplary science, but we cannot overlook instances where scientists are less than exemplary. For every virtue that helps science flourish, there are vices that could do the opposite.

In our study, we asked scientists not only what traits were essential for being an exemplary scientist but also what traits were problematic for doing science. As might be expected, almost everyone simply cited the opposites of the positive traits—being dishonest rather than honest and so on. That makes perfect sense, but even more interesting were other negative traits associated with the least exemplary scientists. Scientists were not shy about their own failings or about bringing up examples of traits that serve science poorly. We will look especially at intellectual arrogance and what might be thought of as intellectual gluttony.

A particularly interesting issue is whether there are vices to which scientists are especially prone. One way that this could occur is if virtues that are especially important for science become twisted or misapplied in some way. As evolutionary science explains, traits that are adaptive in one environment can become deleterious in other circumstances. Having larger antlers may give stags an advantage in competitions between males and a reproductive advantage in sexual display, but only up to a point. The huge antlers of the Irish elk, whose scientific name—*Megaloceros giganteus*—gives a double indication of this giant deer's size, are thought to have become a liability when environmental conditions changed, leading to the species' eventual extinction. Similarly, character traits that serve scientists well in most cases may begin to have the opposite effect if they become exaggerated.

From a virtue-theoretic perspective, one may think of this as a form of teleopathy, a debilitating condition caused by an unbalanced pursuit of goals.[6] Ethicist Kenneth Goodpaster says that signs of teleopathy include fixation on certain goals with a single-mindedness that disregards any possible cost; rationalization of whatever behaviors are taken in the service of that goal; and a social detachment that comes from that single-minded focus. If we can learn to diagnose this kind of condition within science, then we will have a better chance of finding a therapy or, better still, preventing its occurrence in the first place. There are various ways that we could approach this issue, but it may be best to take the hardest case first.

Sweet Temptations

We began with a positive interpretation of the story of scientific curiosity in the Garden of Eden, but a more critical interpretation is that scientists are like Eve not only with regard to their curiosity, but also in their innocence. On this view, scientists are like children with little appreciation of the moral import of their actions. Innocence is sweet but also naive. As one critic put it, "Scientists are like children playing with fire without heeding the disastrous consequences of their games."[7] Even innocent curiosity can be a double-edged sword.

Historian Mark Feige highlighted the positive side of this in his description of what motivated the physicists who cracked open the atom in the Manhattan Project. Professional ambition or desire for prestige may have driven a few, he wrote, "But of all the reasons for studying the atom, the psychological condition of wonder may have been the strongest. Physicists, chemists, and mathematicians studied atoms out of profound curiosity, and when they detected the inner workings of the tiny particles, they experienced awe, amazement, delight, and transcendence."[8] This is the emotional aspect of the scientific mindset that we discussed previously. No one can deny its inspiring power.

However, in pursuing knowledge for its own sake, scientists can be oblivious to other issues. Robert Oppenheimer, the physicist who headed the Manhattan Project, described the scientist's attitude this way: "When you see something that is technically sweet you go ahead and do it and you argue about what to do about it only after you have had your technical success."[9] This aspect of children—when they focus only on the desired candy—is more problematic. The apple of discovery is indeed sweet, but some apples may nevertheless be poisonous.

For scientists, the satisfaction of curiosity is a basic good. However, that does not mean that it is an unmitigated good. The satisfaction of curiosity involves an accomplishment of some sort, but not necessarily an optimization. In an evolutionary sense, it involves being good enough. As we saw, there is an evolutionary benefit to curiosity in variable environments. It provides a basic motivation for learning. It draws us forward into new and unexplored areas. It helps us find the resources we need and ways to overcome challenges we encounter. Curiosity overcomes fear of the unknown and the different, leading us to see these as interesting and attractive rather than as worrisome and scary. In general, this is a highly adaptive trait for beings like ourselves. Without sufficient curiosity to overcome fear, we would not fare nearly as well as we do in the world in which we find ourselves. However, we cannot ignore the fact that sometimes dangers do lurk in the unknown. Curiosity may sometimes lead to trouble, and it does sometimes literally kill the cat. It therefore behooves us to ask not only how to avoid missing important traits, but also how to guard against taking them too far.

Vicious Curiosity

The Vice of Excess Curiosity
Irving Langmuir, the physical chemist who was awarded a Nobel Prize in 1932 for his discoveries in surface chemistry, said in his banquet speech that "The scientist is motivated primarily by curiosity and a desire for truth."[10] In this sense, Langmuir would seem to be an exemplar of the virtues that

we have been examining. But it may be that Langmuir himself was actually an exemplar of taking these into the realm of scientific vice.

The novelist Kurt Vonnegut said he based his character Dr. Felix Hoenikker in the novel *Cat's Cradle* on Langmuir, who worked with Vonnegut's brother at General Electric. In Vonnegut's novel, Hoenikker's scientific curiosity is extreme, which initially serves him well. He is awarded a Nobel Prize and gives a very short acceptance speech saying, "I stand before you now because I never stopped dawdling like an eight-year-old on a spring morning on his way to school. Anything can make me stop and look and wonder, and sometimes learn. I am a very happy man. Thank you."[11] Such an attitude seems to be nothing if not an example of the sort of character that we have offered as a positive model for the scientist. But in Vonnegut's telling, Hoenikker's extreme curiosity has a dark element. He is oblivious to basic social relationships, including those of his own family. Lost in his own research, he rarely speaks to his children, who find him distant and frightening. He becomes one of the creators of the atomic bomb that is dropped on Hiroshima. When one of his fellow atomic scientists says that through their development of the A-bomb scientists have known sin, Hoenikker asks, "What is sin?"

It was actually the leader of the real Manhattan Project who was led to reflect on his sin and that of the other physicists who created the A-bomb. Oppenheimer had an interest in Hinduism, and he later recalled that upon seeing the blast of the first atomic test, a famous scriptural text from the Bhagavad-Gita sprung to his mind: "Now I am become Death, the destroyer of worlds." The line comes from a pivotal scene in the Gita in which the god Krishna reveals his true form, which has "the radiance of a thousand suns," to Prince Arjuna as they converse on a battlefield. Arjuna is wrestling with the question of whether or not he should proceed with the battle, knowing that he will be fighting against soldiers who include his own friends and relatives. Krishna explains that Arjuna should simply fulfill his duty as a soldier and set aside his concerns about the consequences. However, was not this just the problem that led to the bomb—scientists who failed to consider the deadly consequences of their research? Commentators who mention Oppenheimer's reference to the Gita often miss the subtlety of his point. They recognize the existential regret that the quotation expresses, especially when considered in relationship to his other comment about sin, but they overlook the tragic irony in the reference. After all, were not he and the other Manhattan Project scientists just doing their scientific duty?

We may also pose that weighty question in terms of virtues: were not they just expressing their scientific traits in their fullest form? This seems to be Vonnegut's point. He described Langmuir as caring only about finding the truth.[12] It is the tragic disjunct between the sweet technological promise of science and the possible hellish results of taking those good intentions too far that he identified. Vonnegut wrote with some personal knowledge of this, as he was educated as a chemist before he turned to writing. He had viewed science and technology as building a beautiful future but then was pulled up short after seeing the results of the firebombing of Dresden and the dropping of the atomic bombs on Japan. He described feeling sickened by seeing how science created such weapons of mass destruction. He compared it to the feeling of "spiritual horror" that a devout Christian might have in seeing the horrible massacres committed by Christians in war.[13] His portrait of Hoenikker is a stereotype, but Vonnegut believed that the stereotype of the scientist had a kernel of truth, and one cannot deny that the A-bomb was a real manifestation of the issue. Vonnegut's dramatic account may be a science fiction exaggeration, but that is what helps make it function as a cautionary tale of what can go wrong when even a central scientific virtue is taken to an extreme.

Sin or Pathology?

It is worth noting that this kind of literary portrayal is evidence that curiosity is indeed recognized as a core value in science, in that it identifies the characteristic trait that, in excess, becomes the basis of a distinctively scientific kind of villainy and horror. Mary Shelley's *Frankenstein* is an archetype in this vein of cautionary tales; it is the doctor's unfettered curiosity about the spark of life that leads to his overweening scientific ambition to command godlike power to create life from death, resulting in the creation of a monster. Even Alexander Pope, whom we previously saw as someone with a positive view of the scientific temper, warned that there were circumstances where it did not pay to be too inquisitive. "A person who is too nice an observer of the business of the crowd, like one who is too curious in observing the labor of bees," he wrote, "will often be stung for his curiosity."[14] Other writers have made similar points.

The eighteenth-century novelist and playwright Oliver Goldsmith saw curiosity as an aimless waste and no guide to wisdom. "A man who leaves home to mend himself and others is a philosopher," he wrote, "but he who goes from country to country, guided by the blind impulse of curiosity, is a vagabond."[15] At least one philosopher concurred with this assessment. The philosopher Thomas Hobbes is best known for his articulation of the idea of the social contract. He argued that we should rationally agree to give up some individual autonomy to a central government that could protect ourselves from each other and from a life that would otherwise be nasty, brutish, and short. Hobbes judged human beings to be essentially endless in their desires, requiring external powers to save ourselves from our own uncontrollable appetites. Mostly he focused on physical desires, but he thought that mental appetites were no less problematic. Curiosity, he wrote, "is a lust of the mind" that far exceeds "the short vehemence of any carnal pleasure."[16]

In speaking here of curiosity as lust, Hobbes was putting it within the Christian framework of sin. According to Dante, lust lands one in the outermost level of Hell, so while it is not the most serious of deadly sins, it is bad enough. Other sins of the body (gluttony and greed) come next, followed by sins of the passions (sloth and wrath), with the most serious sins that abuse the divine faculty (envy and pride) being at the deepest levels of the inferno. Dante wrote of the seven deadly sins as enumerated by Pope Gregory I, but previous Catholic writers had slightly different lists. We do not need to dig into such details. Indeed, as we saw in the previous chapter, it is not even the case that all Christian denominations would view curiosity in this way, though the worry that curiosity may be a sin remains potent for at least some religious believers.

Psychological pathologies A more modern approach is to see vices not in a religious framework as manifestations of sin but on a medical model as pathologies. Rather than speaking of gluttony or sloth as sins, therapists speak of eating and mood disorders. On this view, lust might be labeled as sexual addiction. Could curiosity be of a kind? Or perhaps it is like a drug that leads one to break bad from its addiction? One does sometimes hear curiosity described in this way, even by scientists themselves. Kevin Mitnick, known as the world's most famous computer hacker, spoke of his own state of mind in these terms to explain the exploits that led to his arrest, conviction, and incarceration: "I was hooked. I was addicted to hacking, more for the intellectual challenge, the curiosity, the seduction of adventure; not for stealing, or causing damage or writing computer viruses."[17] Mitnick's self-diagnosis may not be a completely reliable source of information on this question; however, it is not our purpose to evaluate the strengths and weaknesses of a medical model of vice but just to acknowledge it.

Virtue-theoretic analysis   While seeing these negative traits through the lenses of sin or pathology may be revealing in their own way, I do not adopt either view here. Consistent with the rest of our analysis, we shall look at negative traits and behaviors in science in terms of values and virtues. Virtue theory again provides a useful conceptual framework for this issue. Just as we can understand scientific virtues as those traits that are conducive to doing exemplary science, so too we may analyze scientific vices as traits that corrupt science. This is a pragmatic perspective, for it keeps the focus on the way that epistemic and ethical considerations relate to each other in science in the service of its central, guiding purpose and to our other human goals. As we will see, it allows us to make sense of why traits may lose their virtuous quality when taken to an extreme.

Vonnegut's story in *Cat's Cradle* is a useful character study, serving as an object lesson regarding our question about excessive curiosity. In the novel, Hoenikker becomes the inventor of Ice Nine, which freezes all water and ends up destroying life on earth. Literature in general, and science fiction in particular, allows one to perform philosophical thought experiments and to play out possible ramifications of different conceptual starting positions. It accomplishes this by creating scenarios that bring elements into clearer view, as using different stains to prepare microscope slides may reveal different features of a cell, by means of highlighting and contrast. The creation of Ice Nine in the novel shows the absurdity of extreme curiosity. Causing the end of the world is hardly a benefit to science. Even from the simplest internal perspective, it is clear that the satisfaction of curiosity is not an unmitigated value. Therefore, there must be limits to curiosity in science.

This is a straightforward example, but it shows in outline how virtue-based reasoning can work to help resolve ethical puzzles and correct ethical errors. The import of this should not be missed. Among other advantages, it keeps one from falling into cultural relativism. If something were moral only because a culture said so, one would have no ground for assessing and improving it. But science recognizes that it is not a perfect system and aims to identify its own flaws and weaknesses in order to correct them wherever possible. A virtue-theoretic analysis can help in that it provides a framework not only for understanding scientific culture but also for critiquing it.

Analyzing Scientific Vices

Anyone who has studied the history of science appreciates that the way it is practiced today is not exactly the same as it was in the past. If science were taken to be its distinctive practices as we know them now, one might be forgiven for thinking that it is a very different enterprise than it was in earlier centuries. Theories come and go. Paradigms shift. More significantly, even methods seem to change. Is there even any single, consistent thing that may be called the scientific method? As historians, we like to emphasize what is distinctive in each time period and place, but it is a mistake to portray each as unique and to ignore common threads. What does hold constant are science's core values. Its central guiding purpose has been, and remains, to discover true and useful things about the empirical world, and to do so based on evidence.

Given this purpose, being observant is easily recognized as a virtue, but our virtue-based approach gives us a perspective from which we may assess particular methods of observation in the interest of improving them. To take one example, the creation of scientific instruments is undertaken if there is promise of increasing our powers of observation. The first telescope was a novelty, but Galileo saw that it had revolutionary scientific potential to extend our vision farther into the heavens. Being meticulous is a virtue, stimulating scientists to devise improved telescopes with greater power and resolution. Objectivity is a virtue, so when optical biases are recognized, ways are sought to correct them. And so on. A virtue-theoretic framework provides not only standards for judgment but also the motivation to make needed corrections.

This framework similarly provides a way to judge our own dispositions and to point us in the right direction for improving our character. Aristotle provided a general analysis of this, arguing that to be virtuous is to have the appropriate qualities in moderation: "Moderation in all things." Traits need to be in balance. It is wrong to think that more is always better. Indeed, having a trait in excess often turns a virtue into a vice, as does having too little.

This analysis of vice has a more complex structure than either the religious or medical views.

Aristotle on vice as excess or deficiency   Aristotle analyzed moral virtue on analogy with craft-making skills. Consider what it means to be an excellent cook. The master chef will neither prepare too little food nor prepare too much, will neither overcook nor undercook the meat and vegetables. Put generally, the master of any art avoids both deficiency and excess. A dish with too little salt will be bland, but one with too much will be inedible. Similarly, on Aristotle's view, moral virtue may be characterized as the properly balanced mean. However, finding the golden mean between the extremes is not a simple numerical average; it must be judged, for instance, in relation to other elements of the meal. Seasoned chefs know how to judge the correct amount of seasonings to properly finish the dish.

This view fits with our evolutionary take on the beginnings of virtue in instinct. Remember that instincts are dispositions that are adaptive, and as Darwin explained, what is adaptive is always relative to an environment. Thus, it makes sense that the notions of moderation and balance need to be understood in context. However, we should not necessarily follow Aristotle's approach in every detail. According to Aristotle, there are many ways of going wrong, but only one way of going right.[18] This is too strong, for just as there may often be more than one way to accomplish a task, so too there may be more than one ethical option. Nor should we expect optimality. We are creatures with bounded rationality, and mostly must act with limited time and information, so we should not let perfection stand in the way of the good enough.

That said, Aristotle is certainly correct that there are many more ways to go wrong, so that it is "easy to miss the mark and hard to hit it."[19] Moreover, this provides a useful framework for understanding vice, which is characterized by deficiency and excess.

As with technical skills, Aristotle thought virtues could be learned. However, he rejected Plato's view that virtue is simply a kind of knowledge and vice a lack thereof. Virtue, for Aristotle, also required more than simply having the right dispositions for pertinent action; one must also have the right attitude, which requires having the appropriate emotions and desires. Aristotle took pleasure and fear to be the most basic of human desires. Both are important motivators for action, drawing us toward some things and holding us back from others. It is in relationship to these desires that the character virtues of temperance and courage are especially important in his account. Their purpose is not to repress these desires but to form them so that we fear rightly.[20] As we discussed previously, courage is required when fear needs to be overcome. Individuals who lack fear are not courageous, but reckless. Again, making the point generally, Aristotle said that "to feel these feelings at the right time, on the right occasion, towards the right people, for the right purpose and in the right manner, is to feel the best amount of them, which is the mean amount—and the best amount is of course the mark of virtue."[21]

Curiosity in moderation Aristotle made this point with regard to courage, but it holds equally with regard to curiosity, which we have seen is an instinct with a similar effect in overcoming fear. An Aristotelian analysis is instructive here, in that it explains the patterns we saw in our discussion of curiosity. Lacking curiosity is a vice, especially in science, because that would be a deficiency at the very core of what it means to be a scientist.

There are a couple of terms that one might use to describe individuals with a deficiency of curiosity; the first would be to say simply that they are uninterested in something, which is perhaps not too negative, as it has a limited scope. But someone with a severe deficiency, someone who is uninterested generally, is referred to with the much more cutting term incurious. To call someone incurious is to disparage them in a way that cuts far deeper than the more common insult of calling someone stupid or ignorant. We are all born ignorant and even with our best efforts will remain mostly ignorant for our whole lives. But individuals who are incurious have no excuse; they are ignorant not because they lack intellectual ability but because they lack an intellectual spark. This is a far more damning indictment, as it suggests a dullness of spirit with regard to an essential element of what makes us human.

Interestingly, there is not as clear a term to indicate an excess of curiosity. This might suggest that most people never reach a point of being thought to be too curious, at least in a general sense. There is a suite of terms having to do with excessive curiosity with regard to social relationships—such as being nosy, snooping, and prying—but these may be thought of as negative because they violate expectations of privacy. Vicious curiosity may be uncommon, but we have seen not only how it is possible from a virtue-theoretical standpoint but also at least some plausible cases where it has occurred. And if anyone can be said to be prone to a possible excess of curiosity, scientists are probably more likely candidates than others. Moving to excess is perhaps the most common danger with regard to particular specialized virtues in any vocation, and for scientists it can be hard to know when to curb their curiosity. However, even our short account shows why it is important to be aware of the possibility and to be wary of turning this scientific virtue into a vice. Knowing that the pursuit of knowledge comes with risk, prudence alone should give us reason to pay attention to finding a balance between curiosity and caution.

Excess Objectivity as Indifference?

As a second case, let us turn briefly to the virtue of objectivity. We have already seen how objectivity may be understood in terms of impartiality. Deficiency of this trait is a clear scientific vice; it is the vice of bias and prejudice. It is a bit harder to see how the trait may become vicious in the other extreme. One way would be to see the extreme as a kind of mechanical, robotic logic that makes decisions by a cold, emotionless calculus. We have already discussed why that is the wrong way to think of science, as scientists' passion drives exemplary research. But there is another, related notion that is worth examining in more detail. While impartiality is a good thing in science, in that it eschews bias and looks only to conclusions based on evidence, taking it too far may lead to the vice of indifference. It is one thing to resolve to let the truth be found by evidence without regard to personal preferences but quite another to not care about consequences of any sort. This seems to be the problem that Vonnegut had in mind.

For Vonnegut, the most troubling thing about scientists like Langmuir was that they seemed not to think about, or even care about, the possible applications of their research. "Langmuir," he said, "was absolutely indifferent to the uses that might be made of the truths he dug out of the rock and handed to whomever was around. But any truth he found was beautiful in its own right, and he didn't give a damn who got it next."[22] This is not to disparage either truth or beauty but just to recognize that they are not the only values one needs to care about.

Interestingly, Vonnegut attributed this scientific vice to the literary stereotype of the pure scientist who is so focused on his research as to be absent-minded regarding ordinary matters and indifferent to anything besides his research. He thought that scientists, at least ones from the generation that produced the A-bomb, were all too glad to embrace the stereotype. I am not sure that Vonnegut was right about this; he had a negative and pessimistic view of human nature with regard to such matters, and it is an empirical question whether he was descriptively correct either about scientists in general or Langmuir in particular. What is correct, I submit, is that stories can function as moral portraits and that values can be exemplified and passed along through such depictions. More importantly, from a normative point of view, they provide a way to think through the structure of values of a practice.

Should Scientists Eschew Advocacy?

One brief illustration of an application of this approach involves the question of whether scientists should allow themselves to advocate publicly for or against matters of policy that have a scientific component. It is wrong to be indifferent to the results of one's research, but what is the proper behavior for a scientist in real world circumstances?

A common view is that scientists are right to just stick to the facts. Their job is to investigate the world and to provide science's best assessment of the reality we are faced with and the limits of our ability to intervene to change it. If possible, they should extend those limits by creating technologies that increase our ability to control our fates. But, it is said, they should not engage in advocacy. Virtue theory is clearly relevant in this sort of moral decision. Indeed, the main argument that scientists give for this recommendation is that being seen as an advocate will undermine the scientific reputation for objectivity. How, they ask, can scientists be trusted to provide impartial answers once they begin to be advocates for particular viewpoints?

Moreover, the expertise that we have as scientists is indeed limited. Scientists' virtues, practices, and methods are all structured in relation to science's goal of discovering empirical truths about the world. Their focus on evidence-based reasoning is what provides the justification for expert assessment in the first place. But being an expert on some scientific question does not make one an expert about nonscientific questions, including questions about whether some particular social policy should be implemented or even whether a given amount of scientific research is worth funding compared to other social needs.

However, the fictional Ice Nine and the all too real A-bomb show that such an attitude is deadly if taken to the extreme. Maintaining a reputation for objectivity is of little use at the end of the world. Similarly, the possible loss of a reputation for impartiality is minor compared to the loss of trust that comes when science is seen as to blame for the creation of weapons of mass destruction and the resulting threat of nuclear annihilation. So, let us try to keep things in perspective.

Taking a limiting case like this is a way of demonstrating a general point—here, that objective impartiality is not a categorical virtue. By now, that should be no surprise. The key issue is to examine the application of the virtue in a subtler way and to hone one's judgment so as to know how to weigh it in relation to other virtues in different circumstances. The case suggests that scientists ought not completely reject the possibility of acting as advocates. It does not license being advocates under any conditions, and one must develop moral judgment to sort all this out. However, what is interesting to note here is that one can find cases where advocacy by scientists increases, rather than diminishes, objectivity and trust.

Being disinterested (with regard to the possible results of investigations) is not the same as being uninterested or indifferent. There are many cases of researchers, for instance, who have devoted their scientific lives to the study of some disease because a family member suffered from it. This gives them a keen interest in the subject itself, but their investigation itself is disinterested in that they want evidence to determine the answer to the question—What will cure this disease?—without prejudice for or against some particular answer. Knowing the basis for such scientists' interest in the research question, a third party would likely trust those researchers more, not less. Indeed, it would seem perverse if such scientists, out of a misplaced concern about objectivity, declined to advocate for a discovered cure.

This example is helpful for illustrating two important points. First, it shows how a virtue-theoretic perspective can help resolve particular ethical problems. Scientists have worried about the ethics of advocacy, and many have gone too far in proscribing it. Virtue theory provides a framework that can help one make more nuanced ethical judgments in difficult cases. Second, and this is the more important general point, it shows the danger of taking a trait to an extreme. Objectivity is a core scientific value, pushing scientists to find ways to better avoid bias, but it remains a virtue only when the impartiality it recommends does not turn into the kind of indifference that Vonnegut was calling out. This is Aristotle's point about finding the place between the vice of deficiency and the vice of excess—virtue is the golden mean.

Other Scientific Virtues in Deficiency and Excess?

These two examples show that virtues have a subtler structure than the simplistic bimodal moral concepts that most people start with. Morality is more complex than good vs. evil. On that kind of view, one might think that in science it is a simple matter of identifying virtues and their "opposite," which is a common way that people initially think about these things. Such a view may allow one to quickly identify the most salient negative trait, while obscuring negative traits on the other side. As we saw, this may lead one to contrast being curious with being incurious, which is the more common deficiency case, and to overlook the possible problem of excessive curiosity. It may lead one to contrast objectivity with bias and to overlook possible problems of indifference. However, using an Aristotelian framework as a starting point, it is easier to discern this pattern in other cases. Let us now briefly identify other traits in this more fine-grained manner to distinguish the scientific virtues from their corresponding vices of deficiency and excess.

One might not expect that this pattern will be followed for every trait, but it is a useful heuristic. What about the value of veracity in science? The virtue of honesty is clear, as is the vice of dishonesty, which is its lack. However, one might think that it would not be a moral problem to have an excess of honesty in science. When James Watson published *The Double Helix*, which gave an account of his and Francis Crick's discovery of the structure of DNA, reviewers invariably commented on the fact that he seemed to dispense with any pretense that scientists were paragons of morality, including himself. He presented a "warts and all" picture of scientists. The distinguished evolutionary biologist E. O. Wilson, who was a colleague of his in Harvard's biology department, wrote that Watson fancied himself as "Honest Jim," and that he said whatever came to his mind with a "casual and brutal offhandedness." Wilson said that if it had not been for the magnitude of Watson's discovery, he would have been thought of as just another gifted eccentric and that "much of his honesty would have been publicly dismissed as poor judgement."[23] Wilson acknowledged that Watson probably thought that he was working for the good of science and that "a blunt tool was needed" to move it forward in the way he wanted, but he described as many ways in which Watson's blunt and careless manner harmed science as helped it. He thought of Honest Jim not in a positive way but rather as "the Caligula of biology."[24] We do not need to decide here whether Watson's brutal honesty caused more harm than good, but the example makes it clear that it is indeed possible to be honest to a fault.

We have already examined some of the reasons that skepticism is a virtue in science. A doubting attitude keeps scientists sharp and pushes them to not accept a conclusion before the evidence is in. It is clear why a deficiency of skepticism is a negative trait in science—being gullible makes one more likely to believe something that is not sufficiently supported by evidence. But what about being overly skeptical? Charles Darwin wrote that insufficient skepticism was "a frame of mind which I believe to be injurious to the progress of science."[25] He recognized this trait in himself and guarded against it by making it a habit to always write down any observation or argument he came across that could be a problem for whatever hypothesis he was investigating, so that he would not fail to consider them seriously. It was helpful for a scientist to have a good measure of skepticism, he wrote, so as not to waste time on hypotheses that really were not viable. However, he also pointed out that having too much skepticism could also be problematic. He wrote of knowing quite a few scientists "who, I feel sure, have often thus been deterred from experiment or observations, which would have proved directly or indirectly serviceable."[26] It is important to question oneself, but being too skeptical could make research grind to a halt. Darwin's term "serviceable" here is significant from our theoretical perspective, for it highlights the fact that the correct standard of judgment should be in terms of what works in the service of discovery but also that the standard need not be optimality. As in evolution, science does not expect perfection but rather requires only adaptations that are good enough for the circumstances. This is a useful evolutionary addendum to Aristotle's notion of moderation.

We can see this kind of feature in other virtues as well. For example, with regard to attentiveness, it is easy to recognize inattentiveness as the clear vice on the side of deficiency. To be attentive in excess might be labeled obsessive. Similarly, with regard to meticulousness, the deficiency is being careless or sloppy. An excess of meticulousness might be called being compulsive. Interestingly, a person may have these traits to a higher degree that in other circumstances may be thought of as negative, yet still find them to be virtuous for science. (Recall my colleague with the mild form of obsessive-compulsive personality disorder who thought it helped her research.)

There is much more one could say about the subtleties of this fine structure of virtue, but we must save that for another occasion and here just briefly note the general Aristotelian analysis of the other main scientific virtues. Deficiency of perseverance leads one to give up too easily, while an excess may be called obstinacy. The sort of intellectual humility that is relevant in science—humility to evidence—is similarly clear. Having an excess seems much the same as being too skeptical and might be a form of scientific timidity. Deficiency of intellectual humility is the more serious problem for science, as it leads one to have too much confidence in one's own ideas and preferred hypotheses and to elevate them prematurely or beyond their value. This might be a kind of intellectual pride, which is deleterious to scientific practice for it is a form of authority, one's own, over evidence. One might think that this is a form of scientific arrogance, but that seems to be a different issue and worth separate treatment.

Scientists Behaving Badly

The Arrogant Scientist

We have previously discussed how a basic value in science is that conclusions are to be accepted on the basis of evidence rather than authority. However, dismissal of authority can sometimes become a kind of arrogance, and it is likely this trait that leads some to see scientists as arrogant. Einstein himself was notoriously contemptuous of authority. "A foolish faith in authority," he wrote when he was a student at the University of Zurich, "is the worst enemy of truth."[27] Later in life he realized that he had taken this trait to an extreme and reflected, "To punish me for my contempt for authority, fate made me an authority myself."[28]

The problem is not just that arrogance is an unpleasant trait, which makes arrogant people poor collaborators. This is harmful to science, but it is hardly the worst effect of this trait. Arrogance can also come across as domineering and in some contexts even sexist. An example of this sort of effect may be seen in the way that Russian botanist C. A. Timiriazeff spoke about science in his popular lectures on the life of plants:

> [T]he chief object of the scientist is not to describe but to explain and command Nature; his method must not be that of a passive observer, but rather that of an active experimenter; he must engage in strife with Nature, and by the power of his mind extort from her answers to his questions, so that he may master and subordinate her at will, provoke or arrest the phenomena of life, direct or vary them.[29]

Timiriazeff was right about the importance of experiment and active interventions in science, but one cringes at this way of describing the scientist's methods. Of course, he was writing in an earlier era when the culture as a whole had yet to be sensitized to the need to avoid such an attitude. While culture has progressed in this regard in the intervening century, it cannot be said that the problem has been eliminated.

Arrogance can also breed a feeling of superiority to the degree that someone becomes incapable of considering opposing views or heeding even reasonable words of caution. Enrico Fermi had originally been dismissive of critics of the A-bomb project in just this sort of way. "Don't bother me with your conscientious scruples!" he said. "After all, the thing's superb physics!" But after the test explosion he was not able to drive home because of the shock he felt.[30] Whatever one's view about the morality of the A-bomb project, we can all agree that an arrogance that blinds one to scruples is a serious failing.

This ought to be understandable from a scientific perspective even without bringing in general moral considerations. Arrogance of this sort can blind one to other perspectives and to critical assessment and thereby reduce a scientist's ability to eliminate biases that may stand in the way of finding some truth. It can also sometimes lead scientists to overstep the bounds of their expertise, moving them from science into scientism.

Scientistic Gluttony

The explanatory success of science makes some scientists believe that science can answer every question, leading them to overextend its methods or to push the particular perspective of their discipline into areas where it does not apply. We might think of this lack of restraint and occasional tendency to try to devour whatever it finds in its path as a form of intellectual imperialism or perhaps a kind of gluttony. This can take many forms, but let us look at just a couple of cases by way of example.

Sufficiently accurate for poetry   In our discussion of the virtue of meticulousness, we examined the special importance of measurement and quantification in science. It is for good reason that, in an earlier era, the slide rule and the pocket calculator were icons of the scientist. Wherever possible, scientists attempt to express observations in numerical form so that patterns in the data may be recognized and analyzed statistically. It cannot be denied that quantitative reasoning is extraordinarily powerful; Darwin said that scientists who possessed advanced mathematical abilities seemed to have a sixth sense. But we also noted how some scientists may put too much emphasis on quantification as necessary for scientific knowledge—many still hold to an extreme form of Kelvin's Dictum, which we touched on earlier. When this attitude is taken too far it can lead scientists to disregard other disciplines where quantification is less important. This no doubt accounts for some of the cases of scientists' arrogant behavior, as when they barge in to apply quantitative "rigor" to some area where it is not appropriate.

One example that is sometimes cited to illustrate this involves Charles Babbage, the nineteenth-century polymath best known for his pioneering work in computing and his invention of the first automatic calculator. In a brief letter to the poet Alfred, Lord Tennyson, Babbage wrote:

> In your otherwise beautiful poem "The Vision of Sin" there is a verse which reads—"Every moment dies a man, Every moment one is born." It must be manifest that if this were true, the population of the world would be at a standstill. In truth, the rate of birth is slightly in excess of that of death. I would suggest that in the next edition of your poem you have it read—"Every moment dies a man, Every moment 1 1/16 is born." The actual figure is so long I cannot get it onto a line, but I believe the figure 1 1/16 will be sufficiently accurate for poetry.[31]

Babbage's arrogance was well known in his day, and this may indeed be an example of scientific overreaching, though I am more inclined to read it as being meant humorously—as a bit of good-natured joshing. Whatever the case, ascertaining Babbage's own frame of mind and intent here is not relevant to our current discussion. Even if I am right that the letter was meant in jest, it still exemplifies the point (through self-deprecating humor) that some scientists are prone to overextend the role of quantitative reasoning and other aspects of their methods into areas where they do not belong.

Of course, imperialistic or gluttonous overextension is not exclusive to science. The conflicts with religion and literature we discussed in the previous chapter arise with blame on all sides. However, our task here is not to point fingers or apportion culpability but to do what we can to rein in such tendencies within science. Just as it was problematic when dogmatic religion tried to assert its authority into matters of empirical fact, it is similarly problematic when science tries to push into metaphysical or ethical issues, as though these were resolvable by empirical tests. I have elsewhere defended the distinction between metaphysical naturalism and scientific naturalism (which is methodological rather than ontological) that is applicable here.[32] We have also already seen why one may not "read off" ethics from biology or, more generally, derive normative conclusions (*oughts*) directly from solely factual information (*is's*). The scientist who understands such distinctions will be less likely to overextend the scientific way of knowing into areas that go beyond the limits of its special epistemic methods.[33] Much more could be said about these and other cases, but I will focus here on just one example—a case of overextension of scientific reductionism—to illustrate the problem and how we might begin to address it.

Scientistic reductionism   In scientific contexts, reductionism is often an effective method for explaining phenomena. The term refers to the idea that it is often possible to explain some apparently complicated phenomenon by reducing it to its parts. One example of this from physics is statistical mechanics, which explains the thermodynamics of large physical systems in terms of the motions of their constituent atoms. Mendelian genetics in biology is another example, explaining phenotypic patterns observed in biological organisms in terms of their genes. Showing how the parts of a system cause some aspect of interest about the whole system is pervasive and powerful in science, but it can be taken too far.

Consider this case of overextended reductionism taken from Nobel Laureate Francis Crick, codiscoverer of the structure of DNA. He wrote: "You, your joys and your sorrows, your memories and your ambitions, your sense of personal identity and free will, are in fact no more than the behavior of a vast assembly of nerve cells and their associated molecules."[34] Crick calls this idea "the astonishing hypothesis," and it is the basis of his book by the same name. This is not a benign form of explanatory reductionism but what might more appropriately be called gluttonous reductionism. Crick seems to think that pointing to such biological and biochemical factors is all that is needed to explain these complex human traits, and that they are "no more than" that. This is a form of scientistic gluttony in that it tries to swallow all of human experience into his scientific specialty of molecular biology. But Crick has bitten off more than his science can chew.

Crick seems unaware of how parochial an explanation he has offered. If one is going to be a gluttonous reductionist, why stop at molecular biology? Requiring molecular explanations makes biochemistry the base, but a gluttonous physicist would say that this hardly goes far enough. After all, all those nerve cells and biomolecules are "in fact, no more than" the behavior of atomic and subatomic particles and forces. The absurdity of this sort of reduction ought to be self-evident, but perhaps it is worth pointing out just a few of the problems.

Crick's astonishing hypothesis is really little more than the straightforward claim that things are made of matter. But this is truly simplistic from an explanatory view. He points to human joys and sorrows, but take anything you like, and one could give the same "explanation." What makes plants grow?: "It's atoms and molecules, my dear lad." What makes the sun shine? Atoms and molecules. Why are we here? Atoms and molecules again. Saying this with a knowing, scientific nod of the head does not improve the answer. If this were all there was too it, then we would have a ready-made explanation for anything, and could all head home.

Perhaps this is being unfair to Crick, as he did also speak of the "behavior" of these cells and molecules. That is a bit better, but one could make the same objection here. Saying, "It is all behavior, chap," is not much more satisfying. Think of a parallel example to Crick's "astonishing" hypothesis that you offer after having taken a flight from New York to Amsterdam: This airplane, its jet engines, its wings, ailerons, and trim tabs, its communication and navigation systems, its brand and styling, are in fact no more than the behavior of a vast assembly of metal and plastic parts and their associated molecules. If your listeners had been frozen during the Dark Ages and then been thawed and revived today to witness the "miracle" of flight for the first time, they might have thought that airplanes relied on black magic. Such individuals might indeed find your version of Crick's hypothesis astonishing, but otherwise it is rather unenlightening. Again, there is nothing wrong per se in pointing out that things are made of atoms, molecules, or cells, but it is quite wrong to say that complex phenomena like flight or psychology or personal identity and so on is "no more than" these.

The mistake of gluttonous reductionism is that it fails to recognize the significant processes that operate at higher levels of organization. Explanations in terms of such processes are equally based in natural laws and are no less scientific than ones at lower levels of organization. Indeed, depending on the question, they are often more salient, more revealing, and more useful than lower level ones. If you want to know why this square wooden peg does not fit in this round hole, it would be a laughable explanation to speak of the assembly of cellulose molecules that make up the peg. The more appropriate answer is simply that it is a *square*. The particular molecules are mostly irrelevant for this sort of high level question. This is not to say that the molecular structure does not matter at all. It matters in the sense that the peg needs to be made of matter of some sort. However, as long as the peg is of a certain size it does not matter whether it is made of wood, metal, or plastic; it is the shape, not the matter, that matters.

We may again thank Aristotle for the framework behind this analysis of how explanation works in science. He pointed out that there are different aspects to the explanation of anything. What Crick was talking about falls under what Aristotle identified as the material cause of a thing. However, one may also speak of a thing's efficient cause, which includes the forces that worked to make it so, the formal cause, which involves the shape or form of the thing, and what he called the final cause, which involves what the thing is for. It is especially ironic that Crick ignored this last explanatory element, for that is where evolutionary explanations come in. Darwin wrote a book on the expression of emotion in humans and animals, and for my money his evolutionary explanations are arguably far more revealing and satisfying than simple appeals to nerve cells and molecules. Add psychological, sociological, cultural, literary, philosophical, and other explanations to the mix, and one's understanding of our joys, sorrows, and other mental traits becomes richer still.

Again, the point is not that it is wrong to offer reductive explanations; what is wrong is to arrogantly and ravenously insist that all other forms must give way to one's own preferred reductive level. We would do well to avoid this and other examples of the vice of scientific gluttony.

Extreme Behavior

Curiosity, the instinct for truth, conceptually organizes science and drives scientific practice. It is because this is taken for granted (at least mostly) in scientific culture that scientific results can be trusted (at least mostly). We must include the "mostly" caveat because we know that scientists do not always live up to the ideal and because, as we have seen, it is even possible for a core virtue like curiosity to turn into a vice by overextending it. Just as having virtuous traits and habits disposes one to act virtuously, resulting in excellence and flourishing, so can having a deficiency or excess of the same traits impede or even undermine a practice. Aristotle was right to note that both extremes may be thought of as vices, but there is an interesting asymmetry between the two when it comes to understanding vocational virtues. Someone who is weak in traits that are essential to a practice will be judged a poor practitioner, but if the deficiency is severe they will not be recognized as a practitioner at all. Someone who lacks curiosity and other scientific virtues simply is not a real scientist. On the other hand, someone who displays the requisite traits in excess would still be recognized as a practitioner, despite being judged a poor one. Stereotypes are based upon exaggeration of distinctive characteristics, which is why stereotypical scientists typically are depicted as being honest to the point of bluntness, meticulous to the point of eccentricity, or objective to the point of annoyance.

Kinds of scientific misconduct   There are many ways in which scientists may behave badly in both directions. In this chapter, I have mostly focused on the category of excess, because this is the more neglected and, in many ways, the more interesting case. When does a virtue become a vice by going too far? Rarely is this a simple matter of "crossing a line" from good to evil. More often this is the source of the tragic, when some positive trait is taken to excess from a well-intentioned, single-minded focus, to the neglect of other virtues, resulting in a loss of balance and then some fateful calamity. Much more could be said about scientific tragedy, but I want to turn now briefly to the side of deficiency.

Most of what we think of as scientific misconduct falls under this latter category—they involve cases where individuals fail or fall short with regard to some essential trait. We have already examined one extreme example of this sort—the Schön case—where data were fudged, falsified, and even fabricated. These are cardinal sins in science. Falsification, fabrication and plagiarism, the unholy trinity of scientific misconduct, all involve deficiency of science's basic virtue of honesty and several other key scientific virtues besides. One may analyze various forms of questionable research practices—actions that may not quite rise to the level of outright misconduct, but which are nevertheless problematic because they undermine science in subtle ways—in a similar fashion. These are often cases where meticulousness has been neglected in some way, especially with regard to the strictures of scientific methodology. I would also argue that many general workplace climate issues, including various forms of social bias and prejudice that sometimes escalate to the point of hostility, may be judged in these terms. As we discussed earlier, explicit or implicit bias should be seen as a failure not only of objectivity but also of basic humility to evidence, both essential scientific virtues.

There are some other kinds of scientific misconduct that do not fall as neatly into this virtue-theoretic classification that are important to mention. Research that involves human subjects, for instance, requires that special attention be paid in the design and execution of experiments, so subjects are not harmed. There have been egregious cases of misconduct—Nazi medical experiments, the Tuskegee Untreated Syphilis Study, the Milgram experiment on obedience to authority figures, to name three notorious examples—where the knowledge learned was bought at too high a price. These are now recognized as serious breaches of research ethics because of their disregard for basic duties of care. Research with animal subjects involves its own requirements of care. The possible satisfaction of curiosity that might result from such experiments does not trump such ethical duties.

There is one additional category of misconduct that deserves special treatment. Conflict of interest is typically treated as a distinctive form of misconduct, but it is better understood within a virtue-theoretic analysis. This arises not because of an excess or deficiency of some trait that turns virtue into vice but rather cases where internal scientific values come into conflict with external values. This is a very general and pervasive kind of issue, but I will illustrate it here by repeating an argument I have made elsewhere about the conflict of interest that may arise for scientists and their institutions because of where they get their funding.[35]

Conflicts of interest, conflicts of values   Why should it matter ethically who funds a scientist's research? Scientific research is expensive, and it will not get done at all unless someone pays for it. Viewed practically, it seems that funding is simply a matter of economics. However, a broader perspective reveals that a variety of ethical issues are indeed relevant, and responsible researchers need to take these seriously. The possibility of conflicts of interest that certain funding arrangements can cause is one important area of concern. This has become increasingly significant as universities, which typically are at the forefront of scientific research, struggle to find new sources of funding to supplement or replace traditional sources.

If nothing else, funding from industry or other special interest groups can sway what scientific questions get investigated, which can in turn lead to outcomes that are unjust or ethically problematic in some other way. Long-term basic research can get shortchanged in the pursuit of applied research that promises a quick return. Public research institutions may begin to neglect research that fulfills their founding mandate of serving the public interest in favor of research that benefits the particular interests of industrial sponsors or their own financial health.

These are all important considerations to keep in mind in sorting out the ethics of conflict of interest in scientific research, but here I want to focus on a deeper issue, namely, how financial interests may undermine scientific objectivity. The ethical problem in financial conflicts of interest in research is not, at base, that of divided institutional loyalties but rather that such interests can conflict with the scientist's basic responsibility to investigate the world objectively. This is not just a conflict of interest but a conflict of values. Of course, financial stakes are not the only source of potential conflict; ideological and other confounding interests can be equally dangerous to objectivity, but here we will keep our focus on what many take to be the bottom line.

It may not be impossible for scientists to remain objective in their research, no matter who is providing the funding, but it certainly can be difficult. With their future funding always on the line, researchers may feel great or subtle pressure to get results that will keep their sponsors happy. Even if individuals perform their research perfectly objectively, they may find themselves constrained in reporting their full results if external funders have retained some control over what can be published, since it may not be in the interest of the sponsor to reveal every finding.

This is not to say that research sponsors do not care about scientific objectivity. In most cases, objective research is in their own interest as well and not just because there can be a real economic payoff in the discovery of empirical truths. However, in certain kinds of situations, someone with a financial stake in an outcome might care more about only the appearance of truth. Business advertisers never underestimate the commercial value that scientific support for the efficacy of their product will bring or the blow that a negative finding can have. There are already documented cases in which industry funders removed information from university researchers' reports that revealed health dangers in their products. These are clear violations of the sort of impartiality we should expect in scientific research.

There are also more subtle ways that conflict of interest can undermine objectivity. For instance, there is real evidential value to knowing whether the investigator has a financial stake in the conclusion of a study. When we, as third parties, assess reports of the results of scientific studies, we quite properly consider the trustworthiness of the investigator, because we know that researchers are themselves instruments in any study, just as spectroscopes and survey forms are. Everything from the quality of the data to the accuracy and fairness of the final report can be affected for better or worse by the objectivity (among other characteristics) of the researcher. Moreover, we understand how a researcher's judgment or actions could be influenced to favor the preferred outcomes of those who hold the purse strings. For instance, it was not unreasonable to doubt studies that supposedly refuted health dangers of cigarettes once one learned the research had been funded by the Tobacco Institute.

It should thus be obvious why financial conflict of interest and potential loss of research independence and control ought to be of concern to universities. A major reason why industry and other institutions find it valuable to turn to university researchers is that historically the academy has placed a premium on the value of knowledge for its own sake. Because its primary interest (at least in the ideal, if not always in practice) was in the intrinsic value of learning and research, it was trustworthy (again, at least on the whole, if not in every instance) as a source of objective knowledge. The degree to which this basic value of academic research is compromised will, in like measure, compromise the value of its products.

Another way to look at this is in terms of third-party perceptions. It is probably true that the public tends to overemphasize the degree to which industry funding may be a confounding factor, because most people are never in a position to directly evaluate the quality of the experimental design itself. However, even if a scientific conclusion is properly supported by the evidence, its value will not be appreciated if the public loses trust in the objectivity of scientists and the integrity of the scientific enterprise.

The more that university research is perceived as being beholden to special interest funding, the more likely that its reputation for impartiality will degrade over time.

What is the upshot of such considerations? Do they imply that academic researchers should never accept industry or special interest funding? No, such extreme measures are not necessary. However, it does mean that it should be accepted only when adequate protections are in place. Our task during this transitional period, as research universities begin to forge closer ties to industry, will be to conscientiously devise a system of checks and balances to safeguard the values of open and impartial research that made industry want to benefit from university research in the first place. In the long run, it is in the interest of all parties to uphold the integrity of scientific research. If warranted trust in scientific objectivity is undermined, we will lose much of the fundamental value of science. This is not just a matter of conflict of interest; it is a conflict of values. Thus, from an ethical perspective, it is research integrity, not financial gain, that should be the bottom line.

Cultivating an Ethical Culture

Vice Squad?

So, what is to be done to prevent scientists from behaving badly?[36] The research ethics community has come to a consensus that promoting responsible conduct of research (RCR) cannot be done on a piecemeal basis but will require the cultivation of an ethical scientific culture.[37] One highly cited paper puts it this way: "[A]ll explanations [of research misconduct] seem to share a common denominator—the failure to foster a culture of integrity"[38] Focusing on culture is critical, but ethics and culture interact in complex ways, so fostering an ethical culture is not always straightforward. Science, as C. P. Snow emphasized, has its own distinctive culture and thus its own ways of expressing integrity. Unfortunately, RCR is often framed in ways that are insensitive to how ethical norms are embodied and transmitted culturally in general, let alone in scientific culture.

Whereas deeply rooted cultural norms organically structure a society or a practice from within, RCR literature and training too often theorize and present research ethics in terms of quasi-legalistic external control. The paper quoted above that calls for fostering a culture of integrity will serve as a representative example. It summarizes the issues in this way:

> No regulatory office can hope to catch all research misconduct and we think that the primary deterrent must be at the institutional level. Institutions must establish the culture that promotes the safeguards for whistleblowers and establishes zero tolerance both for those who commit misconduct and for those who turn a blind eye to it.[39]

Such sentences bristle with regulatory and legal terminology. The paper's recommendations for fostering an ethical culture in research are put in the same external, legalistic terms: institute "zero tolerance," whistleblower protections, a clear reporting system, mentor training (specifically so mentors are "more aware of their roles in establishing and maintaining research rules and minimizing opportunities to commit research misconduct"), and alternative oversight mechanisms beyond formal complaints (e.g., institutional auditing of research records). Even the final recommendation to model ethical behavior is formulated in like manner and focuses mostly on "policies," "procedures," and "deterrents."[40] This is not the development of an ethical culture but of an enforcement culture.

Inherent in its name, RCR focuses on behavior—how should scientists conduct their work. Conduct in the RCR literature is typically couched in terms of rule following and rule breaking. Laws are not the only kind of rules, of course, but because the field arose in response to egregious behavior,[41] it is not surprising that RCR rules were originally theorized and are still largely framed in legalistic terms. Putting it bluntly, RCR as currently taught is not so much focused on conduct as misconduct.

A legal framework may be necessary as a way for institutions to deal with misconduct, but this is not the most effective way to foster a culture of integrity. It is not that rules of conduct are problematic in and of themselves, but in understanding cultural dynamics, one must take into account that rules seen as imposed from without are viewed very differently than those that are part of a culture. This is one reason why scientists sometimes see RCR regulations as interfering with science rather than furthering its aims.

Furthermore, a legalistic approach that focuses on misconduct misses an important feature of culture in that it goes beyond behaviors to include attitudes. As we have seen, culture is essentially normative, involving all sorts of values and ideals, including ideals of character. Put another way, culture involves not only what kind of behaviors I should or should not do, but also what kind of person I should or should not be. Thinking in terms of scientific virtues allows one to analyze and promote such values in the culture of science. By better understanding the character traits that make for an exemplary scientist one can acquire a better understanding of the actions that follow. This is directly related to the notion of research itself.

Responsible Conduct of Research

When one speaks of responsible conduct of research, the tacit assumption is that we are dealing with scientific research, which is characterized by its distinctive aims and methods. As we have seen, a scientific virtue-based approach begins here. Aristotle explained how virtues arise in relation to the *telos* or ends of a practice: they are those settled dispositions that are conducive to the achievement of excellence in that practice. The central aim of scientific practice is the discovery of empirical truths about the natural world, and the methods of science reflect its basic epistemic values, such as testability and repeatability. Scientific virtues are thus those character traits that a scientist should try to embody for science to flourish.[42]

The final key term in RCR is "responsible." Typically, this is thought of in this context as a synonym for ethical conduct of research, but it is worth considering what is implied specifically by the notion of responsibility. The primary question one asks in this regard is "responsible for what?" Appropriate answers to this question involve enumeration of one's duties. As previously noted, duty in science is not limited to compliance with laws and rules. But a second question when one speaks of responsibility is to whom or to what is one responsible? This is a more fundamental question, as duties are derivative of it. I argue that the basic responsibility of the scientist is to science itself, in part because science is based on evidence rather than authority. The scientist is not responsible to a scientific leader or any particular person but rather is responsible first to the values that structure science as a practice and then to humanity as a whole, as all practices themselves ultimately aim at human flourishing.[43]

What this means is that scientific integrity is more than research integrity. Integrity involves the notion of a unified wholeness of parts that function together by virtue of the strength of its supporting structure. Scientists are researchers at base, but they are not only that. They are also colleagues and mentors. They interact with other actors in other professions and other walks of life. They are citizens and human beings. Thus, we need to broaden the scope of research ethics in this way, for there is more to science than just the conduct of research.[44]

As a way to speak about this, my own tendency is to retain the traditional sense of RCR with its focus on research integrity and think of that as one core part of a broader category of science ethics, which should be seen as also encompassing the scientific virtues and other topics that may be linked to, but are not directly a part of, basic research. But one does not need to legislate terminology; research ethics is already a rather broad term. As ethicist Kenneth Pimple pointed out, it may even be said to be an "incoherent" field with subject matter that encompasses "ageless moral truths and recent arbitrary conventions; minute details of particular actions and the broad sweep of public policy; life-and-death issues and matters just the other side of simple etiquette."[45] Whether we adopt a new term or further expand the scope of the old one, my point is just that we need a broader notion that incorporates this wider perspective and that explicitly includes the character of the scientist.

The scientific virtue approach does not reject the importance of rules or even of law as a means of supplementing self-regulation. Again, the problem is not with rules and laws per se but rather with whether they are imposed from without or whether they arise as an expression of intrinsic values from within the culture. This book has tried to make the case that science has an inherent moral structure and that the scientific virtues are a promising organizing principle for reconceiving and expanding science education and RCR. To foster a culture of scientific integrity, taking the values already inherent in scientific culture seriously is a good place to begin.

Virtuous Institutions?

Of course, no ethical theory is a panacea and no form of ethics training will work for everyone. In the musical *Camelot*, Mordred is the virtuous King Arthur's moral opposite who brazenly announces, "I like my women married, my willpower weak, my wine strong, and my saints fallen." Arthur's best efforts to reform Mordred are doomed to failure. In his gleefully twisted song about the seven virtues, Mordred portrays them as "deadly" and "ghastly little traps" to be sidestepped and avoided. Courage is but an invitation to death. Humility leaves one not with the earth but the dirt. Fidelity is only for your mate. It may be that some individuals like Mordred really are simply bad apples who will never change their character.

However, even though we cannot ignore the possibility of misconduct by bad apples, we should not let that completely drive RCR training.[46] It would be an even bigger mistake to let it dominate our conception of the issues. Of course, policing and enforcement will always be necessary to deal with such cases, and scientists must take responsibility for this task both individually and corporately. However, focusing narrowly on this as a legalistic matter about catching cheaters is hardly the best way to cultivate a culture of integrity. As we have seen, there can be more general causes for scientific vice and misconduct. If institutions promote RCR in the wrong way, such as through a simplistic compliance model, it can inadvertently undermine the development of an ethical culture and instead engender cynicism. We need to think carefully about how institutional policies, from rules to incentives, shape behaviors. An institutional setting, depending on how it is structured, is an environment that can allow and encourage virtuous behavior, or hinder and discourage it.

This was one reason for mentioning above the case of conflict of interest with regard to funding; it highlights that such ethical misconduct is not simply an individual matter but involves institutions in various ways as well. In that case, the concern was that research universities risk losing trust if they are not cognizant about maintaining independence from outside funders who have different value weights than basic scientists. If that trust erodes, it could also affect individual scientists and eventually science itself. However, research funding is just one example. There are any number of other sorts of cases where the moral (or immoral) behavior of individuals interacts with that of some larger group, ranging from exploitation of graduate students by self-serving researchers to general issues of mutual respect and professional courtesy. In thinking of science as systematic curiosity, one must pay attention not only to the structure of scientific virtues and how they are embodied in individual scientists, but also to the structure of the organizational relationships within which scientists practice. Developing and maintaining an ethical culture of science necessarily involves both.

The idea of culture is usually analyzed in terms of the mores of a group of people—a community—but we should remind ourselves of an important caveat here. While it is a commonplace to suggest that "the community" is the locus of authority in normative matters, we should not do that glibly. The notion of "community standards" is important, but we need to distinguish two ways in which this might be understood. If this is taken in what ethicists call a cultural relativist view—which holds that whatever the community says, goes—then it is problematic. Communities can have systemic biases, including ones that cause injustice, and their standards may need to be improved or replaced. Even setting aside this sort of standard philosophical problem of cultural relativism, it will not do to simply say that the community is the locus of authority, especially for the sciences, as this would reintroduce the mistaken notion that justification is based upon authority. Ethically, the community is not a moral authority any more than individuals are. Of course, we do sometimes speak of particularly wise people as individuals who speak "with moral authority," but if we apply the same distinction between authority and expertise that we used for empirical conclusions, then it might be better to say that such a person has moral expertise. Again, we should not quibble over terminology, but there does seem to be a reasonable conceptual difference between these. Persons with practical wisdom are worth heeding not because they get to legislate by fiat what is (or is not) moral, but because they have developed a more sensitive judgment to discern what is more likely to be (or not be) moral. If there is such a thing as moral expertise, it is something along those lines.

This is the second, and more ethically justifiable, understanding of the notion of community standards. Although morality is not simply what the community says it is, the community certainly ought to be involved in ascertaining appropriate moral standards and in deciding how they should be expressed and enforced. The individual is the locus of moral action, but the group mind may sometimes be more powerful in discerning complexities and working toward solutions. However, and this is the critical point, this must be done with the recognition that community standards must themselves be open to criticism, improvement, and correction. We saw a version of this previously, in our discussion about how scientific methods should themselves be subject to revision when new ways are discovered to reduce biases that interfere with objectivity. The same point extends to other protocols, practices, and policies at all levels of scientific organization. As in the epistemic case, analysis of conduct (and misconduct) should not stop with individuals. Unlike individual moral agents, the "scientific community" is an abstract entity, but it is defined by the same *telos* and so may be said to have a derivative sense of virtue and vice.

We must be sensitive to this difference between what would make me a good (or bad) scientist versus what would make us, communally, better (or worse) scientists, in the same way that, in thinking about justice, we consider both what it means to be a just person and to create just social structures. That scientific institutions have come to recognize the importance of RCR training of individual researchers is all to the good, but it will not do if the broader, structural issues are neglected. RCR training ought to aim to cultivate ethical judgment in individuals, but it cannot stop there. If we allow the derivative notion of virtue, then we may say that, for the sake of the scientific *telos*, some community standards and structures are more virtuous than others. It behooves the scientific community to pay as much attention to the environment and climate of research, including structures of training, collegiality, and reward, for it is in such a culture in which individual virtue will either grow or wither. Bad eggs cannot be blamed for all of scientific vice. For science to flourish, it must be willing to regularly put its own community standards to the test and take appropriate steps to protect cultural practices that support individual virtue and improve those that do not.

The Genie

The story of Eve and the apple is just one of many classic tales that warn against curiosity. Eating the apple brings knowledge, but also death. Death enters the world in another ancient cautionary tale about curiosity, when Pandora cannot resist the urge to open a forbidden jar of "gifts" given to her by the gods, thereby releasing sickness, disease, and other maladies. Pandora herself was created by the gods to punish mankind for Prometheus's having stolen fire from them. In another story—one told by Scheherazade in *One Thousand and One Nights*—a curious fisherman pries open the lid of a strange and marvelous copper jar his nets have pulled up, releasing a genie who has been imprisoned within for centuries. The only wish that this angry genie says he will grant the fisherman is the manner in which the genie will kill him. In some versions of the tale, the genie is identified as the demon Asmodeus, who, by coincidence, represents one of the seven deadly sins; he is the demon of lust and is associated with the second level of hell, which is where the curious are reputed to be condemned to. We clearly need some better stories that depict curiosity in a more positive light!

For science, curiosity is a basic value, and we ought to celebrate it. However, so is honesty, and honesty demands that we acknowledge that curiosity will sometimes open dangerous boxes. It was scientists like Oppenheimer and even Feynman who uncorked the atomic bottle and released the nuclear genie. We tend to remember the Disney version of Aladdin's genie, but not all genies are so benign. It was not by chance that the MB-1 air-to-air nuclear missile, which Boeing called "the most powerful interceptor missile ever deployed," was named the Genie.[47] Ironically, Disney used the story of the Fisherman and the Genie in a 1957 animated educational TV show about *Our Friend the Atom* to illustrate the possible dangers that accompany the benefits of atomic power. As in the original story, the fisherman tricks the genie into returning into the jar and then, having regained control, convinces the genie to work for his benefit. In the Disney version, the fisherman extracts three wishes, that the power of the nuclear genie be put to work for the goals of producing and promoting energy, food and health, and peace. Disney produced the film at the request of the Eisenhower administration as part of its Atoms for Peace initiative. The film, as well as a book and a Tomorrowland ride at Disneyland, was paid for by General Dynamics. It is easy to view *Our Friend the Atom* as nothing more than self-serving propaganda from the military-industrial complex to mollify a skeptical public and justify the expansion of the nuclear weapons program. But that is too cynical and simple a reading.

President Eisenhower's Atoms for Peace speech, which he gave at the United Nations in 1953, must be understood in relation to the Cold War effort, but it should also be recognized as an honest acknowledgment of "the awful arithmetic of the atomic bomb." Confronting an awful situation, it was a good-faith attempt to try to reduce the risks of deadly nuclear warfare by redirecting atomic research and policy toward positive ends. In making "these fateful decisions," Eisenhower pledged the determination of the country "to help solve the fearful atomic dilemma—to devote its entire heart and mind to find the way by which the miraculous inventiveness of man shall not be dedicated to his death, but consecrated to his life."[48] Given the real gravity of the danger, viewing this goal cynically would be a terrible error.

This case of atomic power is a prime example of a general issue—scientific discoveries can bring both promise and peril. More than any other episode, the creation of the A-bomb undermined trust in scientists and caused the public to begin to question science as a consistent force for progress. If scientists ignore or dodge responsibility for possible perils, they further undermine trust. Recognizing such perils requires self-awareness and perhaps an additional measure of intellectual humility—the recognition that science does not answer all questions, that it cannot expect to make wise decisions by itself, and that even its central virtues can turn to vices if taken to an extreme.

This, perhaps, is the way to take these cautionary tales. Rather than reading them as warnings against curiosity in general, they should rather be seen as warnings against taking curiosity too far. Knowledge itself is no sin, but thinking that the little knowledge we have gleaned makes us gods may be. Stealing fire from the gods gives us godlike powers, but that does not come immediately with godlike wisdom. Prying open every jar without due consideration of possible consequences is curiosity without judgment. Virtue ethics does not have the resources by itself to decide every moral choice—consideration of duties and rights, as well as costs and benefits must be brought in to help resolve particular ethical dilemmas—but it provides a framework that helps distinguish virtue from vice and a model for developing the practical judgment needed to find a balance among competing interests and values.

Putting this into practice and defending it in the practice of science requires vigilance, for the scientific community cannot ignore predictable perils, including the possibility of bad apples. Weakness of character may lead some scientists to violate even basic rules and norms. Aristotle was right that both deficiency and excess can turn virtue to vice. Creating an ethical culture in science is therefore essential. RCR training is a step in the right direction, but because this occurs only very late in one's education, it is unlikely to be sufficient to accomplish this task by itself. Scientific values cannot be tacked on at the end but must be cultivated as part of science education from the beginning, and it is to that topic that we now turn. Finding the right balance of traits, developing judgment, and putting our character and practice to use for good ends must be our wish. For the community to neglect this, and to allow scientific values to decay, could indeed mean death—the death of science itself.

9 The Seeds of Science

People like you and me never grow old. We never cease to stand like curious children before the great mystery into which we were born.
—Albert Einstein, letter to Otto Juliusburger[1]

The whole art of teaching is only the art of awakening the natural curiosity of young minds for the purpose of satisfying it afterwards.
—Anatole France[2]

Teach Them Wonders

Goals of Science Education

For society to flourish, it must consider what knowledge science provides and then wisely apply its technological fruits. For science to flourish—for its culture to be propagated and its ways of knowing the world integrated with other forms of reason—society must teach it well, not only for future scientists but also for the many more who will use science in some other way. But what does that entail? Reflecting about how we should we teach people about science, Richard Feynman said, "I think we should teach them wonders and that the purpose of knowledge is to appreciate wonders even more."[3]

Compare Feynman's recommendation to that of one important consensus study report by the National Research Council, the operating arm of the National Academies of Sciences, Engineering, and Medicine, which focused on identifying effective approaches for early education in science, technology, engineering, and mathematics (STEM). Effectiveness, of course, is a relational term that must be judged in relationship to goals, and the report identified three key goals for STEM education:

> 1. Expand the number of students who ultimately pursue advanced degrees and careers in STEM fields and broaden the participation of women and minorities in those fields. ... 2. Expand the STEM-capable workforce and broaden the participation of women and minorities in that workforce. ... 3. Increase STEM literacy for all students, including those who do not pursue STEM-related careers or additional study in the STEM disciplines.[4]

These are laudable aims, but one must admit that they are not especially inspiring on their own. Feynman's recommendation, on the other hand, captured the spirit of the scientific mindset, and its way of thinking about science education is far more expressive and compelling than these bureaucratically articulated goals. And yet the two need to come together. The virtue-theoretic approach that we have been developing may help do just that and provide a useful framework for thinking about science education. Our main focus has been on the values and virtues that underlie excellence in basic research, which applies most directly to the first NRC goal, but this has important implications for science education more broadly as well, including stimulating interest in science, broadening participation, and increasing STEM literacy. To begin to address this, let us first consider what it means to be scientifically literate.

Levels of Scientific Literacy

Civic scientific literacy   Early notions of scientific literacy involved the same sort of things that are commonly taught in basic science classes, namely, the products of scientific research—the discoveries. In most science courses this is what was taken to be the "content" of the class. There are, of course, good reasons for learning the factual content of science. It is also true that someone who does not know that the earth goes around the sun or that genes are coded in DNA or that our world is billions of years old cannot really be said to be scientifically literate in a very basic sense. My colleague Jon Miller has tracked this sense of scientific literacy over many decades, noting that only about 28 percent of American adults have sufficient understanding of basic scientific ideas to read about science at the level it is presented in the *New York Times*.[5] He properly points out that the key here is not thinking of this on a warehouse model of learning, in which pieces of information are to be stored on a mental shelf, but rather in functional terms, namely as tools needed to operate successfully as citizens in society. This notion of civic scientific literacy recognizes the political importance of having functional scientific literacy, and Miller noted with approval that the University of California, Berkeley, changed the name of its introductory physics course from Physics for Poets to Physics for Future Presidents.[6] Amen to that! However, there are other aspects of science that even this broader emphasis on functional content misses, that are at least as important.

Conceptually, literacy involves language—specifically reading and writing—but, as we have seen, for science, language comes later. It is hard for a book lover to admit this, but one cannot learn everything from books. Before the scientific revolution, it was common that scholars looked to the books of earlier generations for knowledge and wisdom. However, seeing books as primary in this way is another example of taking authority over evidence. We now find it amusing to see the odd mistakes that were perpetuated about plants and animals by scholars who took their knowledge about nature from old books when they could (and should) have observed the real objects in nature.

This ought to be relevant even in the earliest stages of science education. To take a simple example, many scientists credit their early experiences in nature as critical in developing their interest in science. The evolutionary biologist E. O. Wilson often spoke with fondness of time spent as a boy, with his nose to the ground, watching ants. With changing living patterns, it is less common for kids to roam free in the woods, even in summers. Children growing up in cities of brick and concrete are especially disadvantaged. The "No child left indoors" science education slogan is, for this reason, very much on target. This is not to denigrate the importance of book learning and acquiring basic facts. There is certainly value in education research about what conceptual scaffolding teachers ought to provide to increase the chance that their students will come away understanding difficult scientific ideas. A student who continues in science will need to know the relevant concepts. But understanding discoveries that scientists have made is not the same as understanding how scientists make discoveries. If children only see nature through books and on screens, it seems less likely that they will develop the kind of firsthand appreciation of nature that inspired Wilson and others.

Thus, especially for teaching and learning science, we might do well to adopt an expanded notion of literacy that recognizes the importance of real observation and experience for understanding what underlies the words.

Robust literacy—reading with understanding—requires appreciation of how those facts were discovered. Put another way, it requires a general understanding of scientific methods and practices.

Scientific methods and practices   This has been a special area of interest and a focus of my own STEM education scholarship. As I have argued, the best science education reveals not only the science of nature but also the nature of science.[7] As we saw, science is natural philosophy, and my argument has been that to teach science properly we ought to include instruction in the methods and practices of science, which is what philosophy of science is about. If a student is to be inspired to think about a career in science, we ought to give them a real sense of the nature of that vocation. This is more important than giving them a longer list of basic scientific facts to study. Moreover, most students will not go on to be scientists, and for such students—possible future presidents—it is more important that they gain an appreciation of how science works so they are better able to understand how scientific conclusions are inductively tested and confirmed. One need only think of the problems caused by counterfeit science—not just creationism but also the anti-vaccination movement and similar cases—to recognize how a general appreciation of evidence-based reasoning would go a long way to helping individuals make better choices both as citizens and in their own lives.

As in the case of basic content, these ought not simply be professed but demonstrated. The problem of merely professing to students from the podium is well known to discipline-based education researchers, and science education is in the midst of a major reform effort that emphasizes active learning. Our philosophical perspective provides another reason we should endorse this kind of reform: the sage on the stage is yet another instance of an authority model, which science rejects on principle. Without understanding the distinctive evidence-based way in which facts are ascertained in science, a student sees little difference between science and any other subject. A science teacher ought to be respected not as an authority but as an expert. In science, someone's say-so is not acceptable by itself but only as a proxy for the evidence.

This sort of recommendation is not unique or even new; Dewey and others after him made similar arguments. What has changed in recent years is that a threshold of agreement has at last been reached among scientists and educators in that science practices are now explicitly mentioned as important learning goals in recent science education standards.[8] Now the challenge is to devise ways to do this effectively for different subjects.

One part of my own interdisciplinary education research has focused on how we can do this for biology, especially with regard to using evolution as an exemplar of scientific reasoning. To demonstrate how to implement this in a practical way in classrooms, I created an educational version of the digital evolution platform Avida that researchers use for experimental evolution. Like the research version, Avida-ED is powerful as a model system because the digital organisms within its environment actually evolve. How is that possible? Because in Avida-ED the core elements of the Darwinian mechanism—random variation, inheritance, and natural selection—are not simulated but instantiated.[9] This is especially compelling for active learning; students can run their own experiments and observe evolution in action. There is no better response to claims that evolution cannot produce novel, complex, functional traits (a common creationist view that some students come in with) than to let students test it themselves and watch evolution do just that.

Since then, together with a great team of colleagues and graduate students, the Avida-ED Project has created a large suite of model exercises and other curricular materials that demonstrate not only basic evolutionary mechanisms but ways that evolutionary hypotheses may be tested and confirmed.[10] Avida-ED can be used for in-class demonstrations, homework or lab exercises, or full open-ended research projects where students work though the entire scientific process from asking questions, formulating hypotheses, designing and implementing experiments, analyzing data, and writing up and presenting results. This experience has convinced us that it is indeed possible and practical for students to get a firsthand sense of the methods and practice of science. Informally, there is also reason to think that this kind of learning environment can help students develop an appreciation of scientific values and virtues, which is the next area that should become a part of science education reform.

Scientific values and virtues Teaching the nature of science—its methods and practices—is a huge step forward, but it does not go far enough. Our survey of scientists revealed a strong consensus that the virtues we have been examining are important to exemplary scientific research. Are we adequately conveying the nature of scientific methods and practices in science courses if we leave out the values that underlie them? Moreover, our respondents said that they look for such qualities when judging who to take into their lab as a graduate student or postdoctoral researcher. If we hope to increase interest and persistence in STEM, especially for individuals who have not considered such a vocation, we ought not neglect this.

The scientific character virtues are a critical part of the scientific mindset, and we would do well to make this explicit in science standards. Of course, there will be predictable resistance at first. Even after people come to understand their philosophical justification and see that these are broadly recognized within the scientific community, some will question whether they can be taught. Are some people just not cut out for science? Or can the scientific mindset (or any virtues, for that matter) be trained? This is an ancient question.

Meno's Question

Plato's dialogue the *Meno* opens with a question to Socrates: "Is virtue something that can be taught?" Or instead, Meno continues his query, "does it come by practice? Or is it neither teaching nor practice that gives it to a man but natural aptitude or something else?"[11] Socrates was a good person to ask, as the Oracle of Delphi had said he was the wisest of Greeks. More importantly from our point of view, Socrates is an exemplar of what it looks like to be a person who always wonders why. Socrates viewed himself as someone who is himself ignorant but who desires to know, and he modeled this attitude of curious inquiry no matter what the topic, helping others do the same.

In the course of their discussion—a quintessential Socratic dialogue—Meno is led to recognize that he does not even have a sure grasp of what virtue is, let alone how it may be learned. This leads him to wonder how one may come to discover anything new at all. "[H]ow will you look for something when you don't in the least know what it is?" Meno asked. Plato's answer, articulated by the character of Socrates, involved his notion of recollection—the idea that we already contain knowledge within our souls and can recall it through disciplined, rational thought—which he demonstrates by showing how an ignorant slave boy can reason to geometrical conclusions. Do not assume that a person's circumstances will prevent such learning. We can certainly agree with that general point even if we are not convinced by Plato's specific epistemological account.

Aristotelian/Darwinian approach   In this book I have taken an approach to virtue that is more inspired by Aristotle, with a heavy infusion of Darwinian evolutionary science that, among other benefits, provides a natural alternative to Plato's doctrine of recollection. The usual empiricist alternative to Plato's rationalist epistemology holds that we are born devoid of knowledge until we are marked by the world—all knowledge comes from experience. My evolutionary account seeks a middle ground. We are not born with innate ideas in the Platonic sense but neither are we blank slates. Rather, we are born with at least some worldly wherewithal inherited from our ancestors and their own selective history, which provides us with functional knowledge of the world, coded in our instincts.

It would take another book to lay out that full argument, but I believe it is a connection of which Darwin would approve. Darwin's Greek was poor, and he never read Aristotle in the original, but near the end of his life the Aristotle scholar William Ogle gifted him with a copy of his translation of Aristotle's *Parts of Animals*. Darwin had previously appreciated Aristotle as "one of the greatest, if not the greatest, observers who ever lived," but Ogle's book opened his eyes to the depth of Aristotle's accomplishment and vision.[12] Though he had already written to thank Ogle for the book when he received it, he was moved to write again after getting further into it to say that he had "rarely read anything which has interested me more" and that it had given him a new appreciation for Aristotle: "From quotations which I had seen I had a high notion of Aristotle's merits, but I had not the most remote notion what a wonderful man he was. Linnaeus and Cuvier have been my two gods, though in very different ways, but they were mere school-boys to old Aristotle."[13]

Aristotle, of course, had no inkling of the Darwinian notion of evolution, but it is not an unreasonable stretch to speak of this in quasi-Aristotelian terms: phenotypic expression by the environment is the actualization of the organism's genetic potential. Moreover, these evolved abilities allow us to then learn about the world during our lifespans. We may make discoveries and pass them along (culturally) to the next generation. The process of discovery is itself Darwinian in form, utilizing the general process of generation of variations that may then be subjected to the test of experience.[14]

I suggest that the same is true of moral virtues. This departs somewhat from the standard view in virtue ethics, which holds that virtues are acquired traits; MacIntyre even built that into his definition of virtue.[15] However, this is an unnecessary and ill-conceived conceptual restriction. Aristotle had no such constraint in his own conception of virtue, and a hypothetical example explains why it is not required. Consider the virtue of compassion or other forms of caring. It would not make sense to disqualify an individual who was born with such compassion in full measure and who naturally acted upon that feeling just because it was innate rather than acquired. Empathy is at least a partly heritable trait, and individuals may have it by nature in different degrees. But suppose, as a thought experiment, that everyone was simply born with saintly compassion so that it did not even need to be cultivated (i.e., everyone is not only disposed to act compassionately, but also thoughtfully chooses to do so when it is called for); that the trait was given, rather than acquired, is not a good reason to say that it is not a virtue or that its expression in the whole population is not virtuous. Thus, the notion of virtue as an acquired trait in the extreme definitional sense is a mistake.

A thought experiment, however, only helps with the conceptual point; the standard view is more realistic. No one is born a saint. Nor is it even reasonable to expect that everyone is born with the instinctual prerequisites for successfully making one's way in the world in equal measure. It is similar to the mistake of an overly simple good-vs.-evil binary morality, if that is taken to think that one either has a trait or does not. Virtues, like truth, come in degrees. With very few exceptions, evolution provides everyone with a sufficient amount (we all come from an unbroken line of evolutionary winners, after all), but each of us is likely to have to work harder to develop one or another trait. Think back to our earlier discussion of Darwin's account of the evolution of the moral sense. Social animals have instincts that conduce to fellow-feelings and helping behaviors, but they will be expressed more or less fully depending on how they are reinforced or discouraged by the natural and social environment. This is the challenge of learning and education—instinctual traits must be developed and given moral force by reason. Put another way, we are born with the seeds of virtue, but these must be cultivated.

To Stand Like Curious Children

The First Emotion

Edmund Burke, the eighteenth-century Irish political philosopher and statesman, wrote that "The first and simplest emotion which we discover in the human mind, is Curiosity."[16] Burke took curiosity to be, at base, the finding of pleasure in all things new, the desire for novelty. Burke may not have been thinking about this biologically, but he was insightful in identifying this mental trait as primitive and pervasive, noting that it is a component of everything that affects the mind and that it blends with all our other passions. However, he judged that it is the most superficial of emotions and thought it to be fickle and quickly exhausted. Samuel Coleridge hailed Burke as "a scientific statesman,"[17] and that is true to a degree, but these latter judgements suggest that Burke did not fully appreciate the depth to which curiosity runs in the scientist.

Born Curious

The physicist Leo Szilard, best known for his idea of the nuclear chain reaction and for his letter with Einstein to President Roosevelt that led to the Manhattan Project, expressed a sentiment that is common among scientists. "I was born a scientist. I believe that many children are born with an inquisitive mind, the mind of a scientist, and I assume that I became a scientist because in some ways I remained a child."[18]

Szilard was identifying an element of his scientific character that he thought went back to childhood and was present at birth. In some sense he underestimated that latter point in thinking only that "many" children are born with an inquisitive mind; we have previously explained the evolutionary reason for thinking that curiosity is an instinct with deep epistemic import, especially for organisms like ourselves who are born into a variable environment and so must be able to learn its features and affordances. We are born curious. As babies we instinctively reach out to touch the world, to bring it to our lips to taste and try. By necessity, infants are natural explorers. Developmental psychologists have long recognized this important fact about human cognition. Piaget, in observing his own infant, saw him as a budding scientist. Robert Emde, following Piaget's ideas about cognitive assimilation, studied how children from birth to age three transform their biologically prepared motives into a basic set of norms, expressed prior to language as procedural knowledge, to "get it right" about the environment. He argued that such processes constitute at least the beginnings a "moral self"[19] and even "pathways to virtues."[20] Emde focused on social virtues rather than the scientific virtues and saw "little to recommend" in the metaphor of "child-as-scientist."[21] I think his negative assessment was premature, coming from a language-first view of science, and that his focus on procedural knowledge actually fits better with the language-later view that I have been advocating. I recommend the research of Alison Gopnik and colleagues, who investigate the notion of "the scientist in the crib" within an evolutionary framework as especially compelling for understanding the instinctual basis of mental traits that are essential for figuring out not only the causal structure of the world but also the problems of language and other minds.[22] Still, it is true that in infancy and early childhood we are only protoscientists. We start with instinctual curiosity in much the same way that Darwin spoke of the social instincts as the basis of the moral sense. The curiosity we are born with becomes scientific in a morally robust manner only later, through reason and training, but its instinctual root and connection to childlike exploration makes for a powerful motivator for many scientists, and one finds this notion regularly mentioned independently by exemplary scientists.

Scientists as children   We have previously mentioned Huxley's famous dictum that, if they are to learn anything of the world, scientists must sit down before fact as a little child. Einstein attributed his discovery of relativity theory to his own childlike view of the world:

> The ordinary adult never bothers his head about the problems of space and time. These are things he has thought of as a child. But I developed so slowly that I began to wonder about space and time only when I was already grown up. Consequently, I probed more deeply into the problem than an ordinary child would have.[23]

One finds this idea—that being a scientist is like remaining a child in one's attitude toward the world—articulated by many scientists as they reflect on their experience. Barbara McClintock recounted her own scientific attitude in similar terms:

> I was just so interested in what I was doing I could hardly wait to get up in the morning and get at it. One of my friends, a geneticist, said I was a child, because only children can't wait to get up in the morning to get at what they want to do.[24]

Speaking about how his interest in science was motivated by the lovely and interesting discussions about nature he had with his father when he was young, Richard Feynman expressed the childlike wonder of discovery he felt as a scientist:

> I've been caught, so to speak—like someone who was given something wonderful when he was a child, and he's always looking for it again. I'm always looking like a child, for the wonders I know I'm going to find—maybe not every time, but every once in a while.[25]

Science and play   Feynman was expressing the same sort of feeling as McClintock, which is that science is fun. Scientific curiosity involves a kind of playfulness. In basic research, scientists experience the childlike enjoyment of exploration—trying things out for the first time, going somewhere they have never been before, seeing things never before seen. That scientific playthings are sometimes abstract does not diminish the enjoyment. Scientists speak of playing around with data, turning problems over and over in their head, like a puzzle toy. Even when the problems become highly abstract, there is always some concrete element to a scientific question. Determining the constraints of a mechanism or trying to solve an empirical problem can be like fiddling with the buttons and knobs of a new gaming device to figure out how it works.

Another element of play that seems relevant to basic science is that there may be no reward for solving a puzzle beyond the intrinsic satisfaction and delight of having figured it out. Nor need there be. This is what we spoke about previously as the internal good of science—what Feynman called the pleasure of finding things out, and what Darwin said was "reason enough" for scientific researches.

The puzzle-solving analogy nicely captures this and other aspects of scientific practice. Thomas Kuhn used the term puzzle solving to refer to a key element of what he identified as "normal science," namely, that between rare periods of revolution, much of what occupied scientists was solving puzzles within paradigms. Of course, every analogy has limitations, and here an obvious one is that many kinds of puzzles are not scientific. In logic puzzles, like Sudoku, for instance, one already knows the answer and the puzzle just involves manipulating the symbols to make it all fit. Theoretical physics might seem superficially like this, in that the work involves searching for patterns and symmetries. The difference, of course, is that it is not enough to be consistent internally; it must also be consistent with nature, which is to say that it must be tested and confirmed experimentally. This was Feynman's point that the wonder of having made an actual discovery does not happen every time, for experiments often prove one's theorizing wrong. But what a deeply felt thrill it is when, every once in a while, the curiosity that drives puzzle-solving is rewarded.

Character is not fixed   Before continuing, it is worth reiterating a point about the instinctual basis of virtues that we ought to be able to take for granted but which often causes concerns because of common misunderstanding about how evolution works. The concern is that in saying traits have a biological basis one is implying that nature predominates over nurture, giving rise to the fear of reductionistic genetic determinism and the social biases that may arise from that kind of view. For some, it brings to mind Freud's claim that anatomy is destiny together with the general worries of biological predestination that accompany the idea, but such ideas were expressed in other ways long before Freud. In his autobiography, Darwin told a story of how Robert FitzRoy, captain of *HMS Beagle* during his famous voyage, had initially thought he would be inadequate to the rigors of the trip. "He was convinced," Darwin wrote, "that he could judge a man's character by the outline of his features; and he doubted whether anyone with my nose could possess sufficient energy and determination for the voyage." Darwin found this naive view more amusing than insulting and joked that, "I think he was afterwards well satisfied that my nose had spoken falsely."[26]

Current evolutionary science makes the proper relationship clear. Biological traits do, of course, have a genetic component, and it would be foolish to ignore that causal factor, but genes are expressed only in relationship to environments, and phenotypic traits are the effect of both. It is thus a mistake to assume that one can judge a person's character simply by who their parents are or how they look. Even more important than basic phenotypic plasticity is that evolution has provided us with great flexibility for dealing with variable environments—we have the ability to learn. As Theodosius Dobzhansky pointed out, educability is the key factor in the evolution of human beings. What this means is that character is not fixed. Innate traits may be developed or diminished.

Curiosity Lost

We are born curious, but all too often education extinguishes that spark. For some, this happens early in childhood from adults who demand unquestioning obedience. Children naturally ask why, but some parents, not always because of weariness, cut that short. Don't ask why; just obey. It is, thus, not surprising that curiosity and authority are regularly at odds with one another. A parent who demands obedience on the basis of authority ("Because I said so") sets a pattern that tells the child to not ask too many questions. It sends a message that they should simply acquiesce and do what they are told. Of course, sometimes simple obedience is called for, as in cases where immediate danger threatens, but if it dominates to the degree that natural curiosity withers, it deprives children of one of their most valuable character assets.

The same dynamic all too often operates in the classroom. In the traditional model, teachers are the authorities and students are expected to not question their pronouncements. The way this was implemented varied, though only slightly, as one ascended through the educational levels. In university, the classic model was the sage on the stage. Professors professed, and students were expected to absorb (memorize) what they offered. The operative metaphor was knowledge as a liquid that "spouted" from the teacher and was funneled into the heads of the pupils. Good students were described as "sponges" for the way they could "soak up" the knowledge that flowed from a teacher. Unfortunately, we move all too quickly from the lovely image of a someone who is a "fountain" of knowledge to the less attractive image of students doing their best to "spit" back what little they had learned.

Certain aspects of the traditional model, even aspects that are regularly mocked, such as the copying of lines, are actually not entirely unreasonable. Part of what copying does, simple though it may be, is help develop a habit modeled on something that presumably has been determined as worth emulating. One would never expect someone to learn to play the piano simply by hearing a lecture about playing the piano, or even by hearing someone play the piano. There is no substitute for practice. What a piano teacher does is help someone learn how to practice. The challenge is to figure out how to do this in a way that develops the needed motivation and skills so that the student can eventually proceed independently. There is no avoiding the fact that this will involve scales and arpeggios, but if learning is no more than endless marching up and down the keyboard the student will quickly lose the natural joy of music.

The same is true of classroom learning. For certain topics and in certain circumstances it is not only efficient but highly valuable to listen to skilled lecturers and to try to take in the knowledge they have distilled. However, in general, this works well only when the listeners have already learned how to learn. Academics are almost always individuals who learned the requisite learning skills that allow them to take something they hear in lecture form (or through "book learning") and process it themselves. They may have become so good at this that they have forgotten how they acquired the skills in the first place, let alone know how to impart them to others. Too often such instructors teach as though they are speaking to others who are like themselves, rather than to students who are still acquiring the needed skills. Anyone who recalls sitting through dry lectures knows how deadly and uninspiring this can be.

Darwin's Mentors

Much of Charles Darwin's postsecondary education was like this, even in the areas for which he already had an interest. Reflecting upon his university experience later in life, Darwin wrote that his courses at University of Edinburgh and Cambridge University had been mostly useless for his development as a scientist. It was his experiences outside of the classroom that made him a scientist. Competitive beetle collecting, geological field trips with his science professors, and especially his experience on the *Beagle* were what he most credited for developing the mindset and habits of a scientist. We have already mentioned how he called out the importance of the *Beagle* voyage for developing the virtues of industry and attentiveness—habits of mind that allowed him to draw connections between what he read, observed, and thought about. In his autobiography, he wrote that the most important factor was "the habit of energetic industry and of concentrated attention to whatever I was engaged in, which I then acquired. Everything about which I thought or read was made to bear directly on what I had seen or was likely to see; and this habit of mind was continued during the five years of the voyage. I feel sure that it was this training which has enabled me to do whatever I have done in science."[27]

How did Darwin develop these traits? Mostly through practice, which established them as habits. This was fostered by the influence of a few of his professors who departed from the traditional model, such as the Reverend Professor John Stevens Henslow, the Regius Professor of Botany at Cambridge, who engaged his students in what educators now call active learning. Henslow led his students on regular fields trips and had them work in the college gardens, always encouraging them to make their own observations. Darwin became a regular member of Henslow's scientific circle and recalled an example of his gentle form of mentoring:

> Whilst examining some pollen-grains on a damp surface I saw the tubes exserted, and instantly rushed off to communicate my surprising discovery to him. Now I do not suppose that any other Professor of Biology could have helped laughing

at my coming in such a hurry to make such a communication. But he agreed how interesting the phenomenon was and explained it's meaning, but made me clearly understand how well it was known; so I left him not in the least mortified, but well pleased of having discovered for myself so remarkable a fact, but determined not to be in such a hurry again to communicate my discoveries.[28]

Henslow was certainly his most important scientific mentor and colleague, and Darwin included this story to illustrate his "kind consideration" as a teacher. However, this is more than an example of consideration and kindness. Darwin drew the lesson that he should not be overly quick to announce some discovery, but this was more than a lesson in patience. Another teacher, operating as an authority, might have pointed out that the fact was already well known and thereby diminished the young Darwin's pleasure at his discovery. That sort of response could easily make one feel ridiculous and thereby squash a student's excitement in this personal discovery. What a different experience it is to have a professor affirm the interest and scientific significance of the observation. Rather than making Darwin feel small and insignificant about something he didn't know that everyone else already did know, Henslow's approach made him feel like a member of the community. He had, on his own, discovered something that the scientific community already saw as worthy. In this way, making a rediscovery was not an indication of inferiority but rather a sign that one might have what it took to be a scientist oneself. This is one way that a scientist with practical wisdom acts as a role model.

Character Education

Role Models/Exemplars/*Phronomoi*

Thinking of mentors as roles models is especially apropos for our discussion in that the term was coined by Robert Merton, who used it in a study of how medical students were socialized into the behavioral norms of medical practice. Their professional socialization occurred in relation to the reference group of medical professionals—a group to which they did not yet belong but hoped to be accepted into—by modeling themselves on the social-behavioral expectations of the role. As we discussed before, Merton was mostly focused on institutional norms rather than values and character virtues, but his concept is easily adapted for our question about character education and to an evolutionary framework.

One may think of a model as a template that can be copied. In general, copying successful models is a good approach for any kind of learning. Moreover, in that copying process, a model can be modified and so become more (or less) adapted to circumstances. While it is important to pass on the wisdom of past experience, teachers should not make the mistake of trying to make their students into clones of themselves; the evolutionary point about educability is that it allows organisms to adjust to environmental variations.

The notion of a "role" is also useful, though potentially misleading and problematic from our perspective, if it is taken to suggest that such models are defined in the way that a character is defined within a play or some other story. This would take us back to the problem of a language-first view. Character, in the sense that we are interested in it, is not restricted to the notion of character roles in dramas or other narratives. Narratives are powerful for expressing character, and one of the problems for science education is that there are relatively few well-known narratives in which scientists are depicted in more than stereotypical fashion—caricatures rather than character. Perhaps for that reason, very few scientists in our survey identified books or movies, whether fiction or nonfiction, as significant inspirations. Science certainly needs better stories. However, as we saw in our discussion of postmodernism, narratives are not primary; they must be created in relation to the foundational values of a vocation as embodied in practice. That is the sense in which the term role is appropriate for us; in evolutionary terms, it is the notion of *functional role*.

This also fits nicely with our Aristotelian virtue-theoretic framework, as the character virtues are defined functionally in relation to some purpose (*telos*). In science, the character virtues we have seen to be important are justified, similarly, in relation to the central, guiding purpose of discovering truths about the natural world. Character education in science ought to, therefore, involve modeling oneself after individuals who are exemplars for doing this well. Aristotle had a term for such individuals—*phronimoi*. Aristotle's concept prefigures Merton's and in many ways is more precise and relevant for us. Aristotle used the term to refer to individuals who, because they have the practical wisdom of experience and have developed a good character, may be entrusted with the habituation of others. Aristotle's notion is especially pertinent because it is not primarily about institutional norms and social reference groups but comes with an implicit moral framework—a virtue-theoretic account of the sort we have been developing for science—and because it emphasizes know-how and not just theoretical knowledge.

We should note in passing that, in Aristotle's view, practical wisdom is different from theoretical wisdom, which is also important for full understanding but which includes "scientific" knowledge combined with "intuitive reason" of those things that are "highest by nature."[29] The *phronimos* embodies virtue but has also reflected on the nature of virtue and is able to thoughtfully explain it. One does not expect working scientists, even exemplary ones, to necessarily have theoretical wisdom about their practice, as many have studied only their subject matter without having to think about the philosophy behind it. We expect more from science teachers, but for scientists who are researchers, it is less important that they profess virtue than that they embody and demonstrate it.

Another important point about practical wisdom is that it involves knowledge and practice that has been tested by real-life experience. In this sense, the Aristotelian moral exemplar should be thought of, using a distinction introduced previously, not as a moral authority but as a moral expert. The idea of moral expertise seems foreign in contemporary culture, where relativism, or at least moral skepticism, predominates. Other philosophers have discussed this general issue of moral expertise in ways that are relevant, but I will not review the arguments here since, for our immediate purposes, a narrower notion is sufficient. Scientific virtue, as we have seen, is a form of vocational expertise. The notion of morality here involves character traits, methods, and practices that are conducive of scientific excellence. Our empirical study was an attempt to collect some of this practical wisdom from exemplary scientists. *Phronimos* is unlikely to ever become a common term, so I will mostly use exemplar or role model interchangeably, with the hope that Merton would not seriously object to my Darwinian-Aristotelian interpretation of his term.

Character Development

We are born curious, but instinctual inquisitiveness is only the beginning. Character in the moral sense involves rational intention. It is not just given; it is shaped by experience and developed by practice. It is not the same thing as personality. One might be fearful (a personality trait) but learn to manage one's fear to become brave (a character trait), whereas cowards are unable to control their fear and give in to it. This is one reason that the concept of virtue is connected to the ideas of both moral justification and power. To say "by virtue of" means both "because of" and "on the causal basis of." For example, one might say, "The sapling grew into a magnificent tree, by virtue of its strong root system." In teaching virtue, we are empowering our students.

As we have seen, the basic process of learning a virtue involves modeling, which does not require language. But we do, of course, want to be able to talk about the process, especially if we are to theorize about it, so it is worth considering what terms to use.

It is fine to speak generically of teaching and learning virtue, but what verbs best describe the nature of that process? Are virtues instilled? The term develop is neutral, and I often use that. However, with a nod to my late father, who was an agronomist, permit me to suggest an agricultural verb: cultivate. There are a variety of advantages to using cultivate when talking about developing virtues in the ways we have been discussing. The agricultural metaphor emphasizes the biological basis of character traits as well as the work that each of us must do to bring them to fruition. As Voltaire put it, "Il faut cultiver notre jardin" (We must cultivate our garden). It expresses the idea that the virtues grow (from within, but always in some environment) and that their growth is nurtured by the care and guidance of gardeners, including parents, teachers, trainers, mentors, and other role models.

Of course, no metaphor is perfect. Although the notion of cultivating virtues nicely captures important aspects of the process, it fails to capture other aspects. In particular, it underemphasizes the fact that one needs to practice virtuous behavior to strengthen the disposition. Philosopher Julia Annas, taking up a point in Plato's early Socratic dialogues, argues persuasively that virtue may often be usefully thought of as a skill.[30] If it is to become habitual, virtue must be exercised. Thinking of virtue in this way will strike some as odd, if only because of the good-vs.-evil way that morality is often dramatically portrayed, as though one is virtuous or not. However, if moral education is to be meaningful, we need to take seriously the idea that one can become a better person through practice. It may even be possible to do this methodically.

Franklin's art of virtue   For a prime example of how this may be accomplished, there is no better model than Benjamin Franklin, who took what may truly be called a scientific approach to the development of his own moral character. In his autobiography, the prototype of the American self-help book, Franklin described how he improved himself through self-reflection and practice. For instance, he explained how as a youth he had been fond of disputation, aiming to confute and defeat every interlocutor. He came to recognize this as a bad habit that made one disagreeable in company, soured conversations, and produced disgust and enmity where there might have been occasion for friendship.[31] Later, upon reading about Socrates, he recognized the advantage of the indirect method and worked to switch from contradiction and argumentation to pretending to take on the Socratic role of "humble inquirer and doubter" with an artful way of extracting concessions in a conversation that people did not foresee and thereby "obtaining victories that neither myself nor my cause always deserved."[32] Eventually, he came to realize that pretending to be like Socrates was still problematic, and he adopted a truly Socratic form of intellectual modesty. Franklin explained this personal transformation in considerable detail as an example of how it helped him not only persuade others, but also be persuaded himself by others, to his own benefit.

Working to develop intellectual humility was the most difficult of the virtues that Franklin set for himself. He had originally drawn up a list of a dozen he aimed to perfect, ranging from temperance in the consumption of food and drink, to chastity, sincerity, justice, and industry ("lose no time; be always employ'd in something useful; cut off all unnecessary actions"). Though Franklin did not use science to determine what virtues to strive for, the process he used to train himself in virtuous habits was scientific in the geekiest of senses. He made a chart, recorded data of his daily successes and failures in field notebooks, and used these measures to track his improvement over time.

What practice does is take conscious intention and make it habitual. Franklin did this with regard to general moral virtues, but there is no reason to think that similar sorts of training would not work as well for the sorts of mental traits that make for exemplary science. Put another way, we should try to cultivate scientific habits of mind.

Dewey's scientific habits of mind The notion of scientific habits of mind is most associated with the educational philosophy of John Dewey, who was a pioneer in arguing for a reformation in science education to focus on science as a way of thinking. In one especially influential article, he distinguished science as subject matter and as method, and he argued for the importance of the latter in science education, writing: "[S]cience has been taught too much as an accumulation of ready-made material with which students are to be made familiar, not enough as a method of thinking, an attitude of mind, after the pattern of which mental habits are to be transformed."[33] Again, the point is not that scientific subject-matter content is unimportant, but rather that it has been overemphasized in relation to the scientific methods, practices, and mental habits that make the discovery of such facts possible in the first place. What our survey of exemplary scientists did was reveal for the first time some of the unappreciated value elements of the characteristic scientific mindset and the broad agreement about their importance in science. The question and challenge for science education is the same as Plato's character Meno's, namely, how these virtues may be learned.

Learning Scientific Virtues

I have suggested that combining elements from Aristotle and Darwin along the lines that we have been developing may help answer Meno's question about how virtue is learned. Evolution has already done the work of providing the species with the basic instinctual dispositions. Beyond that, it has also provided us with the capacities to learn, and that is where education comes in. We may cultivate our virtuous instincts and train them by copying the behaviors of virtuous people. In the basic sense this could be just a mechanical imitation, but at a more sophisticated level it also involves cultivating the motivations and emotions that go along with the behaviors.

To a certain extent, it is possible to cultivate virtue on one's own by observing individuals with practical wisdom—scientific exemplars, in our case—and trying to emulate them. Such natural copying begins in childhood, which may be why scientists mostly credit their parents, especially if one was a scientist, for inspiring their interest in science. Habits form early. The other major influence is teachers. A small proportion of exemplary scientists did not develop an interest in science until college, but most were inspired much earlier, citing not just high school teachers but elementary and middle school teachers. We will need to look at more than initial inspiration to understand how the scientific virtues are cultivated, but roots are especially important, and a common thread among these stories is how some teachers' directed passions were infectious.

Developing a Scientific Mindset

Did You Ask a Good Question?

Nobel Prize–winning physicist Isidor Rabi was asked what had made him become a scientist rather than a doctor, lawyer, or businessman like the other immigrant kids in his New York City neighborhood. Rabi credited his mother, but in a specific way, and his answer is instructive: "My mother made me a scientist without ever intending it. Every other Jewish mother in Brooklyn would ask her child after school: 'So? Did you learn anything today?' But not my mother. She always asked me a different question. 'Izzy,' she would say, 'did you ask a good question today?' That difference—asking good questions—made me become a scientist!"[34]

The difference that Rabi identified in his simple anecdote gets to the heart of the scientific mindset. It is the difference between learning facts and developing an inquisitive frame of mind. Most often when parents ask what their child learned, they want to hear about what knowledge they have absorbed. Hearing about science is better than not hearing about it, but if children come to think of learning science only as memorizing what some teacher said about it, then the true lesson of science has been undermined. Asking good questions, on the other hand, is the essence of a curious mind.

This is just one example of the general point about developing a scientific mindset. Other scientists have similar examples of how they acquired a scientific way of thinking from their parents. Feynman often recounted stories of how he learned science from his father, who took him for walks in the Catskill Mountains to see interesting things going on in the forest. He said that other children would later tease him when he could not give the name of some bird they saw, saying that his father did not teach him anything. Feynman did not say how this affected him as a child, but as an adult he recognized what important lesson his father had taught him. He said that his father would point to a bird and say, "It's a brown-throated thrush—but in Portuguese it's a—, in Italian a—, and so on." But then his father continued, "Now you know all the languages, you want to know what the name of that bird is and when you've finished with all that you'll know absolutely nothing whatever about the bird. You only know about humans in different places and what they call the bird. Now, let's look at the bird and what it's doing."[35] In this simple and memorable way, Feynman's father introduced the fundamental idea we have discussed that science begins not in words but in wonder.

Couldn't one do this sort of thing as a classroom teacher? There is no good reason that one should not be able to develop curiosity and other scientific virtues as part of one's regular education, at appropriate levels.

Virtue is not developed only in childhood, but that is where the process begins. It does not help that curiosity is often viewed as a problematic trait at that stage. Consider the best-known stories for young children where inquisitiveness figures prominently—the tales of Curious George, the monkey. The tacit message of these otherwise endearing stories is that curiosity invariably leads to trouble. Would not it be better if we had more examples that showed the fun and the positive rewards of curiosity? Even when curiosity is valued, circumstances too often make it hard to encourage. In classes with a high student/teacher ratio, simple classroom management often mitigates against allowing children to explore, as harried teachers have to worry first about keeping kids from hurting themselves or others. There are exceptions, of course, where teachers and programs find ways to make this part of the learning process. Montessori schools, for example, have developed wonderful materials that encourage children to practice skills and explore subject matter in ways that support independent learning. When a child's natural curiosity is engaged, the difference between schoolwork and play is erased.

Technology may be able to help, as in the work of AI and computer-based learning pioneer Seymore Papert. Founder of the Epistemology and Learning Research Group at MIT, Papert had majored in philosophy in college and then went on to get two PhDs in mathematics. He became well known for his ideas about how to use technology to create learning environments that encouraged hands-on, experiential learning. He is justly famous for developing the computer language Logo as a learning environment that would let children play in "mathland" by programming a "turtle" to draw. As he explained, "[I]n teaching the computer how to think, children embark on an exploration about how they think. The experience can be heady: thinking about thinking turns the child into an epistemologist."[36]

My digital evolution software Avida-ED, which I described above, attempts to do something similar for students at higher levels.[37] Because digital evolution instantiates the evolutionary causal mechanism, it allows students to perform real experiments to test their hypotheses about how evolution works. This means that they can learn not only about evolution but also about the scientific process itself. And because digital evolution can happen as fast as one's computer, it is easy to run quick tests and learn by doing, asking questions, and trying things out to see what happens. It is especially exciting to see students run with such a tool, following their curiosity where it leads them. When students say, "This is the first time I have really felt like a scientist" and are eager to play around beyond their assigned tasks, then they have begun to know what being a scientist can be like.

There are many other ways that a scientific mindset can be cultivated. One may even take a page from Benjamin Franklin's method. Have students spend a day or a week attempting to embody some particular virtue and keeping track of their results. I have been invariably pleased by the results that students report after I have them spend a week being intentionally attentive about something that interests them. Nearly everyone expresses surprise and delight at having made some unexpected discovery. Sometimes it is a discovery about themselves. Sometimes it is a discovery for themselves. Sometimes it is both. One senior wondered why she had never seen any snakes on campus. She eagerly discussed her experience at the end of the week, having seen one for the first time. They were there, but she had never noticed any until she started to pay attention. Realizing that she had that power was also a revelation. This is the kind of experience that makes one think of continuing on to graduate school, which is the time to start to develop virtues in a professional sense.

The transition one must make from college to graduate school in science is from gaining a general understanding of a field and having an interest, to a much more specialized familiarity and, eventually, expertise. Undergraduate textbook science focuses mostly on what we already know, so it is not surprising that many undergraduates think they know it all. The wisdom to be learned in graduate school is (or ought to be) the Socratic wisdom that develops from delving into a subject to such a depth (really, beyond one's depth) that one comes to appreciate how much one does not know. That is, after all, what qualifies one for a title of doctor of philosophy. Because curiosity involves new things that we do not understand, most of one's research career is spent in a state of confusion, struggling to make sense of difficult puzzles. Graduate school requires learning to be comfortable with being confused and coming to see confusion as a clue to discovery.

At this level, one sees curiosity expressed through deeper and deeper dives into the subject matter. It means coming to know the literature in one's area of research well enough to appreciate what previous researchers have discovered and where the open questions are. Participating in research group journal clubs, where members read and discuss the latest publications and preprints, is not seen as a chore but as a pleasure. By the end, of course, one must make a novel contribution—some discovery of one's own that is recognized by other experts as having advanced the community's understanding of their special, beloved part of the world.

The hidden value curriculum   For the sake of simplicity, in this chapter I am mostly focusing on curiosity as a central example of how scientific habits of mind may be taught, but scientific character education ought to involve all the scientific virtues that we have been discussing. Many instructors already recognize the importance of these traits, if only when noting their absence. Allow me to give a few examples of complaints I have overheard in a variety of collegiate settings. Speaking about introductory biology courses, one professor noted how students come in thinking that science proves things with certainty without appreciating how evidence works: "I'm trying to change that student mindset. We really want them to be observant and objective and learn how to evaluate evidence and use evidence to make conclusions." Another science instructor lamented how his students were unable to deal with the everyday tedium that is part of scientific practice: "Data collection is boring. Students don't have the patience." Others speak of virtues in relation to skills. Many science educators have concerns about students' lack of quantitative skills, but that is not the most critical lack. As one professor put it in commenting about the problem he saw with many students: "They can't add. They don't know how to cancel. And most important, they don't have a curiosity about problem solving."

So, if teachers see this lack, why don't they do something to remedy it? Some do, of course. Scientists speak with great fondness of teachers who sparked their curiosity in science. But the more common view of science classes as boring suggests that this is the exception. Too many science teachers follow a traditional model that inadvertently stifles rather than fans curiosity. The reform movement in STEM education is a step in the right direction in that it engages students in the practices of science in tandem with the learning of content. What I am arguing here is that this does not go far enough, in that the values that underlie scientific methods remain hidden. They are not hidden by design, but by negligence.

It may be that part of the reason for this neglect is an inherent tension in the way we think about teaching. The issue involves the subjects we teach. On the one hand, we think of ourselves as teachers of subject matter. When asked, teachers usually default to self-identification in those terms: "I am an English teacher." "I am a philosophy professor." "I am a biology instructor." However, there is another sense of subjects: those individuals who are the subjects of our instruction, namely, our students. It is usually only in early grades that one hears teachers self-identify in these terms: "I am a preschool instructor." "I am an elementary school teacher." Around middle school one is more likely to hear a combination: "I am an eighth-grade science teacher."

Education swings from one to the other. Dewey was working to correct the one extreme that focused just on subject-matter content. At the other end, it is equally a problem when teachers are assigned science classes to teach even though they have no background in science. Ideally, STEM education needs to unite these, with an appropriate degree of balance, at every level of instruction. Carefully designed materials and thoughtful pedagogical methods are needed, but they will likely not be sufficient for most students without the guiding hand of a practiced teacher who recognizes the importance of developing these traits and, for each level of development, can appropriately model them.

Curious Teachers Spark Learners

The virtue-theoretic approach that I have been developing here to help understand the values that structure science may be profitably applied to other vocations, including that of teaching. If we wish to improve STEM education, or education of any sort, we ought to first be sure that we understand the *telos* of the teacher. There are various ways that one could try to articulate this, but to maintain our Aristotelian line, I suggest the following: the purpose of education is to actualize the potential of our students, guiding them toward excellence in their subject or practice. Identifying the goal of education in this way unites the two notions of subject. At each level of education, from elementary school through graduate school, instructors serve as guides and exemplars. In science teaching, this means that one needs to model what it means to be a scientist. This suggests that to succeed in teaching their subjects, in both senses of the term, science teachers should themselves be curious. Putting this generally, teachers who themselves exhibit the relevant virtues will be better at cultivating them in students.

This follows from the Aristotelian notion that we learn virtue by following the example of those who are themselves virtuous. If the goal is to cultivate virtue, one must do more than study what the virtuous have accomplished. You may know the stats of every player on a winning sports team, but if you never got to watch them play, it would leave you with an impoverished notion of the nature of the game. Watching amateurs gives a basic sense of how a game is played, but watching champions provides an appreciation of what excellence means. And yet, even that would never make you a player. Players know the rules of the game, but they have also practiced the moves. To improve, they train. The best players know the theory, the history, and the stories of past champions and championships. They study the details of the best games of the best players and their methods for working out, diet, lifestyle, and so on. Importantly, they try to mimic these behaviors as a way of improving their own practice. They try to play with others who are better than they are to learn from the experience of higher-level play. For the best training, they work with coaches who have the abilities to a sufficient degree themselves and the added special skill of being able to cultivate those traits in others.

There is an insulting adage that those who can, do, and those who can't, teach. This betrays a lack of understanding of what teaching requires, and it does a disservice to those who dedicate themselves to education. The vocation of teaching has its own set of virtues that deserve to be treated separately, but the basic point is straightforward—teachers must have not only subject-matter knowledge, but also the relevant traits that will allow them to be effective models for their students, because achieving the *telos* of the teacher involves the latter as much as the former.

When scientific literacy is defined only in terms of content knowledge, then it is not surprising that science education focuses on teaching scientific facts. Of course, there are good reasons to learn about important scientific facts, not the least of which is that they provide cultural touchstones and an appreciation of some of the wonders of the natural world. Such encounters can stimulate interest. However, if the goal is to encourage budding scientists and develop in them the traits that will help them one day make discoveries of their own, then more is required. It is not enough to learn about discoveries made by the curious, one must be curious oneself.

This point is general. To develop a scientific mindset, it is not enough to be told that science is based on careful, repeatable observations; one must learn to be meticulous, persistent and attentive. This is an application of the lesson about how virtue is learned by modeling those who have practical wisdom. Just as it would be foolish to try to learn bravery by following the model of a coward, so is it hopeless to expect to learn the habits of curiosity from a teacher who is incurious.

Teaching to the test Despite the importance of this insight, we are only now beginning to think about what it looks like in practice. Indeed, educational policy has kept us looking in a different direction. When implemented, the laudable mandate to *leave no student behind* looked much more like *leave no student untested*. In the name of accountability, incentives (and disincentives) too easily shift such that schools and teachers prioritize test-taking skills over subject skills. Again, this is not to dismiss the importance of basic subject-matter knowledge, but that is only the first level of literacy. For students to understand the justification for this knowledge, we also must teach about the methods and practices of science. Moreover, even appreciating the methods is unlikely by itself to make a student care to persist in a STEM field. Hume's point about passion applies here; knowledge alone is insufficient to motivate action. A teacher who diligently marches through a curriculum may help students pass a test, as students are indeed motivated by grades, but such external goods are at best indirect incentives. They completely miss the internal goods of scientific practice.

Teaching to the test is a misguided and dangerous pedagogical practice if the goal is to develop lifelong learners. It may help students graduate, but it will not transform their lives as it never encourages them to value the subject and the material beyond its utility to produce the external good of a passing grade. The students who will persist in science are those who have internalized its values. The students who will persist in science are those who have had their natural curiosity stimulated and fueled, so that they eagerly turn their lights to the as-yet-unexplored areas of the world and revel in the possibility of finding answers to the questions that excite them. This is what it means to develop an identity as a scientist.

Teaching to the test is bad for students, but it is also bad for teachers. It bends their vision and forces them to aim for external goods as well. It judges their success as teachers in terms of their students' scores on standardized tests instead of seeing success in terms of helping their students acquire and improve their abilities (including knowledge and skills), which is the internal good for a teacher. If we want STEM education to encourage student persistence in science, then we need teachers who are focused on the internal goods of science and who can model the traits—the virtues—that a science student ought to develop, for it is by developing those habits of mind that dispose one (i.e., provide one with the appropriate passions) to scientific research.

Developing a trait means training oneself so that one is more disposed to reliably act in the appropriate situations. Having the trait means that one is more likely to be able to consistently reproduce the action or behavior. A formal test setting is only one possible situation, and a necessarily artificial one at that, so it is an imprecise metric. Educational tests to assess student learning are necessary and important, but as in any other case of testing, one must be aware of their limitations. Diligent instructors know that they need to assess their test questions for reliability and validity—technical terms in testing methodology—to increase confidence that they can consistently assess what they mean to assess. More importantly, one must not forget that it is a test *of* something, namely, of the trait that one aims to develop. Minimally, this tells us that we need better tests.

The paradox of metrics   One does, of course, need standards of measurement in order to gauge progress toward success. Success is the development of appropriate dispositions. But passing a test is evidence of success, not success itself. Too often we mistake the test result for the thing being tested for. This is similar to the problem of fitness in evolutionary biology. We measure fitness in terms of reproductive success, but fitness is not actual reproductive success itself but rather the adaptive traits that are dispositional for reproductive success. Consider two identical individual organisms—clones that, for the sake of argument, are the same in both their genotype and phenotype—that are tested in the same environment. One successfully produces a hundred viable offspring. The other is accidentally killed before it can reproduce, and so has none. It would be a mistake to say that the first was more fit than the second on the basis of their actual reproductive success. Their fitness was equal, as their dispositional reproductive potential was equal. In this case, it was the measurement that failed, as measurements always have a chance of doing. The paradox of metrics is that they are needed to identify the dispositional trait, and they are expressions of the trait, and they may even partly define the meaning of the trait, but they are not the trait themselves.

Mistaking the test for the goal reverses the relationship and so easily leads to skewed pedagogical methods and incentives. Think of this in terms of the work of practicing research scientists. Their primary task is not to show what they know but rather to figure out something they do not know how to do. (The situation may be slightly different for applied scientists or for scientific technicians, who use scientific knowledge without necessarily aiming to add to it, but even they will not generally be taking tests; they will be making them.) Misunderstanding this can lead teachers to inadvertently undermine scientific integrity. For example, consider the common case of "cookbook" science labs where grades are based on getting the right answer. If this is the only metric, then it signals to students that the goal is getting a known answer, when it really should be getting good data. When taught in the traditional way, students are inadvertently being incentivized to fudge or fake. Moreover, it is unrealistic; in real research one does not have the answer in advance. By rewarding the wrong thing, one undermines the *telos* we hope to achieve.

One could argue that cookbook labs where a particular result is expected are teaching meticulousness, as careful technique should indeed result in students getting the same result. Fair enough. So, have a lab or two where that is the announced learning goal. Making that explicit and explaining why it is important could help students appreciate the point. However, if the goal of science education is to foster scientists rather than lab technicians, one cannot stop there. Meticulousness is only one of many important scientific virtues, so would not it be better to also have grades that reflect and incentivize objectivity and self-skepticism? The significance of a lab experience, after all, is to come to appreciate how conclusions in science are based on empirical evidence. In real science, experiments fail regularly, so why should students fail if their experiments do? The lab experience ought to demonstrate that it is the possibility of failure that makes an experiment a real test. If the data go against a hypothesis, then we have learned something. Acknowledging both sorts of failure ought to be part of the process. Teaching students to always accept what their teacher says rather than what the data say is to instill obedience to authority where we ought instead to be fostering humility to evidence. Learning to be attentive to the evidence and how to follow where it leads is the path to the satisfaction of curiosity.

Everyone is born with the capacity to develop these virtues, and thoughtful science education ought to create fertile environments where they can blossom.

Capacity for Original Work

The problem is that not all environments have been equally hospitable for all would-be scientists. Evolution has provided us all with the biological wherewithal to thrive, but environmental circumstances can prevent the expression of traits. The pioneering astronomer Maria Mitchell spoke about this issue in her 1876 presidential address to the Association for the Advancement of Women (AAW), challenging critics who doubted whether women really were capable of making scientific discoveries.

> Does anyone suppose that any woman in all the ages has had a fair chance to show what she could do in science? … The laws of nature are not discovered by accident; theories do not come by chance even to the greatest minds; they are not born of the hurry and worry of daily toil; they are diligently sought; they are patiently waited for, they are received with cautious reserve, they are accepted with reverence and awe. And until able women have given their lives to investigation, it is idle to discuss their capacity for original work.[38]

Mitchell's point, of course, was that women had not had a fair chance. They had these capacities as well as men, but social circumstances had systematically prevented their having adequate opportunities to develop and express them. In such inhospitable circumstances, they needed to have a special sort of resilience and tenacity.

They persisted Mitchell said of her own character, "I was born of only ordinary capacity, but of extraordinary persistency."[39] She was being modest about her intellect but not about her mental toughness. We have previously discussed how scientists cite perseverance as an important trait for exemplary scientific practice. Mitchell needed it not only for the long, late nights she spent patiently sweeping the sky with her telescope but also in her quiet but persistent work for women's equality. This was true even in her position as the first faculty member hired for the newly established Vassar Female College.

Taking it for granted that women, having "the same intellectual condition as man," should also have the same right to education, intellectual culture, and development, Matthew Vassar founded the college expressly "to give one sex all the advantages too long monopolized by the other."[40] Vassar College was not the first to admit women, but it was widely seen as an important experiment in higher education for women, and journalists pointed to its first faculty member, the "famed and honored" professor Mitchell, as "a perpetual illustration of the accomplishment which is supposed to be peculiar to the other sex."[41]

Mitchell said that Vassar was the best college of its kind but that as a female college, "it is not of the right kind."[42] Separate but equal was not true equality. She lobbied for women to be admitted at Harvard and advised Ezra Cornell to take the lead in his new college by making his institution coeducational from the beginning. Though well ahead of its time in many ways, Vassar was in other ways still very conservative. Mitchell chafed against many of the college's policies and battled with administrators over inequalities in her accommodations and salary compared to her male faculty colleagues. That women did not receive equal pay for equal work was another social constraint that they had to overcome.

Such examples make a general point. Women and scientists of color have had to persevere not only against the usual intransigence of nature to give up its secrets but also against a resistant social environment. The pioneering minority scientists who broke new ground had the same core scientific values that make for exemplary research but needed additional virtues—resilience and an extra measure of perseverance—because of the exceptional additional challenges of an environment that has not been as hospitable as it should be.

A fair chance in science   Mitchell's own classroom reflected her pioneering spirit and her desire to provide an environment where her charges had the opportunity to develop a genuine scientific mindset. "I do not expect to make of you philosophers like Newton, astronomers like Kepler or mathematicians like Laplace," she said to her students, "but I do hope to open your eyes and your understanding to the wonderful lessons of the Universe."[43] What this meant was that she wanted them to learn the way that scientists do their work and discover new knowledge. She wanted them to know of previous scientists but did not want students to just accept their views on authority. She wrote:

> There is this great danger in student life. Now, we rest all upon what Socrates said, or what Copernicus taught; how can we dispute authority which has come down to us, all established, for ages? We must at least question it; we cannot accept anything as granted, beyond the first mathematical formulae. Question everything else.[44]

This is what we have discussed as the virtue of skepticism. Mitchell saw it as critical for developing as a scientist, especially for women who were raised to be obedient.

> Until women throw off reverence for authority they will not develop. When they do this, when they come to truth through their own investigations, when doubts lead them to discovery, the truth they get will be theirs, and their minds will go on unfettered.[45]

Mitchell's approach to science teaching reflected this attitude. Eschewing traditional classroom pedagogy, she experimented with what we today recognize and advocate as active learning. "We must have a different kind of teaching," she said. "It must not be text-book teaching. I doubt if science can be taught in school-rooms at all. Certainly, it cannot be taught by hearing recitations."[46] Her students learned how to operate the telescope and other instruments that had been installed in the specially built observatory, and they regularly observed the night skies. She hosted what became her famous annual "Dome Party," recounting stories over strawberries and cake of another pioneering astronomer—Caroline Herschel—while the stars shone through the open dome. She included her undergraduates in her own astronomical research, even taking a group of eight students on a 1500-mile trip to Iowa to take part in observations of a total eclipse of the sun.

As astronomers and others congregated in Burlington for the event, one Iowa newspaper reported that Mitchell's group was not only the most interesting but the most interested parties of observers. They were not sightseers but scientists. However, other local papers that saw the group in a rather less serious manner, speaking of them as "schooled in ladylike demeanor" and lending "a touch of romance to the occasion," recommended that the "beaux at Des Moines … pack their carpet bags and make their appearance in Burlington or miss their opportunity."[47] Such subtler forms of sexism are another impediment that female scientists had to overcome. Mitchell saw to it that annoying visitors did not interfere with their days of preparation. To collect data, every student had an assigned task during the eclipse. Afterward, Mitchell wrote of how well they acquitted themselves. Astronomers acknowledge how the powerful effect of seeing a total eclipse often leads observers to forget their tasks and look around, but her young student assistants "would not have turned from the narrow line of observation assigned to them if the earth had quaked beneath them.... Was it because they were women?"[48]

The real answer, of course, was that it was not because they were young women but because they were young scientists, who just happened to be women. Mitchell preferred to be seen just as a scientist. This was a common attitude among pioneering female scientists. Barbara McClintock, for example, had no desire to be or be seen as a "lady scientist." That would suggest that science was male, which she thought was absurd. McClintock recognized and resented the disparity between opportunities for women in science and those for men in science and the intrusion of sexual stereotypes. Her wish was to transcend gender and to be treated "person-to-person." She rejected female conventions and was often described as boyish, but she also rejected being seen as "more boy than girl." Why should her way not be a way of a woman? None of that should matter to the science; she insisted on "her right to be evaluated by the very same standards as her male colleagues."[49]

That was Mitchell's attitude as well, a hundred years earlier. She thought her female students could and did do at least as well as male undergraduates. At least some reporters recognized and made a point of highlighting that as well, noting that Vassar's examination standards were "as high as that at any college for young men," and commended Mitchell's students for "the enterprise and spirit they manifest."[50]

In her presidential address to the AAW Congress years later, Mitchell specifically made the point that the education system still mostly failed women, but her recommendations for how to improve science teaching are worth heeding for the sake of all science students.

> It is a feeble kind of science which can be put upon a blackboard, placed in array upon a table, or arranged upon shelves, the facts of science may be taught by such means—the spirit of science, which is the love of investigation, they cannot

arouse. If science can be developed at all in class rooms, it must be by debate; free thought, and free inquiry are the very first steps in the path of science.[51]

Her insistence on free inquiry refers to what we have seen as the basis of the scientific mindset—the passion for exploration and discovery, coupled with the rejection of authority in favor of reason and evidence. What she modestly omitted from this was her own importance as a scientific role model. It was seeing Mitchell's own scientific excellence that helped her students themselves begin to embody the virtues that make for exemplary research. It was Mitchell's example and the hospitable environment she fostered that opened her students' eyes to the wonders of the universe. One of her graduating students from 1879 elegantly described that feeling of an endless frontier to explore that being in Mitchell's classes engendered:

> Our Astronomy classes were small, but full of enthusiasm. We seemed to enter into a new world, and to have presented to us, in all their sublimity, new ideas of the boundless expanse of the heavens. What scenes compare with those which we have witnessed from the Observatory, on clear winter evenings, when the stars were brightly twinkling, or meteors flashed, or auroras tinged the northern skies with their soft crimson glow! With what feelings of wonder and awe did we for the first time sweep the heavens with the large telescope, and behold Jupiter with his satellites and mottled belts, or examine the apparently smooth surface of the moon, and notice its deep caverns and lofty mountains.[52]

How far might students, of all backgrounds, go if science teachers embodied the scientific spirit as Mitchell did, so as to inspire such an attitude.

Cultivating Curiosity

To conclude, let us return to the three goals for STEM education laid out by the National Research Council, which our discussion may now let us view more clearly. Scientific literacy ought to be seen as involving more than being able to read some basic scientific facts. One cannot be said to understand scientific facts without some appreciation of the inductive methods and practices that evidentially validated them. Expanding the STEM-capable workforce is socially and economically important because so many jobs in our technologically advanced society require some training in science. To be ready to fill such jobs, students need to learn more than scientific facts; they must learn how to think like a scientist, using scientific methods to weigh evidence, sort out irrelevant factors, and draw inductive conclusions to solve empirical problems. However, if we also want more individuals to pursue advanced degrees in science, we will likely fail if we do not recognize that basic research science is not a job, but a vocation.

Thus, beyond learning basic facts and methods, one ought to come away from one's science classes with a feel for the scientific mindset. I use that phrase intentionally, for it is not enough for instructors to tell their students about the values that structure science. Stories about scientific heroes and inspiring tales of discovery are a good first step. However, to really get across what it means to be a scientist, they ought to attempt to have their students experience those feelings and develop those traits in themselves.

Central to this is making curiosity a part of science education. I do not mean that the goal is to teach curiosity for, as already seen, we are born with that. On an evolutionary account we know that we all begin with a suite of instinctual traits in various degrees, like any biological variation, but we also know that phenotypic expression depends not just on genes but also on environment. Traits can be fostered, and encouraged so they are actualized and developed, or they may encounter an inhospitable environment where they are discouraged. Minimally, we ought to teach in a way so that students do not lose their natural curiosity. More positively, the goal ought to be to nurture and develop our curiosity and other traits—to cultivate the virtues.

There are stages to this process. The seeds of science are already naturally in place, so the early stage involves simply providing a conducive environment where they may take root and grow. This is an environment of exploration, play, and discovery that exemplary scientists say they never lose the feeling of. At a later stage, after someone has already identified an interest in science and a desire to go into the vocation, cultivation can become more focused. At this stage, usually in college, though sometimes earlier, we may teach curiosity and other virtues in the sense of training. This term also has an appropriate biological root, as in training a plant to grow in a certain direction. Thinking of education in this way is different from preaching and professing. Instructors cannot simply utter virtuous platitudes; they must demonstrate the virtues.

This is perhaps our most important take-home lesson: to be effective, science teachers ought themselves be curious. The scientific exemplar (the equivalent of the Aristotelian *phronimos*) is someone who embodies the scientific virtues and is able to guide their development in others. Exemplary scientists who think back to who or what inspired their interest in science cite the stimulation of their curiosity by parents and teachers more than any other factor, by a wide margin. If we make no other change in science education, the best thing we could do would be to hire and support teachers who are role models in this sense.

There is one final point, which is that we do not just learn virtues and then have them. If one thinks of virtue as a skill, then it becomes clearer that it must not only be trained but continually practiced. Virtues must be exercised. Curiosity stretches the mind. That is a good thing in itself. Our goal should not be to turn everyone into a scientist but rather to nurture the scientist within everyone. If we do this well—fanning the spark rather than snuffing it—the goal of increasing and broadening the range of individuals who pursue advanced degrees in science should come naturally. For those whose interests take them elsewhere, they should still be left having had their eyes opened to the wonders of this world and the joys of discovery. This is the root of lifelong learning, and, as we shall see, it is part of what it means to be human.

10 We Are Scientists Second

We are scientists second, and human beings first. We become politically involved because knowledge implies responsibility.

—Freeman Dyson[1]

The Scientific Identity

There Is No National Science

One sometimes hears people speak of science as "Western," but we intuitively know that there is something very wrong about viewing science with a geopolitical adjective attached as though it was somehow located and proprietary. As the Russian playwright Anton Chekov put it, "There is no national science, just as there is no national multiplication table; what is national is no longer science."[2] Trust a writer to make the point so clearly. Science may have come into its own historically in Western Europe, but the telescopic observations of the moons of Jupiter could have been made as well in Russia as in Italy. The pioneering chemist Robert Boyle complained that his home country of Ireland was barbarously ill-suited to chemical research, but the experiments that confirmed the law describing the pressure of a gas in relation to the volume of its container could have been carried out equally well in Lismore as in London. Evolution operated as efficiently in Borneo as on the Galapagos Islands, and data that led to its discovery were not limited to either locale. Polonium may have been named after Marie Curie's homeland, but she isolated it in France and could have done so as well in Argentina. More importantly, much as we rightfully laud Galileo, Boyle, Darwin, Wallace, and Curie for their observations and discoveries, it was not that they were European that made their findings true, as scientists from India, Japan, Kenya, or elsewhere could equally well test and confirm them. With the requisite mindset and methods, anyone may make or confirm a scientific discovery anywhere.

There are, of course, cases where geopolitical ideology has overridden scientific findings in some nationalistic way. We have already mentioned Lysenkoism in the Soviet Union, which rejected Mendelian genetics and Darwinian evolution as a "bourgeois perversion" in favor of a Lamarckian view of the inheritance of *acquired* characteristics that the party thought was more properly in line with Marxism, as a classic example. Russian biologists who challenged the party authority were fired, imprisoned, or even executed.[3] It should be clear why this kind of power-enforced construction, socially effective though it was in its historical time and place, was not real science.

Scientists see science as inherently international. One might put this even more strongly—they see themselves as constituting a community of their own. That is part of what it means to identify as a member of the scientific culture and embody its norms. We have been exploring what those norms are and how one becomes acculturated into the scientific community. As we have seen, this involves more than acquiring scientific knowledge. Get scientists to talk about what drew them to science and most will have a personal story to tell. In our study, we asked scientists when they first thought about becoming scientists and what inspired their interest. Many spoke of being inspired by their father or mother or by a teacher. Many talked about some experience they had in nature or at a science center or in a science course. But many also spoke of their own mindset, and it was very common for scientists to say that they began to view themselves as a scientist from the time they were very young. "I was always curious as a child," many explained. What they are saying here is that they saw in themselves a core trait that stood out in a noticeable way, perhaps compared to other interests or to the usual interests of others at a comparable age, that in hindsight they recognized as central to what it means for them to be a scientist. Being a scientist implies that one cares about certain things, asks certain kinds of questions, sees the world in a certain way, and places a special value upon these.

Identity Theory

As we saw in our discussion about science education, in addition to the importance of learning content is the significance of how the cultivation of this distinctive mindset is critical to developing an identity as a scientist. The institutionalization of science has brought about a greater focus on credentialing—academic degrees in science from recognized universities, participation in professional societies, publications in scientific journals, and so on. Merton's work on the ethos of science showed that, even at an institutional level, there is more to being a scientist than one's credentials, but as with the paradox of metrics it is too easy to mistake the degree for the qualities—the qualifications—they are supposed to certify. Obituaries of scientists today tend to focus just on their affiliations (e.g., where they studied and worked) and accomplishments (e.g., what they discovered and what prizes they received), whereas in the past it was common to speak of scientists' character and the traits that were recognized by the scientific community as laudable. It would behoove us all to swing the pendulum back in that other direction, away from credentials to return toward character, for that is far more important to the normative notion of scientific identity.

The notion of personal identity has long been a central topic within philosophy, as it has important implications for a wide range of epistemological and ethical questions. Historically, the soul was taken to be the locus of personal identity, but philosophers pursued conceptual questions about personal identity that were independent of what metaphysical substrate might be the guarantor of its continuity. David Hume argued that personal identity was not a simple thing at all, noting that when reflecting on "what I call myself"[4] one finds a bundle of perceptions that are given a "union in the imagination."[5] To investigate the concept of personal identity, philosophers often conduct thought experiments. One classic kind involves "mind switch" scenarios of a kind made popular in science fiction. Suppose that all your memories and thoughts were transferred into an android equivalent. Now consider two scenarios that might occur right after the mind transfer. In the first, a phaser blast destroys your original body; in the second, it destroys the android body. The common intuition is that the death of your original body is not the death of your personal identity, whereas the second scenario, in which your original body survives without your mind, would be.

Extending this line of reasoning, I suggest that personal identity (in the sense of who we are, what we would wish to live on, and what we would identify as ourselves in a classic body-switch thought experiment) involves not just our memories and thoughts but our values and dispositions—our character. Consider a value-switch thought experiment in which one retains one's memories and knowledge but replaces one's character and everything that one cared about. In such a case, one is arguably no longer the same person. One's identity as a scientist may be considered in a similar manner. Following a value-switch experiment, one may retain all one's credentials, but if one no longer cares about making discoveries, curiosity changing to apathy, honesty to deceit, and so on, one is no longer a scientist in anything but a superficial sense. Without the qualifications, in the sense of the relevant normative qualities, the credentials are hollow. To have an identity as a scientist, or any vocation for that matter, is to embrace the guiding purpose and embody the values of that practice.

Thought experiments of this sort should alert us to another issue. Consider a value-switch thought experiment in which one's scientific knowledge is transferred into the android body and also all the relevant scientific values and character dispositions, but nothing else. Such a science fiction character lacks something that we take for granted but which cannot be overlooked if we are to make sense of scientific moral aspirations. This book has focused on the virtues and ideals that make for an exemplary scientist. Our hypothetical android could be a scientist in that sense. But we are scientists second. Our first identity is as human beings. How do these identities and ideals fit together? It is to that question that we now turn at the end.

Does Science Make Us Better People?

The Surest Guide to Virtue?

In his Two Cultures address, C. P. Snow opined that scientists were "the soundest group of intellectuals we have."[6] Literary intellectuals were suspect in his view, especially in light of the fascist tendencies among notable writers in his day. There is little to say in defense of the literary fascists Snow had in mind. One may even agree that science's antiauthoritarianism does somewhat inoculate scientists to fascist views. However, I will not defend Snow's assessment if taken as a generalization, as I know of no data that support a claim for the general moral superiority of scientists over nonscientists. However, I do want to argue for a weaker claim that science does contribute to ethics and makes us better people, though not perhaps in the way that one might initially have thought.

Fanny Wright, Orestes Brownson, and Benjamin Jowett   The nineteenth-century social reformer Frances "Fanny" Wright argued that science was the surest guide to virtue. She was precise about her reasons for this assessment:

> The sciences have ever been the surest guides to virtue, because, demanding calm observation, obtaining all their results by means of dispassionate investigation, they bring into action our noblest faculty, the judgment, and submit the imagination to its guidance; dispose us by the previous accurate observation of things to an equally accurate observation of men, and, confirming us in the habit of tracing effects to causes in the world without, prepare us to follow attentively the train of occurrences in the world within.[7]

The favorite student of the Utilitarian philosopher Jeremy Bentham, Wright was also an admirer of Benjamin Franklin, so it may be that she drew some inspiration for this view about virtue from his writings. An early feminist and freethinker, Wright wrote and lectured tirelessly for the rights of women, the abolition of slavery, and universal, free, secular education that emphasized scientific knowledge.[8] Her activism for social equality and justice was closely tied to her scientific philosophy.

However, not everyone had so positive an opinion of science as any sort of guide to practice. Orestes Brownson, who had at one point collaborated with Wright, broke with her views upon his conversion to Roman Catholicism after a twisting spiritual journey that had led him from Presbyterianism through Universalism and Unitarianism, as part of the Transcendentalist movement. Brownson argued that her vision of an education based solely on science would lead to a nation of "infidels, libertines, sharpers and rogues" because "not natural training but grace alone can elevate our fallen nature to genuine virtue."[9]

The British theologian Benjamin Jowett, writing somewhat later than Wright, took an even dimmer view of science: "Research! A mere excuse for idleness; it has never achieved, and will never achieve any results of the slightest value."[10] This was not just a claim about the worthlessness of scientific knowledge but a moral indictment of science and scientists generally; with his reference to idleness, it is clear who Jowett thought ran the scientific workshop. Logan Pearsall Smith, who recounted the story of Jowett's statement, wrote that Jowett challenged him to cite any results of research that were of any value. The only example that he could think of on the spot was a recent discovery he had read about that striking a patient's knee would produce an involuntary kick that was a useful indication of health. "I don't believe a word of it," Jowett said, and told Smith to give his knee a tap. Smith obeyed and reported that "the little leg reacted with a vigor which almost alarmed me, and must, I think, have considerably disconcerted that elderly and eminent opponent of research."[11]

The bane of modernism   However, even if we set such religion-based objections aside, there remain other reasons to doubt that science is an all-purpose savior. As we have seen, science has specific goals, and its practices and methods are tailored to its special function. For instance, its ideal of evidence-based knowledge requires that conclusions be drawn on the basis of observations and reasoning that others could check for themselves—in principle, anyone ought to be able to repeat an experiment to reconfirm the conclusion. Scientific methodology thus involves interchangeability of observers. As we noted previously, this is one of the reasons that scientific publications are formulaic, so as to, as far as possible, remove the individuality of the investigators. This focus on sameness over difference is a boon for evidential testing, but its useful blandness is a poor recipe for other aspects of human life. The bane of modernism was its zealots who thought that they could indiscriminately apply this formula everywhere.

To give just one example, think of the sorry aesthetic effect of this view on much of modern architecture. Le Corbusier's modern ideal, summarized in his slogan that "A house is a machine for living in," resulted in many cost-effective, but mostly dreary, homogeneous structures. Le Corbusier's idea was that human inhabitants would activate these efficient containers. The best buildings of the modernist period are sublime, but the nondescript skyscrapers of many cities are a sad testimony to the deadening result of taking this form of modernism too far. Who wants to live or work as an interchangeable part of a machine? This is one aspect of the postmodern critique of modernism that was worth taking seriously. Human differences are as important as human commonalities. To his credit, Le Corbusier was not dogmatic about his view. On being told how the inhabitants of a housing project he had designed had altered it, he responded philosophically: "You know, it is life that is right and the architect who is wrong."[12]

For such reasons, I cannot fully endorse either Snow's or Wright's sense of the claim, but I do still think it is reasonable to say that science makes us better people. Why? Because curiosity is one of the core elements of what it is to be a human being, and science is a practice that seeks excellence in the pursuit of that satisfaction.

Curiosity Is More Than a Vocational Virtue

In this book I have focused on curiosity and other traits mostly as scientific virtues…as vocational virtues. They provide a moral structure to the practice of science through their expression in its exemplary practitioners. But as Darwin revealed, curiosity has deep roots. As we saw, scientific curiosity stems from an instinctual, biological form of curiosity that helps organisms like us thrive by driving us to explore and learn about the variable world into which we are born. Human beings live in a complex and changing environment, and our inquisitive nature is one critical trait that helps us find our way around it. What this means is that curiosity is not just a vocational virtue; it is a basic human virtue. As such, it is important not just for scientific flourishing but for human flourishing as well.

Although instinctual curiosity may have evolved originally by virtue of its selective value for resource discovery and reproductive fitness, in complex social environments it has new and more subtle effects. It does not take a scientist to see this. The British-American journalist Alistair Cooke noted its general beneficial effects for human beings: "Curiosity endows the people who have it with a generosity in argument and a serenity in their own mode of life which springs from their cheerful willingness to let life take the form it will."[13] As a student of American history, it may be that Cooke was consciously echoing the statesman William Samuel Johnson, one of the original signers of the U.S. Constitution, who said that "Curiosity is, in great and generous minds, the first passion and the last.…nor perhaps can there be proposed to those who have been accustomed to the pleasures of thought, a more powerful incitement to any undertaking, than the hope of filling their fancy with new images, of clearing their doubts, and enlightening their reason."[14] One could easily multiply such examples; we readily acknowledge the value of curiosity in the general mental and social lives of human beings. A similar argument may be made with regard to the other scientific virtues that are linked to curiosity and the goal of discovery. This is the sense that, in the ideal, science makes us better people—it is the expression of deep elements of human nature in their most excellent form.

Curiosity and Integrity

Because curiosity is part of what it means to be human, it is also a part of human flourishing. However, it is not the only part. So, to get a more complete picture we need to look at how it fits in relation to other aspects of human flourishing. Scientific understanding is not the only worthy goal in life; for human beings, the good life involves a variety of life projects. If a life of virtue is to be a coherent one—if it is to be a life of full integrity—scientific values will have to be integrated with the values of our other pursuits, including our diverse social, religious, and political ones.

We have discussed this previously, but it is worth reiterating that the ethical notion of integrity is complex. It is not just a synonym for honesty, which is only one of the virtues that a person with integrity should exemplify. Integrity is also sometimes used just to mean being ethical or being principled. This is fine, so long as it is not assumed that ethics is just a matter of following a set of absolute rules. Indeed, as we have seen, even honesty can be taken too far; a person of integrity will recognize, for instance, that brutal honesty is no virtue.

A more sophisticated notion of integrity involves the notion of the integration of values. This notion connotes a unified whole, not an ad hoc collection. Throughout this book I have reserved the term for the latter notion to refer to the unity of scientific virtue. Integrity involves a balanced interrelationship among the particular virtues.

Scientific virtues hang together because of their foundation in the special goals of science. Scientists with integrity will embody these virtues in a unified balance, expressing them in their research practices. This has been the main focus of this book, but this cannot be the whole story for the reasons outlined above. We are, as physicist Freeman Dyson put it, scientists second and human beings first. Accordingly, we now need to expand our view and look at how to integrate scientific virtues with more general human virtues. We need to seek a broader integrity.

Human Integrity and Scientific Responsibility

Trust in Science

Part of what it means to say that a person is principled or has integrity is that they are not inconstant in their character and that they can be expected to behave steadily even in the face of countervailing pressures. They cannot be bought. One thing that we value about such individuals is that they are trustworthy. This perspective is helpful for thinking about science.

We previously looked at how values structure scientific practice so that its conclusions may be trusted by other researchers. Researchers interacting with each other are already a part of the scientific culture and share its goal, so this is an internal issue. This is trust within science. We have also seen how this allows scientific conclusions to be trusted for practical purposes, whether in engineering, medicine, or other applied science. But now that we have pulled back to look at science as part of the overarching human project, we need to consider what trust in science means in that larger context. If science led to nothing but safe bridges and cures for disease, few would question it, but it is not surprising that people begin to question the trustworthiness of science when they see it giving rise to nuclear weapons and eugenics. We might distinguish a narrow from a broad notion of trustworthiness that tracks a narrow versus a broad sense of scientific purpose; the former involves trust to fulfill a vocational *telos*, while the latter involves trust to serve the human condition.

Failure of trust in science in this broader sense will be the inevitable result if science is not integrated with other human values. As we saw in our analysis of the scientific vices, even central virtues can be perverted if taken to the extreme. The mad scientist trope in literature is characterized by the domination of some scientific virtue over all others. This is the paradox of single-mindedness—it often leads to efficiency and success but at the expense of everything else around it. Such is the case of anyone who fails to see how particular vocational virtues fit within the overall human project. One should not assume that scientific standards and approaches may be applied to all human endeavors. We should not think that a technological fix is always wise. When the scientific mindset becomes single-minded, it may no longer be trustworthy in the broad sense.

So, what is to be done? In what ways and within what limits should science and scientists be trusted when it comes to not just scientific flourishing but human flourishing? Any answer to this difficult question will have to involve understanding the relationship between the virtues of the scientist and the virtues of the human being generally.

Classical Analyses of Human Virtues

There is a large and ancient body of literature on the human virtues. In the *Republic* and elsewhere, Plato highlighted what are called the four classical cardinal virtues: prudence or practical wisdom (*phronēsis*), courage (*andreia*), moderation or temperance (*sōphrosynē*), and justice (*dikaiosynē*). Aristotle included these traits in his own, much more detailed and systematic analysis of human virtue in various works, especially the *Eudemian Ethics* and *Nichomachean Ethics*. Aristotle listed courage, temperance, and justice among what he called the moral virtues—*ethike*, from which we get the term *ethics*—which also include virtues such as friendship, humility, generosity, and honesty. He included practical wisdom under a separate heading of what he called the intellectual virtues, which also included intuitive reason (*nous*), technical knowledge (*techne*), scientific knowledge (*episteme*), and philosophical wisdom (*sophia*). As we have previously discussed, I concur with Aristotle's view of virtue as a state of character but add an evolutionary basis; in the case of science, certain adaptive instincts are then cultivated by habit and systematized in scientific methods and practices.

Aristotle's lists, overlapping as they are in several ways with the scientific virtues we have been discussing, suggest that there are important connections between the vocational virtue analysis in terms of the scientific *telos* and the more general analysis in terms of human *telos*. For Aristotle, understanding the human *telos* involved understanding our human essence. Because he took us to be rational animals, the actualization of our distinctively human potential necessarily involves the exercise of reason. Moreover, because scientific knowledge is one of the intellectual virtues, the scientific virtues should figure among the human virtues as well, just as we have posited.

However, we must be careful not to conflate the two. At first glance, one might think that Aristotle's inclusion of scientific knowledge in his list covers the scientific virtues that we have been discussing, but that would be a mistake. Our analysis focuses not on the state of having scientific knowledge but rather on the character traits that we should cultivate to make it more likely that we can discover such knowledge. In this sense, the scientific virtues have both an epistemic and an ethical character. Moreover, our account of the scientific virtues as vocational virtues was, by intention, more restricted than this general account of human virtue. In the end, however, it must fit within the larger framework of virtue theory and the ultimate goal of human flourishing.

Virtue ethics in the broad sense has itself seen a revitalization of interest among ethicists, but this is not the place to review the important advances by contemporary philosophers—thinkers from Anscombe, Foot, and MacIntyre to Hursthouse, McDowell, Nussbaum, Annas, and others—who have been exploring general issues of virtue and human flourishing, though I hope that connections to these views are clear. These philosophers were not addressing virtue in science but rather virtue for human beings generally. Some might think that connecting virtue to science is simply a matter of applying this general analysis to the particular case, but I have taken a different approach.

The major difference is that I do not think it is helpful to begin with classical notions of human virtue and ask how to apply or instill them in science. Rather, my approach has been to start with the more limited notion of vocational virtues and explore virtues in the scientific context starting from its own specific *telos*. Because of its narrower focus, it is easier to assess the relative importance of possible virtues for this particular practice in a more rigorous and empirically supportable manner. Moreover, it provides a basis for comparisons, as the same process may be carried out for other vocations. This also helps make better sense of conflicts of interest, which loom large in any ethical deliberation about responsible conduct. Moreover, because the satisfaction of curiosity—i.e., making empirical discoveries—is not just an instrumental but also a constitutive good for human flourishing, I suggest that this approach will turn out to be useful for the larger project as well.

However, that is not our current task. For the moment, I want to keep the focus on what ought to be the ideals for the scientist and for scientific practice. My point here is that scientists may not stop at the borders of their practice; they must recognize that virtuous scientific practice is not exhausted by the simple satisfaction of scientific curiosity. Nonscientific values must also be recognized and accommodated. Putting this in Aristotelian terms, the scientific *telos*, like the *teloi* of other vocations, is not an ultimate and independent end; rather, it participates as a constituent part of the larger human *telos*. This means that scientists must recognize that not everything can be analyzed just in scientific terms; as sociologist William Cameron put it, "[N]ot everything that can be counted counts, and not everything that counts can be counted."[15]

Social Responsibility

As I have noted previously, it is a basic evolutionary fact that every trait that is a strength in one environment can be a weakness in another. Scientists' passion in their vocation and the disciplined focus that it gives them can sometimes lead to myopia with regard to the bigger picture. Robert Merton talked about this in terms of what he called the "imperious immediacy of interest," where the scientists' primary concern with the furtherance of knowledge leads them to disregard possible consequences that lie outside that area of immediate interest.[16] What may seem rational from the point of view of that immediate interest may actually be irrational if it inadvertently undermines other social values with which they ought to be integrated. Merton warned scientists that this kind of shortsighted perspective and behavior could have grave repercussions for science, in that society will blame scientists for the negative consequences of scientific research.

> Precisely because scientific research is not conducted in a social vacuum, its effects ramify into other spheres of value and interest. Insofar as these effects are deemed socially undesirable, science is charged with responsibility. The goods of science are no longer considered an un-qualified blessing. Examined from this perspective, the tenet of pure science and disinterestedness has helped to prepare its own epitaph.[17]

In speaking of "pure science," Merton was referring to what is also called "basic science" in contrast to "applied science." While the former aims simply to discover truths about the world, the latter aims to use confirmed scientific theories for practical purposes (e.g., development of technology and techniques to harness natural processes). It is common to see these two at odds with one another, especially where there is political pressure for science to focus on applications. Basic research is often ridiculed, especially by politicians hoping to score some easy points. Senator William Proxmire often picked on basic science with his Golden Fleece Award. Why are we wasting taxpayer money investigating questions that seem silly or pointless? Scientists have rightly pointed out that one cannot predict what basic research will result in new applications, but experience shows that it will. The Golden Goose Award is the clever rebuttal to critics like Proxmire, highlighting examples of seemingly "pointless" basic research that led unexpectedly to significant applications.

In speaking of "disinterestedness," Merton was referring to the institutional virtue he had identified, namely, impartiality with regard to possible outcomes. While a virtue in the narrow scientific setting, disinterestedness may become problematic in a broader social context if it leads to a "not my job" attitude, whereby scientists deny responsibility for the uses to which their discoveries may be put. Merton's point was that the two traits in combination may set up conditions for the death of science. Scientists have a tendency, he thought, to conflate truth and social utility and assume that the social effects of science must always be beneficial in the long run, but that may not always be the case. Golden geese notwithstanding, what about technologies, like nuclear weapons, that put humanity at risk? Society will cut down the tree of knowledge if it bears evil fruit, he warned.[18]

This connects to our earlier discussion about scientific vices and responsible conduct of research. If science tilts out of balance, it may become vicious. Merton was not imposing values from without here; rather, he was trying to warn scientists about the dangers in terms of the possible effect on their practice. However, my goal is to make the argument about the broader notion of scientific responsibility from a virtue-theoretic perspective, which hopefully shows why scientists have social responsibilities in fully internal terms. On my view, to focus on the furtherance of scientific knowledge to the exclusion of other social values is to miss the full sense in which science is and should be integrated as part of the larger human project.

Scientists are insistent about the primacy of basic research. In one sense, this is understandable, for basic research does provide the foundation for other results. However, some scientists can get uppity about this to the degree that they devalue applied research generally. That goes too far. Darwin did not make that mistake; his point was that following the instinct for truth is reason enough for scientific researchers, not that it is the only acceptable reason. Making discoveries that satisfy our innate and cultivated curiosity is a good in itself, but that is not the only thing that they are good for. Because curiosity is a constituent part of what it means to be human, its satisfaction also is a social good. However, as we have seen, that is not a complete end in itself, but rather a component of the general human goal. What this means is that science does not have a responsibility only to itself; it has a shared responsibility with other vocations to foster human well-being generally.

Curiosity is but one human value in this broader conception of the purpose of science. The unity of virtues implies that virtues ought to work together to that common end. Indeed, science is instrumentally useful for a wide variety of other human goals. Consider the simplest human needs, such as for food and shelter. Science has discovered ways to satisfy these with increasingly sophisticated technologies, but that relationship has it roots in the original instinctual utility of curiosity and other scientific virtues. As we saw, the selective value of these traits from an evolutionary point of view was that they served to increase organisms' chances of surviving and reproducing in variable environments. Basic science is sometimes called "curiosity-driven science," but that is a misnomer, as curiosity covers both basic and applied cases. I may be curious to know something for its intrinsic interest or for its instrumental value. For instance, I may be curious about how cell division works for its own sake or because I think it may help yield a cure for cancer. Instinctual curiosity is based on its adaptive utility, so applications have been there from the beginning. The pursuit of empirical truth is the first rule of science, but in this sense, we may say that the pursuit of human flourishing is its zeroth rule.

For human organisms, this evolutionary imperative became a complex social imperative. While we still have basic needs for food and shelter, we now have many additional individual and corporate goals, and science retains its role in serving them. Scientific professional institutions have highlighted this in various ways. "Advancing Science" is the first half of the motto of the American Association for the Advancement of Science (AAAS), but the second half is "Serving Society," and AAAS undertakes a wide variety of efforts to be not only a voice but a force for science in both regards. The National Science Foundation funds both basic and applied research, but in making decisions in either area its first review criterion is scientific merit, and its second involves the broader impacts of the research.

The general idea is that the internal goods of science should not be sought in isolation from consideration of the broader social goods. It is a mistake to dismiss either of these goods. What this means is that "curiosity-driven" ought not be viewed as a term of derision. And neither should "applied."

Confluence of interest   Scientists participate in and should contribute to the larger task of advancing human flourishing. So do and should practitioners of other vocations. Engineers emphasize innovation. Physicians and other health-care-givers emphasize care. Practitioners of the various arts emphasize creativity. From the broader perspective, we must recognize that there are many more than two cultures that will need to be connected if we are to solve the wicked social problems that impede human flourishing.

The larger moral task cannot be accomplished alone; the human project is too big and too complex. This is not only a matter of our limited mental bandwidth. Human flourishing also requires relationships, as these constitute some aspects of flourishing for us as social creatures. Thus, we must divide the task and take special responsibility for particular elements of the process, while at the same time cooperating to see that the elements are properly integrated. Institutions have a special role to play in this task, as they provide a forum for cooperative deliberation and action.

Individual scientists do not always recognize this broader responsibility beyond their specific vocational imperative to the degree that they should; for example, some scientists groused about the NSF broader impact requirement. This is not surprising, as that imperative for someone qua scientist still operates in relation to the vocational *telos*, which can obscure the ultimate goal. This was Merton's point about the imperious immediacy of interest. Professional societies may have a better chance of avoiding this problem. They still recognize the central purpose of their vocations (indeed, they play an important role in formalizing them and, when needed, enforcing them), but as representatives of the profession, they are more likely to see those values in relation to those of other vocations because it is their task to represent them in broader contexts.

Still, we must acknowledge that meeting the full challenge posed by the goal of unifying the virtues is no simple task. Indeed, it is beyond the scope of professional institutions by themselves. Our previous discussion about conflicts of interest showed some of the problems even in simple cases of scientific integrity, and the difficulties loom larger as one considers vocations in relation to one another. A division of moral labor is necessary because of human limitations, but it will fail in its purpose if it only divides rather than serving to unite us. The general solution is to see all of these in relation to human integrity. The positive goal here must be to understand and achieve a *confluence* of interests. But how is that done? Aristotle argued that integration of the parts is the task of politics.

A Well-Ordered State

Considering what the integration of human values ought to look like, Aristotle concluded: "In general it belongs to excellence to make the condition of the soul good, using quiet and ordered motions and in agreement with itself throughout all its parts; whence the condition of a good soul seems a pattern of a good political constitution."[19] The recommendation here was that moral excellence involves the smooth working of its parts in the ordered service of their common purpose. In the *Politics*, Aristotle used the analogy of sailors on a ship, who each have their appointed tasks. Just as they all must work properly together to preserve their common vessel, so must citizens work together to preserve the ship of state. Ethics fails if politics is not considered; the possibility of good conduct and well-being is undermined in an unethical culture. In times of political turmoil and distrust, however, when quiet and ordered motions are in short supply and the parts of the body politic are in conflict, Aristotle's comparison of a well-ordered soul to a well-ordered state seems a pipe dream. It is, therefore, especially important for us as citizens to consider the role of science in political life.

One sometimes hears the claim, especially in academic circles, that all science is political. But what does this mean? Often, it is just an example of a general view that everything is political and has political ramifications. In a straightforward sense, this is uncontroversial, not only in cases where scientific findings may have political implications but also in the sense that science as an institution is a political stakeholder. Other times it is meant to call attention to the way that science is, and perhaps always has been, a political football, tossed about in the struggle between the right and left, with each side trying to use it for their own purposes to score a partisan win over their political opponents. It will be important to pay attention to how science may become politicized, in this negative sense, which is quite at odds with the notion of a well-ordered state. However, sometimes the claim that all science is political is made within the sort of extreme postmodern philosophy that we have previously discussed, with the implication that scientific claims are equivalent to all other claims in a political game of power. Whoever has the football gets to determine what scientific knowledge is. We have already noted some of the problems with that dogmatic view, but it will behoove us to revisit those lessons in this broader context. We need to pay attention to all these possibilities if we are to understand the proper role of science and the responsibilities of scientists in a well-ordered state.

This is not a book of political philosophy, so I will not have the space to try to do justice to the complex questions involved in reconciling a vision of the good life and a good political constitution that will serve it. For instance, what are political prerequisites to human flourishing? How do we balance the individual freedom needed to be able to act as autonomous moral agents and pursue our vision of the good with the corporate constraints of justice that are needed to ensure that we have the opportunity to do so? I do not pretend to solve or even address the difficulties in articulating such a vision. My goal here is more limited. I want simply to defend the critical role that scientific truth has in making any such vision possible. My thesis is that neither freedom nor justice can be successfully conceptualized or defended in practice without the help of science and a basis in truth. *Liberty, Equality, Reality* must be the hendiatris of a well-ordered democratic state.

Liberty, Equality, Reality

Liberty—Freedom vs. Truth?

Let us begin with the role that science plays in advancing freedom. The simplest sense of this involves the various freedoms delivered by the technological fruits of science: Freedom from the harshness of the elements. Freedom from hunger. Freedom from disease. Freedom from toil. We mostly take these for granted, so they rarely are discussed except in cases where they are still lacking because science is insufficiently developed or deployed, or in cases where science and its technological fruits may go too far, so that freedom becomes a burden or an evil. The displacement of workers by machines liberates some from toil, but it "liberates" others of their means of employment. Technology is, thus, not an unadulterated blessing. We will need to consider how this affects issues of justice later, but judged on a simple Utilitarian scale, it seems clear that science has produced overall benefit when one considers freedom in this way. Nor are these only base-level freedoms; after all, freedom from hunger also makes possible freedom to create, to question, and to explore. It will not do to forget about these benefits and this very practical sense of freedom, but let us set them to the side for the moment.

Another way that science gets credit for advancing freedom is in the way it frees the mind. As we have seen, the key concept behind science is the rejection of authority in favor of evidence. It overcomes dogma and dispels superstition. Darwin expressed this liberating aspect of science in general terms: "[F]reedom of thought is best promoted by the gradual illumination of men's minds, which follows from the advance of science."[20] It is significant that Darwin wrote this in a letter to Edward Aveling, politely declining Aveling's request to dedicate a book on evolution and religion to him. Darwin explained that he saw no good reason to dispute Christianity and theism; better to just allow free thought to develop on its own with the advancement of science generally. Truth shall set one free even if (or, perhaps, especially because) it may not come in the form one originally expected. That is part of the gift of the Enlightenment. While there may be some nostalgia for the lost enchantment of the world, the dispelling of occult superstitions was surely a good thing overall. One may complain that science led to discounting the herbal remedies of witches while dispelling the supernatural fantasy of witchcraft, but at least it helped stop these women from being burned at the stake.

One could easily extend this review of ways that science is liberating for both the body and mind, but it will be more revealing if we look instead at those who reject this standard view. Who opposes scientific truth? Why might someone say that it limits, rather than advances, freedom?

The simplest cases involve those who see that it may curb some action they want to take. Someone may want to smoke and then science says that it will harm their health and that of others. Or, more commonly, someone may want to sell you cigarettes, but science says that…you get the picture. Science may limit someone's advantage over someone else; they want to be able to do something that others cannot. Science can also limit someone's freedom of action if it reveals their doing something they are not allowed to do. Forensic science may lead to someone's being found guilty and locked up. It is also easy to see why acknowledging scientific truth may be problematic for someone who is trying to gain an economic, personal, or political advantage, so we must admit that there are various obstacles to the idea of a well-ordered state. What is really going on here is that such people simply want truth for themselves, to enhance their own freedom over someone else's. Such examples do not negate the value of truth or of science's role in discovering truths but rather support it. Truth is valuable, so there is incentive for some to try to control it for their own purposes.

However, there is a stronger argument that has been made against science and the relationship between scientific truth and freedom by the philosopher Paul Feyerabend, whom we encountered previously in our discussion of postmodernism. Feyerabend claimed that science was just another myth that had no special method or claim upon the truth over any other ideology—it was best read as a "fairytale." Feyerabend saw himself as defending society "from all ideologies, science included," and he used the term fairytale to indicate that ideologies should not be taken too seriously. They may have "lots of interesting things to say, but which also contain wicked lies."[21] Feyerabend acknowledged that science at one time served the cause of freedom, liberating us from previous ideologies that had bound us but said that there was no reason to think that it would remain an instrument of liberation.

> A truth that reigns without checks and balances is a tyrant who must be overthrown, and any falsehood that can aid us in the over throw of this tyrant is to be welcomed.... There is nothing inherent in science or in any other ideology that makes it *essentially* liberating. Ideologies can deteriorate and become stupid religions.[22]

Feyerabend argued that science had become such a religion; scientific truth, he said, had become a tyrant: "My criticism of modern science is that it inhibits freedom of thought. If the reason is that it has found the truth and now follows it, then I would say that there are better things than first finding, and then following such a monster."[23] Thus, quite at odds with the idea that science is an essential contributor to a well-ordered democratic state, Feyerabend argued for "a formal separation of science and state just as there is now a formal separation between state and church."[24]

As noted previously, Feyerabend was probably exaggerating his case in his role as devil's advocate. *Against Method* included in its academic index a label for "wicked comments," and there is quite a long list of entries. In his introduction, he noted that the plan had been for Imre Lakatos to give an equally wicked rebuttal and that his untimely death made the book incomplete. With regard to arguing that science is a religion—a line that he often repeated in print and in presentations—Feyerabend explained that he originally made that claim to earn a paycheck as part of an anthology on science and religion where he thought it would make the book sell better if he was as provocative as possible and figured this was the way to do it. Turned out that it was an easy argument to make, he said. Feyerabend was being intentionally crude and superficial to make his point, but many scholars took his criticisms and viewpoint seriously, so it is worth considering to what degree we should as well.

Is scientific truth really a tyrant and a threat to freedom? What was Feyerabend's beef? The first thing to note is that he was not starting from a particular objection to science but rather from a general objection to any "comprehensive system of thought," which later postmodernists would speak of as a "master narrative." What he objected to was "a truth that reigns without checks and balances." That is to say, he opposed any form of intellectual totalitarianism. How does this threaten freedom? Because "once we have discovered the truth—what else can we do but follow it?"

Feyerabend thought that this was how someone would defend the privilege of science and that it was equivalent to what any ideology would say "to reinforce the faith of its followers." It may sound neutral to speak of truth, he wrote, but "it is easy to twist matters and to change allegiance to truth in one's everyday affairs into allegiance to the Truth of an ideology which is nothing but the dogmatic defense of that ideology."[25]

We have already shown that some scientists can be prone to intellectual gluttony, so this is not a merely hypothetical concern. However, Feyerabend provided little evidence that this purported slippery slope is inevitable, and there is reason to think that it is not. Indeed, it flies in the face of what we have seen is a fundamental feature of scientific reasoning: science explicitly rejects the notion that its truths are absolute. Scientific conclusions are inductive, so they always come with some degree of probability attached. What this means is that when making decisions and taking action based on scientific conclusions, one is placing a bet of sorts. In a cost-benefit analysis, science helps determine the associated odds of success, but it does not restrict one's freedom of choice. The odds may strongly favor one assessment of the facts over another, and our confidence may sometimes be close to certainty, but one remains free to reject a related course of action even in such cases. Indeed, if the likely costs outweigh the likely benefits, it can be quite reasonable to do so. Thus, when it comes time to make a policy decision, science makes a necessary but necessarily incomplete contribution. In the well-ordered state, it is the job of wise leaders to weigh the relevant factors, make an informed judgment, and then make the best choice that they can under conditions of partial knowledge.

Despite his provocative rhetoric and radical reputation, Feyerabend's specific model for the role science should play in politics is not very far from this. He wrote that:

> Science may influence society but only to the extent to which any political or other pressure group is permitted to influence society. Scientists may be consulted on important projects but the final judgement must be left to the democratically elected consulting bodies.[26]

For the most part, this is just what we have recommended—the role of politics is to weigh the relevant factors and seek a balance when making policy decisions, and science is not the only factor to consider. In another way, however, Feyerabend's view really was extreme in that he did not differentiate among inputs that might guide political judgment. While it is reasonable for democratically elected bodies to consult a wide variety of consultants and stakeholders, it is not reasonable to take their judgment equivalently on all questions. In an earlier era, it may have been acceptable to consult one's barber if one needed surgery, but today that would be irrational. Similarly, one ought to turn to a barbershop quartet if one needs a demonstration of expertise in four-part harmony and to a team of geologists for expertise about the effect of fracking on groundwater and not the other way round. To say that a cappella singers have some expertise about the first topic and earth scientists have it for the second is not to say that either are oppressors or that our freedom is compromised because of their earned knowledge.

It is wrong to think of truth as a tyrant in this way. That a bridge is tested and found to be truly capable of bearing a load is not tyrannical. What that truth does is provide not tyranny but trust. It would be debilitating, not liberating, if every time one needed to cross a bridge one never had a sound basis for confidence that it would not collapse. Moreover, such useful expert knowledge does not constrain one's freedom; one may still choose not to cross it. By my lights, freedom is far less threatened by truth than by those who would deny it.

For the sake of argument, however, let us grant that in some cases scientific truth does reduce our freedom. However, lack of scientific truth may do so as well. One may freely choose to deny gravity and ignore the bridge entirely, but stepping out from the cliff's edge will quickly curtail any further exercise of freedom. In the end, the world will claim its due. Given the alternatives, is it not better to be constrained by truth than by ignorance?

Equality—Social Justice vs. Truth?

What about equality and justice? Many of the same points noted in our discussion of freedom apply here as well. We have already mentioned how the technological fruits of science have broadly raised the standard of living. Reduction in infant and childhood mortality and the concomitant increased human life expectancy, for instance, is largely due to advances in the science of public health, medicine, and diet. Disparities remain, but these are not due to failure in the science so much as political and economic failure to distribute the fruits of science equitably. However, we also mentioned the displacement of workers caused by development of more efficient machines; the industrial revolution was the classic example of this, but we are likely to see an even more significant disruption as machine learning and artificial intelligence do the same thing for an even broader range of jobs, exacerbating social inequalities as its new efficiencies are injected into the market. Science cannot expect to be held free from blame.

In "Science and the Social Order," Robert Merton noted a hostility to science that arose from such circumstances. He mentioned British industrialist Josiah Stamp and others as an example of a general "anti-science" movement that was based on concerns about economic justice.

> [T]he rapid development of science and related technology has led to an implic-
> itly anti-science movement by vested interests and by those whose sense of
> "economic justice" is offended. The eminent Sir Josiah Stamp and a host of less
> illustrious folk have proposed a moratorium on invention and discovery, in order
> that man may have a breathing spell in which to adjust his social and economic
> structure to the constantly changing environment with which he is presented by
> the "embarrassing fecundity of technology."[27]

Those who object to machines and their efficiencies are often shrugged off as mere Luddites, but it is a mistake to dismiss such concerns too blithely. Stamp was not saying that technological progress should be stopped but that there ought to be some response to the social disruption it caused. As we saw, it is incumbent upon scientists to be part of any decision about whether possible technologies should be actualized in particular cases, and possible economic injustice is a real effect that they ought to provide warning of and be ready to help alleviate.

Merton pointed out an aspect of the scientific ethos that contributes to justice in a more general way. The scientific value of universalism takes human equality, in the sense of requiring observational interchangeability, as a cornerstone of its methodology. This is a clear and powerful position that serves as a bulwark against prejudice and politicized privilege. He illustrated this with a contemporary case that showed just how relevant this seemingly abstract value was in relation to a real case of political authoritarianism:

> It is a basic assumption of modern science that scientific propositions "are invari-
> ant with respect to the individual" and group. But in a completely politicized
> society—where as one Nazi theorist put it, "the universal meaning of the political
> is recognized"—this assumption is impugned. Scientific "findings" are held to be
> merely the expression of race or class or nation.[28]

It is ironic that this politicized view of science by the fascist right became a core element of the radical postmodern left. Such a standpoint is to be resisted; in rejecting the possibility of objective truth and adopting this kind of identity-based view of knowledge, it affirmed what it had set out to overturn, inadvertently undermining social justice in the process. Without a regulative notion of truth, there is nothing but power and no basis for establishing the occurrence of oppression, let alone for alleviating it. Social justice requires truth in the face of the tyrant. This is a critical benefit of science.

However, just as we considered how scientific virtues may become vices if taken to the extreme, we ought to consider the possibility that science could become the oppressor. This takes us back to Feyerabend's critique. He thought that science itself, though a stimulus to freedom and equality in the past, had indeed become a tyrant. Feyerabend thereby pronounced a pox on all houses, science included. His solution was to cut down all traditions to the same level; all should be equal in a free society. Let them all battle equally. He referred positively to a Darwinian notion where ideologies fight it out on the political stage.

> Considering the sizeable chauvinism of the scientific establishment we can say: the more Lysenko affairs, the better....Three cheers to the fundamentalists in California who succeeded in having a dogmatic formulation of the theory of evolution removed from the text books and an account of Genesis included.[29]

With regard to the Lysenko affair, Feyerabend did note that "(it is not the *interference* of the state that is objectionable in the case of Lysenko, but the *totalitarian* interference which kills the opponent rather than just neglecting his advice)."[30] One would think that something more than a parenthetical caveat would be warranted, given the scientists who opposed Lysenko who really were killed.

Again, it is ironic that this kind of view was embraced by, and became associated with, the radical left, for it is equally an instrument of the far right. Feyerabend appealed to both extremes, as his example of creationists shows. Indeed, the ID creationists adopted this kind of postmodern theory as the basis of what they called their wedge strategy to break apart not only science but the whole liberal, modernist edifice in law, social policy, education, and ethics, and replace it with a conservative, Bible-based worldview. To be fair, Feyerabend did include a parenthetical caveat about creationists as well, predicting that "they would become as chauvinistic and totalitarian as scientists are today when given the chance to run society all by themselves."[31] This is consistent with his general opposition to totalitarianism, but his specific objection to ideologies that removed their opponents was that thereafter they became "boring and doctrinaire." This kind of attitude is considerably more elitist and superior than science ever was.

Once again, it may be that Feyerabend intentionally overstated his view as a provocation. The generous reading is that Feyerabend was simply trying to issue a corrective to the situation as he saw it at the time in which, "In society at large the judgement of the scientist is received with the same reverence as the judgement of bishops and cardinals was accepted not too long ago."[32] If that were really so, I would readily concur that a correction was in order, for science neither deserves nor desires such reverence. However, was he just a court jester, cutting down the king to keep him humble, or something more dangerous?

It seems that he may actually be more of a Loki, a shapeshifter who closes the loop and ties the far left to the far right. Perhaps because of the anarchist label he gave himself, he was embraced as a provocateur of the left and by the left. But this is a misunderstanding. He dismissed Marxism almost in passing. One might think that is because anarchism is even farther to the left. However, as is often the case, the far left circles back and closes ranks with the far right, for anarchism is little different from an extreme libertarianism.

Feyerabend considered the objection that his view was a luxurious elitism and that his flippancy and self-indulgence came at the expense of the starving, enslaved, and downtrodden. Well, of course we need to try to act to help the oppressed, he said, but only so long as that does not curtail as much freedom as we can achieve for ourselves. His way of making a virtue of selfishness could have come straight out of Ayn Rand.

> We are supposed to give up our selfish inclinations and dedicate ourselves to the liberation of the oppressed. And selfish inclinations are what? They are our wish for maximum liberty of thought in the society in which we live now, maximum liberty not only of an abstract kind, but expressed in appropriate institutions and methods of teaching.[33]

Feyerabend was disinclined to put aside concrete intellectual or physical liberty in the service of overcoming injustice. Rather than making the oppressed "succumb" to our attempts to liberate them, we should let them realize their own wishes and liberate themselves in their own way. Feyerabend admitted that this may slow down social progress, but should we proceed just because science tells us something that may explain the problem and offer a solution? If we accept that science has special expertise in testing competing hypotheses and discovering empirical truths, he said, then doesn't that just make following science simply another form of slavery? Better to reject the "self-righteous and narrow-minded" would-be scientific liberators because their "systematic" slavery is worse than the "sloppy" slavery they replace.[34]

Unlike the radical postmodernists who followed and extended his view, Feyerabend did not dismiss truth per se. For example, he was quite adamant that Kuhn's view of scientific change was false. Indeed, he put the term in italics, so he clearly thought that one can make truth claims emphatically. What he rejected was Truth, with a capital T, as a totalitarian ideology. But science, properly understood as an inductive enterprise, eschews such an authoritarian ideology as inimical to its core values, so his attack was misdirected. Recently, Feyerabend explained that he never meant to reject science: "Of course I go to extremes, but not to the extremes people accuse me of, namely, throw[ing] out science. Throw out the idea science is first. That's all right. It has to be science from case to case."[35] If his provocations were meant only as a correction, then it is clear that it was an overcorrection that threw out the baby with the bathwater.

Going on half a century later, I judge that Feyerabend and the radical postmodernists succeeded all too well. We see the frightening effects of a post-truth world when real tyrants and bullies come to power. Asserting their own "alternative facts" may seem amusing when it is applied to disputes about inauguration ceremony crowd size, but it is dangerous when it comes to public health issues like vaccines and climate change. Forget the ideal of a press corps that seeks to ascertain the truth; if facts exist only in the eyes of the beholder and everyone has their own set, then calling "fake news" cannot be criticized as unjust. The danger is real even when considering more garden-variety politicians who may become trapped in some ideology that makes it hard for them to face facts; we should all desire a leader who takes reality seriously when making life-and-death judgments about military versus diplomatic options during international crises.

Reality—a Democracy Prerequisite

During the administration of George W. Bush, many people were concerned about what one Republican advisor to Ronald Reagan called President Bush's "weird, Messianic idea of what he thinks God has told him to do" to defeat Islamic Fundamentalism: "This is why he dispenses with people who confront him with inconvenient facts.... He truly believes he's on a mission from God. Absolute faith like that overwhelms a need for analysis. The whole thing about faith is to believe things for which there is no empirical evidence." Ron Suskind, the journalist who related this in a *New York Times Magazine* article, had a range of such examples where the administration ignored or dismissed facts in favor of faith, including one story of his own meeting with a high-ranking Bush aide, later identified as Karl Rove.

> The aide said that guys like me were "in what we call the reality-based community," which he defined as people who "believe that solutions emerge from your judicious study of discernible reality." I nodded and murmured something about enlightenment principles and empiricism. He cut me off. "That's not the way the world really works anymore," he continued. "We're an empire now, and when we act, we create our own reality. And while you're studying that reality—judiciously, as you will—we'll act again, creating other new realities, which you can study too, and that's how things will sort out. We're history's actors...and you, all of you, will be left to just study what we do.[36]

Political writer Mark Danner identified this sneering dismissal of reality as a full-bore expression of the postmodern critical theory we have previously discussed: "Though we in the 'reality-based community' may just now be discovering it, you have known for years the presiding truth of our age, which is that the object has become subject and we have a fanatical follower of [Michel] Foucault in the Oval Office."[37]

If truth is nothing but the imperialistic assertion of power, it can be fairly created by anyone powerful enough to do so. In this case, it appeared that an extreme religious faith was behind the dismissal of facts, but there are any number of reasons, whether based in ideology or bare self-interest, that facts may be rejected and alternative realities created in political settings. In such cases, it is not unusual to blame the messenger. It is very common for partisan opponents of scientific findings (whether in the service of tobacco executives denying that smoking is a threat to health or energy companies denying climate change) to try to cast doubt on the data, but that is not the end of it. It is also common to attack the character of scientists who conducted the studies that identify politically inconvenient facts; scientists, they say, are elitists who are defending their power or who are in it for the money or the fame. But epidemiology and climate modeling are hardly positions of power, and as we have seen, scientists do not choose their vocation for the money. Neither is science a path to fame. Stephen Jay Gould achieved some degree of general fame after he appeared as an animated cameo character in an episode of *The Simpsons*, but otherwise few people can name or even recognize any scientist save Einstein. Rather, individuals choose science primarily because it gives them a chance to exercise their capacity for wonder in the hope of discovering some truth about the world that no one had ever known before. Scientists need courage to maintain their integrity and the integrity of science in the face of such opposition to truth and reality. That is a mindset that should be recognized by the public (even if not by autocrats or con artists) as valuable and ought to be defended as essential in any working democracy.

There have, of course, been scientists of insufficient courage who have allowed partisan politics to prevail. Some have been co-opted into the service of tyrants, or even joined them willingly. In general, however, authoritarian regimes are antithetical to the scientific mindset, which rejects submission to authority in favor of humility to evidence. A "science" that bows to authority rather than to empirical evidence is counterfeit. The aforementioned philosopher Karl Popper made a similar point in *The Open Society and Its Enemies*,[38] which argued against the falsely "scientific" historicism of Marxism that he thought led to totalitarianism. Popper would have rejected elements of my view, as I reject many of his, but we share the general view that science provides part of our best defense against oppressive ideologies. Another philosopher, Philip Kitcher, has more recently examined the role of science and truth and their role within democracies.[39] Kitcher's analysis is subtle in that he is keenly aware that not all truths will be welcomed and that many are not worth exploring in the first place, so we should not assume that it is always better for us to know the truth. Neither should we assume that truth is "higher" than other human concerns, especially when it comes to the possibility of what he calls "subversive" truths that may undermine presuppositions upon which social preferences are founded.[40] This is a different way of making the point discussed above that even honesty can be taken too far. We must be aware of such possibilities and guard against them, but recognizing that virtue in the extreme may become vicious only supports the value of a virtue-theoretic analysis for thinking about the value of science for a just government.

A democratic political system, in contrast to an authoritarian one, is dependent on a shared method of determining what is real, what works, and what does not work. Science provides knowledge of the situation we are in and the constraints within which we must operate. Making decisions without knowing the facts on the ground is a recipe for disaster. This is so for ordinary policy decisions and even more so for extreme circumstances. No general would want to send troops into avoidable danger. One needs to know not only the positions of opposing forces but also the lay of the land with its various obstacles and affordances. A social justice warrior needs no less. Science must also come into play after we have made policy decisions. How else can we know whether our action had the desired effect? How else do we know the degree to which we came close to or missed our target? Again, a story will not do; if one is to make progress one needs to be able to ascertain the facts. That is the difference between a created narrative and a discovered reality.

This is not to say that we may just turn over all decision-making to a government of scientists, though some have tried to argue for technocracies. In a world of perfect science, some say, we could simply use science to decide the best policy choices—its understanding of cause-effect relationships and ability to predict outcomes of different courses of action would help us know in advance which action to take. However, science is far from attaining such a level of perfection. Moreover, even a hypothetically perfect science would be insufficient for making such decisions, for scientific knowledge is only one factor that must be taken into account as we weigh possible decisions. For example, science may confirm that, for some particular problem, choice A would produce a better result than choice B, but if A is too expensive or if its achievement would force the loss of something else of significant value, it may nevertheless be reasonable to go with B. Ultimate judgments about the relative importance of economic or other values are not subject to experimental test and so are not matters where scientific expertise is applicable. It would thus be a mistake to think that we could ever turn over such decisions to technocrats. However, the imperfection and incompleteness of science do not give politicians license to ignore it. Perfection should not be the enemy of the good, or the better.

Although we have been considering this issue in relation to democracy, the point holds equally for autocratic governments. One would think that even a tyrant would have an interest in getting true information about, say, whether rising sea levels were going to threaten a nation's coastal cities. Very often, however, one sees the opposite. Facts are dismissed as irrelevant or fake. Autocrats expect the world to bow to their will. But natural laws bend to no king. How can we seriously allow ourselves to be subjects of a government that does not recognize that we are all subject to gravity and other natural laws? A politician may deny evolution, but Darwin's law will continue to operate just the same. Natural facts are denied at our common peril.

Politicization of Science

Of course, a more common reason that autocrats deny facts is simply that they are lying. They know that truth matters, but to advance their own personal and political interest, it is more important to them that they appear to be right than that they are. While some may come to believe their own lies, others are more cold-blooded and calculating in their manipulative mendacity. That autocrats and others in power often use their power to try to create or "construct" a reality that serves their own purposes is an additional reason to reject the radical postmodern view. It is the possibility of a common grounding in tested reality that keeps us from being at the mercy of such narratives of would-be masters. The way to respond to those who use power to assert their own reality is not to theorize that there is nothing to truth but power. If truth is nothing more than a power-based social construction, we have no basis for calling out the lies of real tyrants. Indeed, in a post-truth world, the notion of a lie is meaningless. This ought to be the real worry about master narratives.

It is a harm to all when one or the other side of the aisle politicizes science. Such behavior turns political ideology and identity into blinders that obscure what ought to be recognized as our common interest in getting the facts right. Again, this is not to say that scientists have always been on the side of freedom and justice, and we have seen that there is always a danger that even a scientific virtue like curiosity can be taken too far. The lesson thus applies to scientists themselves, who must be watchful that they not slip from speaking as experts to dictating as authorities.

In his acceptance speech upon receiving the Nobel Prize, Swedish biochemist Sune Bergström emphasized that scientists should not politicize science. Starting with the same point as Chekov that science "does not know any national borders," Bergström noted that the scientists of the world were forming "an invisible network" by virtue of their free exchange of scientific information, a network of freedom that was accepted across the nations, irrespective of political systems or religions. This network provided a unifying global bulwark against divisive ideologies. "Great care must be taken that the scientific network is utilized only for scientific purposes," he advised, because "if it gets involved in political questions it loses its special status and utility as a nonpolitical force for development."[41]

In citing science's role in development, Bergström was highlighting its instrumental value. Applied science is its most obvious and commonly acknowledged social benefit. But this should not be thought of as limited to technology, which is not always an unalloyed good. More significant is its value in testing our assumptions and determining the likely effects of possible courses of action. Moreover, as we have seen, in addition to its instrumental value is its intrinsic value—the human value in discovery for its own sake. Robert Wilson, founder of the National Accelerator Laboratory (Fermilab), planted the flag for this during his testimony before Congress to get funds for the project when he was asked by a senator, John Pastore, what the effect of building the lab would have for the security of the country.

Pastore: Is there anything connected in the hopes of this accelerator that in any way involves the security of this country?

Wilson: No sir; I do not believe so.

Pastore: Nothing at all?

Wilson: Nothing at all.

Pastore: It has no value in that respect?

Wilson: It only has to do with the respect with which we regard one another, the dignity of men, our love of culture.

Wilson went on to compare the value of the new scientific knowledge that the new lab may help researchers discover to the value of painting, sculpture, and poetry. Science and literature alike, he said, are "all the things that we really venerate and honor in our country and are patriotic about." In this sense, Fermilab "has nothing to do directly with defending our country, except to make it worth defending."[42]

Wilson's testimony is widely quoted as an example of scientific integrity—standing up for the intrinsic value of pure science against the opposition of crass politicians. One cannot help but admire Wilson's refusal to pander. However, cynically seeing this in contrast to the attitude of politicians is unfair, even in this particular instance. In fact, Senator Pastore was one of the biggest supporters of the Fermilab project, and the unedited transcript of the exchange makes it clear that he was looking for practical arguments that would help it pass when it came to a vote in the House and make the decision agreeable to taxpayers.[43] When conducted with its own form of integrity, the political process of crafting coalitions of support to advance the public good in ways that combine intrinsic and instrumental goods is exactly what we should expect of our governmental representatives. But we need to insist that political balancing always be done in light of our best methods for understanding the real circumstances in which we make our choices, and the real effects of those choices.

During dark times when it seems that politicians care little for facts or the help science can offer in providing a basis for shared empirical knowledge, it is worth reminding ourselves that disdain of these is not a universal political attitude. In the very first State of the Union address to the U.S. Congress, President George Washington expressed a sentiment that politicians would do well to recall and emulate. "There is nothing which can better deserve your patronage," he said, "than the promotion of Science and Literature. Knowledge is in every country the surest basis of public happiness."[44]

Full Circle

The Scientific Temper

Human integrity involves well-rounded wholeness. Human flourishing requires actualizing the whole human potential. Science may not be the most important aspect of human flourishing, but it is a fundamental constituent of it and essential to our well-being. However, it is one of the more neglected parts when it comes to being recognized and articulated in value terms. I hope that this book, by demonstrating how curiosity and other virtues help provide a moral structure to science, has taken a step to correcting that. Science is not just about the facts; it is also the mindset that is involved in getting the facts right and caring to do so. Facts have value. If we are to gain scientific knowledge, we should aspire to cultivating the traits that will help us do so. Moreover, it is a fact that certain virtues are more helpful than others in that quest; epistemological anarchism is not a viable option. Such knowledge feeds back into the system, potentially helping us change environmental circumstances or even adjust intermediate values in light of their effect on more ultimate goals. These are just a few points along the fact-value continuum. Integrating these is no simple task, but we are compelled to try. This, as Aristotle said, is the role of politics and we ought to work to support that process. As Freeman Dyson observed, "We become politically involved because knowledge implies responsibility."[45]

Science, united with the humanities, ought to figure in a democratic political vision. Washington was by no means the only statesman to emphasize its value. The most integrated example that I know of came from Jawaharlal Nehru, a leader of India's independence movement and its first Prime Minister, who articulated this kind of vision in *The Discovery of India*, written while he was a political prisoner during the struggle for independence. Ostensibly a history of the country, the book looked to the future as much as to the past and offered a deeply philosophical perspective on how an ancient culture may renew itself. Nehru spoke of science throughout the book not only in terms of its technological import but also as a frame of mind. Central to this, he recognized, was its utility in uncovering truth—not necessarily the eternal Truth of ultimate reality but aspects of truth as far as our limited minds may grasp it: "a living truth for humanity, supplying the essential need for which it craves, and offering guidance in the present and for the future."[46]

Nehru put C. P. Snow to shame in that he sought to bridge not just the two cultures of science and the humanities but the many cultures that politics must attempt to accommodate. He had no illusions about the difficulty of uniting "the diversities and divisions of Indian life, of classes, castes, religions, races, different degrees of cultural development."[47] He did this with a wisdom that acknowledged the limitations but also the strengths of each. Of religions, for example, he said that "Instead of encouraging curiosity and thought, they have preached a philosophy of submission to nature, to established churches, to the prevailing Social order, and to everything that is."[48] This kind of religion brings comfort and stability, but it can stand in the way of human progress. Philosophy, he wrote, avoids such pitfalls and encourages thought and inquiry, but it too often remains isolated in its ivory tower, exploring logic and reason but mostly failing to link these to matters of fact and to the daily problems of human life. Art, poetry, and other intuitive mental experiences are likewise necessary, but for these and all other methods "we must hold to our anchor of precise objective knowledge tested by reason, and even more so by experiment and practice, and always we must beware of losing ourselves in a sea of speculation unconnected with the day-to-day problems of life and the needs of men and women."[49] Science is thus a critical part of the broader human project. Nehru concluded that "The scientific approach and temper are, or should be, a way of life, a process of thinking, a method of acting and associating with our fellowmen."[50] One can only marvel at the audacity of enshrining this in the country's founding document; the Constitution of India states that it is a fundamental duty of all its citizens "to develop the scientific temper, humanism and the spirit of inquiry and reform."[51]

Nehru was articulating a political ideal that recognized the value of the scientific mindset—his notion of "the scientific temper" was central to his political vision. What did he mean by it? In what is probably his more compact statement of this he gave a description that will seem very familiar:

> [What is needed] is the scientific approach, the adventurous and yet critical temper of science, the search for truth and new knowledge, the refusal to accept anything without testing and trial, the capacity to change previous conclusions in the face of new evidence, the reliance on observed fact and not on pre-conceived theory, the hard discipline of the mind—all this is necessary, not merely for the application of science but for life itself and the solution of its many problems.[52]

Even in this short passage Nehru touched on many of the scientific virtues that we have identified. He highlighted not only the core virtues of curiosity and truth seeking ("the search for truth and new knowledge") but the "adventurous" spirit of the quest, as curiosity takes one into unknown realms of its endless frontier. We find the special virtue of being attentive and observant ("reliance on observed fact") and also the general discipline of perseverance and meticulousness ("hard discipline of the mind"). We find the skeptical attitude ("critical temper") that keeps us honest in that quest for truth. We also see the special form of intellectual humility that is important in science—humility to evidence ("capacity to change previous conclusions in the face of new evidence"). Do not accept the word of authorities, even the authority of religion and historical tradition, but seek the evidence that will lead forward to the truth and to the future.

The insistence on looking for evidence, rather than acquiescing to authority, is one general lesson that we would do well to learn from science as we attempt to integrate our various human goals to form a just political and social system. We fail when we allow power to predominate. Science is emancipatory because its basic loyalty is to the truth, and truth is needed if we are to be free.

Final Flourish

In this book I have aimed to uncover, in broad outline, the values that structure science. The search for truths about our natural world has normative import. The scientific *telos* provides a moral touchstone for scientific practice. Curiosity is a core virtue and from it radiates other virtues. To truly satisfy our curiosity requires that we be attentive and observant to notice clues to the workings of the world; meticulous in our measurements to increase accuracy and reliability; skeptical of received dogma and of ourselves to guard against negligent acquiescence to authority and dangerous complacency; objective in our observations and judgments to fairly evaluate the evidence and reduce the skews of bias; and courageously humble to accept what the evidence indicates, even if it goes against our immediate preferences.

Discovering the hidden truths of nature is the special aim of science, so scientists have a responsibility to truth, but not only that. Scientific goals and science itself must be integrated into the greater framework of the human project. Politicians and the public ought to recognize science's critical role in our common natural aim. Science is required for human flourishing in the instrumental sense of providing needed materials through technology, but not only in that way. It is our best method for understanding the affordances and constraints of the world we live in. It provides the theory and practice that can cure diseases and advance health; the knowledge of physical laws that lets us extract power from the winds and from the atom; the understanding of forces that can direct rockets with precision to the planets and beyond.

Even more significant than its practical benefits, scientific practice is partly constitutive of human flourishing in that satisfaction of curiosity is a basic human good. Science is a philosophy that springs from the deep desires of evolved instincts. Satisfying the instinct for truth that drives us is, as Darwin thought, reason enough for scientific research. But the scientific mindset has broader benefits as well. It challenges our prejudices. It fuels our imaginations. It answers our natural questions and leads to new questions, still unanswered. It points to areas ready to explore. Curiosity takes us back to the start. Science ends where it began—in wonder.

Notes

Preface

1. Robert T. Pennock, *Tower of Babel: The Evidence against the New Creationism* (Cambridge, MA: MIT Press, 1999).

2. Aaron M. McCright and Riley E. Dunlap, "Defeating Kyoto: The Conservative Movement's Impact on U.S. Climate Change Policy," *Social Problems* 5, no. 3 (2003): 348–373.

3. John Timmer, "Science Education Group Decides It's Time to Tackle Climate Change," *Ars Technica*, January 16, 2012, https://arstechnica.com/science/2012/01/science-education-group-decides-its-time-to-tackle-climate-change/?comments=1.

4. Louis Jacobson, "Yes, Donald Trump Did Call Climate Change a Chinese Hoax," *Politifact*, June 3, 2016, http://www.politifact.com/truth-o-meter/statements/2016/jun/03/hillary-clinton/yes-donald-trump-did-call-climate-change-chinese-h/.

5. Oxford Dictionaries, "Word of the Year 2016," https://en.oxforddictionaries.com/word-of-the-year/word-of-the-year-2016.

6. Robert T. Pennock and Jon D. Miller, "The Values of Exemplary and Promising Younger Scientists: A Preliminary Analysis" (American Association for the Advancement of Science Symposium paper, February 13, 2016).

7. Imre Lakatos, "History of Science and Its Rational Reconstructions," in *PSA: Proceedings of the Biennial Meeting of the Philosophy of Science Association 1970* (Chicago: University of Chicago Press on behalf of the Philosophy of Science Association, 1970), 91–136.

8. Noretta Koertge, *Scientific Values and Civic Virtues* (Oxford: Oxford University Press, 2005), v.

9. Thomas S. Kuhn, "Objectivity, Values and Theory Choice," in *The Essential Tension: Selected Studies in Scientific Tradition and Change* (Chicago: University of Chicago Press, 1977), 320–339.

10. Nancy E. Snow, *Virtue as Social Intelligence: An Empirically Grounded Theory* (New York: Routledge, 2010).

11. See, for example, Ernest Sosa, *Knowledge in Perspective: Selected Essays in Epistemology* (Cambridge: Cambridge University Press, 1991); Jonathan L. Kvanvig, *The Intellectual Virtues and the Life of the Mind: On the Place of the Virtues in Contemporary Epistemology* (Savage, MD: Rowman & Littlefield Publishers, 1992); Linda Trinkaus Zagzebski, *Virtues of the Mind: An Inquiry into the Nature of Virtue and the Ethical Foundations of Knowledge* (New York: Cambridge University Press, 1996); Jason Baehr, The Inquiring Mind: On Intellectual Virtues & Virtue Epistemology (Oxford: Oxford University Press, 2011).

12. Rosalind Hursthouse and Glen Pettigrove, "Virtue Ethics," in *The Stanford Encyclopedia of Philosophy*, ed. Edward N. Zalta (Winter 2016 Edition), https://plato.stanford.edu/archives/win2016/entries/ethics-virtue/.

13. Kenneth Boulding, "The 'Two Cultures,'" in *Technology in Western Civilization*, vol. 2, ed. Melvin Kranzberg and Carroll W. Pursell (London: Oxford University Press, 1967), 686–695.

14. C. P. Snow, *The Two Cultures and The Scientific Revolution* (New York: Cambridge University Press, 1959), 14.

15. Charles Darwin, "Letter to John Stephens Henslow," April 1, 1848. Darwin Correspondence Project, "Letter no. 1167," http://www.darwinproject.ac.uk/DCP -LETT-1167.

Chapter 1

1. Charles Darwin, "Letter to John Stephens Henslow," April 1, 1848. Darwin Correspondence Project, "Letter no. 1167," http://www.darwinproject.ac.uk/DCP-LETT -1167.

2. Think of this as a methodological form of naturalism that only allows appeal to natural laws and processes in explanations of phenomena.

3. Plato, "Theaetetus," trans. F. M. Cornford, in *Plato: The Collected Dialogues*, ed. Edith Hamilton and Huntington Cairns (Princeton, NJ: Princeton University Press, 1961), 860.

4. Charles Darwin, *The Descent of Man, and Selection in Relation to Sex* (1871), in *From So Simple a Beginning: The Four Great Books of Charles Darwin*, ed. Edward O. Wilson (New York: W. W. Norton, 1958), chaps. 3–5.

5. Darwin, "Letter to John Stephens Henslow."

6. Frederick Suppe, "The Search for Philosophic Understanding of Scientific Theories," in *The Structure of Scientific Theories*, 2nd ed., ed. Frederick Suppe (Urbana: University of Illinois Press, 1977).

7. Lewis Thomas, "Natural Science," in *The Lives of a Cell: Notes of a Biology Watcher* (New York: Viking Press, 1974), 118.

8. Thomas, "Natural Science," 119–120.

9. Charles Darwin, *On the Origin of Species* (1859), in *From So Simple a Beginning: The Four Great Books of Charles Darwin*, ed. Edward O. Wilson (New York: W. W. Norton, 1958), 759.

10. Darwin, *Descent of Man*, 767–1248.

11. Darwin, *Descent of Man*, 800.

12. Darwin, *Descent of Man*, 800.

13. Darwin, "Letter to John Stephens Henslow."

14. Darwin, *Descent of Man*, 819.

15. Robert T. Pennock, "Moral Darwinism: Ethical Evidence for the Descent of Man," *Biology & Philosophy* 10 (1995): 287–307.

16. Alistair McIntyre, *After Virtue: A Study in Moral Theory*, 2nd ed. (Notre Dame, IN: University of Notre Dame Press, 1997), 148.

17. Moliére, *The Imaginary Invalid* (1673), act III, sc. iii.

18. Aristotle, *Nicomachean Ethics*, in *The Complete Works of Aristotle*, trans. Ross Urmson, ed. Jonathan Barnes (Princeton, NJ: Princeton University Press, 1985), 2.6; 1106b36–1107a2.

19. David Hume, *Enquiry concerning the Principles of Morals* §9, Pt. I (Oxford: Oxford University Press, 1777), 268.

20. Hume, *Enquiry*, 172–173.

21. Peter Drucker, quoted in Jack Trout, "Peter Drucker on Marketing," *Forbes*, July 3, 2006, https://www.forbes.com/2006/06/30/jack-trout-on-marketing-cx_jt_0703drucker.html#184a57eb555c.

22. Although the notion that maximizing profit is commonly taken to be a basic theorem of business, Drucker himself held that it was not only not the primary purpose of a business, but that it could be detrimental to society and even sometimes dangerous to the organization's health. See William A. Cohen, *Drucker on Marketing: Lessons from the World's Most Influential Business Thinker* (New York: McGraw Hill, 2013).

23. C. P. Snow, *The Two Cultures and the Scientific Revolution* (New York: Cambridge University Press, 1959).

24. David Armitage and Jo Guldi, "Bonfire of the Humanities," *Aeon Magazine*, October 2, 2014.

25. Kurt Baier, *The Moral Point of View* (Ithaca, NY: Cornell University Press, 1958).

26. We will take this up in detail in chapter 7.

27. Vannevar Bush, *Science: The Endless Frontier* (Washington, DC: U.S. Government Printing Office, 1945).

Chapter 2

1. Isaac Newton, "Quaestiones quaedam philosophicae" [Certain philosophical questions], *The Newton Project*, 2003, http://www.newtonproject.ox.ac.uk/view/texts /normalized/THEM00092.

2. Rudolph Carnap, *Der logische Aufbau der Welt* [The logical structure of the world], 1928; Rudolph Carnap, "The Two Concepts of Probability," *Philosophy and Phenomenological Research* 5, no. 4 (June 1945): 513–532. For an in-depth discussion, see Michael Friedman, "Rational Reconstruction, Explication, and the Rejection of Metaphysics," in *Carnap's Ideal of Explication and Naturalism*, ed. P. Wagner and Michael Beaney (London: Palgrave Macmillan, 2012), 190–204.

3. Richard Feynman, "Cargo Cult Science" (Caltech commencement address, 1974), in *Classic Feynman: All the Adventures of a Curious Character*, ed. Ralph Leighton (New York: W. W. Norton, 2006), 490–497, 492–493.

4. NOVA, *The Best Mind since Einstein*, December 21, 1993.

5. Richard Feynman, *Surely You're Joking, Mr. Feynman? (Adventures of a Curious Character)*, ed. Edward Hutchings (New York: W. W. Norton, 1985).

6. Alasdair MacIntyre, *After Virtue: A Study in Moral Theory* (Notre Dame, IN: University of Notre Dame Press, 1997), 187.

7. MacIntyre has reasons for building in some specific elements in his definition that make it too narrow for our purposes. For instance, cooperation and social sanction are important parts of *social* practices, but I will argue that they do not belong as part of the basic definition of a practice per se, which, for example, may start as habitual individual procedures and only later involve others.

8. MacIntyre, *After Virtue*, 188.

9. We will look in detail at the role of habits in chapter 5.

10. Emotions and motivations are a critical element in virtue ethics, and we will examine their role in science in detail in chapter 4.

11. Richard Feynman, "The Pleasure of Finding Things Out," interview, *NOVA*, WGBH Educational Foundation, 1983.

12. I discussed curiosity as an instinct as such an example in chapter 1 and will discuss attentiveness in chapter 4. But these discussions give just an outline of the view. I will

have to save my detailed account of how to move from evolutionary know-how to a philosophically recognizable form of scientific knowledge for another occasion.

13. My informal evidence for this comes from experience reading nomination letters while serving as president of a Sigma Xi chapter and letters of recommendations for candidates for faculty positions.

14. Walter Isaacson, *Einstein: His Life and Universe* (New York: Simon & Schuster, 2008), 311.

15. Isaacson, *Einstein*, 313.

16. Richard P. Feynman, "The Making of a Scientist," in *Classic Feynman: All the Adventures of a Curious Character*, ed. Ralph Leighton (New York: W. W. Norton, 2006), 18.

17. Richard P. Feynman, "Banquet Speech," The Nobel Foundation, 1965, https://www.nobelprize.org/nobel_prizes/physics/laureates/1965/feynman-speech.html.

18. See Robert T. Pennock, *Tower of Babel: The Evidence against the New Creationism* (Cambridge, MA: MIT Press, 1999), chap. 4.

19. Roger Highfield, "Al-Qa'eda's Atom Plans Were Spoof Science," *The Telegraph*, November 20, 2001.

20. U.S. Department of Health and Human Services, "United States Government Policy for Oversight of Life Sciences Dual Use Research of Concern," http://www.phe .gov/s3/dualuse/Documents/us-policy-durc-032812.pdf.

21. White House Office of Science and Technology Policy and Department of Health and Human Services, "Doing Diligence to Assess the Risks and Benefits of Life Sciences Gain-of-Function Research," October 17, 2014, https://obamawhitehouse.archives .gov/blog/2014/10/17/doing-diligence-assess-risks-and-benefits-life-sciences-gain -function-research; National Institutes of Health, "Notice Announcing the Removal of the Funding Pause for Gain-of-Function Research Projects," December 19, 2017, https://grants.nih.gov/grants/guide/notice-files/NOT-OD-17-071.html.

22. This topic will be addressed in chapter 8.

23. Eugenie Samuel Reich, *Plastic Fantastic: How the Biggest Fraud in Physics Shook the Scientific World* (New York: Palgrave Macmillan, 2009).

24. Malcolm R. Beasley et al., "Report of the Investigation Committee on the Possibility of Scientific Misconduct in the Work of Hendrik Schon and Coauthors," Lucent Technologies, September 2002.

25. The figures go up to about 2 percent and 14 percent, respectively, if one also includes wording about "modifying" or "altering" data. The latter number is harder to interpret, as individuals may be reporting the same cases. In one study where this was controlled for, the number was lower—5.2 percent. All these figures come with various caveats but given the nature of the topic are likely to underestimate the real

incidence of misconduct. See Daniele Fanelli, "How Many Scientists Fabricate and Falsify Research? A Systematic Review and Meta-analysis of Survey Data," *PLOS ONE* 4, no. 5 (2009): e5738, https://doi.org10.1371/journal.pone.0005738.

26. Schön had also published fraudulent papers in *Physical Review*, *Applied Physical Letters*, and *Advanced Materials*. All in all, twenty-eight papers were withdrawn by these journals.

27. Steven L. Salzberg, "Creationism Slips Into a Peer-Reviewed Journal," *Reports of the National Center for Science Education* 28, no. 3 (May–June 2008): 12–14, 19, https://ncse.com/library-resource/creationism-slips-into-peer-reviewed-journal; Daniel Cressey, "Paper That Says Human Hand Was 'Designed by Creator' Sparks Concern," *Nature* 531 (March 3, 2016), https://www.nature.com/news/paper-that-says-human-hand-was-designed-by-creator-sparks-concern-1.19499\.

28. Matthew Mailes, Andrew G. Lyne, and S. L. Shemar, "A Planet Orbiting the Neutron Star PSR1829–10," *Nature* 352 (July 25, 1991): 311–313, https://doi.org10.1038/352311a0.

29. Andrew Lyne, in "Hunt for Alien Worlds," *NOVA*, February 18, 1997, http://www.pbs.org/wgbh/nova/transcripts/2407alien.html.

30. Geoff Marcy, in "Hunt for Alien Worlds," *NOVA*, February 18, 1997, http://www.pbs.org/wgbh/nova/transcripts/2407alien.html.

31. Feynman, "Cargo Cult Science," 497.

Chapter 3

1. John W. Cornforth, "Scientists as Citizens," *Australian Journal of Chemistry* 46 (1993): 266.

2. Robert K. Merton, "The Normative Structure of Science" (1942), in *The Sociology of Science: Theoretical and Empirical Investigations* (Chicago: University of Chicago Press, 1973), 268–269.

3. Merton, "The Normative Structure of Science," 268–269.

4. Robert K. Merton, "Science and the Social Order," *Philosophy of Science* 5 (1938): 321–337.

5. Ian I. Mitroff, "Norms and Counter-Norms in a Select Group of the Apollo Moon Scientists: A Case Study of the Ambivalence of Scientists," *American Sociological Review* 39, no. 4 (August 1974): 579–595; Stephen Cole, "Merton's Contribution to the Sociology of Science," *Social Studies of Science* 34 (December 2004): 829–844.

6. Some substituted "originality" for the O in CUDOS following John Ziman's proposal in *Real Science: What It Is and What It Means* (Cambridge: Cambridge University

Press, 2000). See "Scientific Norms," http://sociology.iresearchnet.com/sociology-of -science/scientific-norms.

7. Pliny the Elder, *Naturalis Historia*, ed. John Bostock and H. T. Riley (London: Taylor and Francis, 1855).

8. Stephan Lewandowsky et al., "Misinformation and the 'War on Terror': When Memory Turns Fiction into Fact," in *Terrorism and Torture: An Interdisciplinary Perspective*, ed. W. G. K. Stritzke et al. (Cambridge: Cambridge University Press, 2009), 179–203.

9. Marcello Truzzi, "Editorial," *The Zetetic* 1, no. 1 (Fall/Winter 1976): 4.

10. Thomas Henry Huxley, "Darwinian Hypothesis," in *Darwiniana: Essays*, ed. T. H. Huxley (New York: D. Appleton, 1896), 20.

11. We will examine this point in more detail in chapter 8.

12. Merton, "Science and the Social Order," 334.

13. Merton, "Science and the Social Order," 334.

14. Merton, "Science and the Social Order," 334.

15. Oren Harman and Michael Dietrich, *Rebels, Mavericks and Heretics in Biology* (New Haven, CT: Yale University Press, 2008).

16. Edwin C. Kemble, *Physical Science, Its Structure and Development*, vol. 1 (Cambridge, MA: MIT Press, 1966), 2.

17. Kemble, *Physical Science*, 2.

18. Thomas Henry Huxley, *Lay Sermons, Addresses and Reviews* (London: MacMillan, 1880).

19. Albert Einstein, "Letter to Jost Winteler," July 8, 1901, in *The Collected Papers of Albert Einstein*, vol. 1, *The Early Years, 1879–1902*, ed. John Stachel, David C. Cassidy, and Robert Schulmann (Princeton, NJ: Princeton University Press, 1987), 310.

20. Albert Einstein, foreword to Galileo Galilei, *Dialogue concerning the Two Chief World Systems* (Berkeley: University of California Press, 1967), vi–xix, xvii.

21. Blaise Pascal, quoted in H. Floris Cohen, *The Scientific Revolution: A Historiographic Analysis* (Chicago: University of Chicago Press. 1995), 159.

22. MacIntyre, *After Virtue*, 92.

23. Kitzmiller v. Dover Area School District, 400 F. Supp. 2d 707 (W.D. Pa. 2005).

24. Francis Bacon, *The New Organon, or True Directions concerning the Interpretation of Nature*, book 1, aphorism 58, in *The Philosophical Works of Francis Bacon*, ed. John M. Robertson (London: George Routledge and Sons, 1905), 269.

25. Francis Bacon, *The Advancement of Learning* (London: MacMillan, 1898), 38.

26. Konrad Lorenz, *On Aggression*, trans. Marjorie Kerr Wilson (London: Routledge, 1963), 9.

27. Merton, "Normative Structure of Science," 277.

28. Stanford University, "Openness in Research," October 10, 2017, https://doresearch .stanford.edu/policies/research-policy-handbook/conduct-research/openness-research.

29. Edward O. Wilson, *Consilience: The Unity of Knowledge* (New York: Vintage Books, 1998), 66.

30. Helen Longino distinguished two senses of attribution of objectivity. First is "to claim that the view provided by science is an accurate description of the facts of the natural world as they are" and second is "to claim that the view provided by science is one achieved by reliance upon nonarbitrary and nonsubjective criteria for developing, accepting and rejecting the hypotheses and theories that make up the view" (Helen Longino, *Science as Social Knowledge: Values and Objectivity in Scientific Inquiry* [Princeton, NJ: Princeton University Press, 1990], 62). Elisabeth Lloyd, in a paper on the concept as it relates to feminist epistemology, distinguished four distinct basic types of meanings of "objective" and "objectivity" (Elisabeth A. Lloyd, "Objectivity and the Double Standard for Feminist Epistemologies," *Synthese* 104, no. 3, Feminism and Science [September 1995]: 351–381). More recently, Heather Douglas's careful analysis distinguished eight varieties of objectivity, arguing that none are reducible to the others (Heather Douglas, "The Irreducible Complexity of Objectivity," *Synthese* 138, no. 3 [February 2004]: 453–473). If that is not enough, the historians Lorraine Daston and Peter Galison have a 500-page book on the history of the concept (Lorraine Daston and Peter Galison, *Objectivity* [New York: Zone Books, 2007]).

31. Israel Scheffler, *Science and Subjectivity* (Indianapolis: Bobbs-Merrill, 1967).

32. John M. Ziman, *Public Knowledge: An Essay concerning the Social Dimensions of Science* (New York: Cambridge University Press, 1968).

33. Karl Popper, *The Logic of Scientific Discovery* (London: Routledge, 2002), 44.

34. Albert Einstein, "Motives for Research," in *Zu Max Plancks sechzigstem Geburtstag. Ansprachen, gehalten am April 26, 1918 in der Deutschen Physikalischen Gesellschaft, von E. Warburg, M. v. Laue, A. Sommerfeld und A. Einstein* (Karlsruhe: C. F. Müllersche Hofbuchhandlung, 1918), 29–32, http://alberteinstein.info/vufind1/images/einstein /ear01/view/1/4009_000000616.pdf.

35. Robert T. Pennock, "Research Funding and the Virtue of Scientific Objectivity," *Academic Integrity* 5, no. 2 (Spring 2002): 3–6.

36. Barbara McClintock, quoted in Evelyn Fox Keller, *A Feeling for the Organism: The Life and Work of Barbara McClintock* (New York: Henry Holt, 1983), 117.

37. Merton, "Normative Structure of Science," 270.

38. Merton, "Normative Structure of Science," 277–278; Merton, "Science and the Social Order," 334.

39. John Rawls, *A Theory of Justice* (Cambridge, MA: Harvard University Press, 1971), chap. 3.

40. As with the large literature on objectivity, we do not have space here to do justice to the debates in this area, but see Sandra Harding, ed., *The Feminist Standpoint Theory Reader* (New York: Routledge, 2004) for an overview of the approach.

41. Merton, "Normative Structure of Science," 276.

42. Merton, "Normative Structure of Science," 276.

43. Robert T. Pennock, "Fostering a Culture of Scientific Integrity: Legalistic vs. Scientific Virtue-Based Approaches," *Professional Ethics Report* 28, no. 2 (2015): 1–3.

44. Merton, "Science and the Social Order," 332.

45. Merton, "Science and the Social Order," 330.

46. Merton, "Science and the Social Order," 332.

47. Reinhold Niebuhr, "Ideology and the Scientific Method," in *The Essential Reinhold Niebuhr: Selected Essays and Addresses*, ed. Robert Mcafee Brown (New Haven, CT: Yale University Press, 1986), 215–216.

48. Merton, "Normative Structure of Science, 273.

49. Merton, "Normative Structure of Science, 273.

50. James D. Watson, quoted in John Parrington, *The Deeper Genome: Why There Is More to the Human Genome Than Meets the Eye* (Oxford: Oxford University Press, 2015), 69.

51. Jim Warren, "Triumph of the Nerds," interview by Robert X. Cringley, *The Television Program Transcripts: Part 1*, PBS, 1996, http://www.pbs.org/nerds/part1.html.

52. We can't pursue this now, but this also suggests that universities should reconsider some aspects of the academic reward structure, which sometimes uses simplistic metrics for judging merit.

53. Bas Van Fraassen, *The Scientific Image* (Oxford: Clarendon, 1980), 40.

54. Darwin, *Descent of Man*, 1236.

55. Daniel E. Koshland, Jr., "Crazy, but Correct: How a Non-Conformist Theory Beat Scepticism and Got Into the Textbooks," *Nature* 432 (November 25, 2004): 447, https://doi.org/10.1038/432447a.

56. Roald Hoffmann, "McPherson Lecture," Michigan State University, September 25, 2002.

Chapter 4

1. Darwin, *Descent of Man*, 803.

2. George Bernard Shaw, *John Bull's Other Island*, https://www.gutenberg.org/files/3612/3612-h/3612-h.htm.

3. Sigma Xi, "Constitution 1886," https://www.sigmaxi.org/about/history.

4. Aristotle, *Nichomachean Ethics*, 1177a39.

5. Aristotle, *Nichomachean Ethics*, 1172a1–3.

6. Darwin, *Descent of Man*, 837.

7. I discuss Darwin's account of the evolution of the moral sense in detail in Robert T. Pennock, "Moral Darwinism: Ethical Evidence for the Descent of Man," *Biology & Philosophy* 10 (1995): 287–307.

8. Charles Darwin, *The Autobiography of Charles Darwin*, ed. Nora Barlow (New York: W. W. Norton, 1958), 139.

9. Charles Darwin, "Letter to T. H. Huxley," July 9, 1857. Darwin Correspondence Project, "Letter no. 2122," https://www.darwinproject.ac.uk/letter/DCP-LETT-2122.xml.

10. Leon Lederman, *The God Particle: If the Universe Is the Answer, What Is the Question?* (Boston: Houghton Mifflin, 1993), 7.

11. Bush, *Science: The Endless Frontier*, 1945.

12. Charles Darwin, "Letter to J. D. Hooker," May 10, 1848. Darwin Correspondence Project, "Letter no. 1174," http://www.darwinproject.ac.uk/DCP-LETT-1174.

13. Charles Darwin, "Letter to H. E. Strickland," February 4, 1849. Darwin Correspondence Project, "Letter no. 1221," http://www.darwinproject.ac.uk/DCP-LETT-1221.

14. Charles Darwin, "Letter to W. D. Fox." October 24, 1852. Darwin Correspondence Project, "Letter no. 1489," http://www.darwinproject.ac.uk/DCP-LETT-1489.

15. Janet Browne, *Charles Darwin: The Power of Place* (Princeton, NJ: Princeton University Press, 2002), 412.

16. Darwin, *Descent of Man*, 803.

17. Darwin, *Descent of Man*, 803.

18. Patrick M. S. Blackett, quoted in Barbara Lovett Cline, *Men Who Made a New Physics: Physicists and the Quantum Theory* (Chicago: University of Chicago Press, 1987), 21.

19. Zachary D. Blount, Christina Z. Borland, and Richard E. Lenski, "Historical Contingency and the Evolution of a Key Innovation in an Experimental Population of *Escherichia coli*," *Proceedings of the National Academy of Sciences* 105, no. 23 (June 2008): 7899–7906, https://doi.org/10.1073/pnas.0803151105.

20. Zachary D. Blount and Richard E. Lenski, "Ecological Divergence and Incipient Speciation in an Experimental Population of E. coli," Forthcoming.

21. Keller, *Feeling for the Organism*, 69.

22. Darwin, *Autobiography*, 78.

23. Darwin, *Autobiography*, 79.

24. Darwin, *Autobiography*, 79.

25. Albert Einstein, quoted in Abraham Pais, *Subtle Is the Lord: The Science and Life of Albert Einstein* (Oxford: Oxford University Press, 2005), 257.

26. Edward Eickstead Lane, quoted in W. W. Bartley III, "What Was Wrong with Darwin?," *New York Review of Books*, September 15, 1977, 37.

27. Alfred Russel Wallace, quoted in James Marchant, *Alfred Russel Wallace: Letters and Reminiscences*, vol. 1 (London: Cassell, 1916), http://www.gutenberg.org/files/15997/15997-h/15997-h.htm.

28. Albert Einstein, quoted in Alice Calaprice, *The Ultimate Quotable Einstein* (Princeton, NJ: Princeton University Press, 2011), 20.

29. Robert Oppenheimer, quoted in *Gaither's Dictionary of Scientific Quotations*, ed. Carl C. Gaither and Alma E. Cavazos-Gaither (New York: Springer, 2012), 267.

30. David Hume, *Treatise of Human Nature*, ed. L. A. Selby-Bigge (Oxford: Oxford University Press, 1978 [1888]), 416.

31. Hume, *Treatise of Human Nature*, 415.

32. John Nobel Wilford, "Mary Leakey, 83, Dies; Traced Human Dawn, *New York Times*, December 10, 1966 (emphasis added), http://www.nytimes.com/1996/12/10/world/mary-leakey-83-dies-traced-human-dawn.html.

33. David Hume, "Of the Standard of Taste," in *The Philosophical Works of David Hume*, vol. 3, ed. T. H. Green and T. H. Grose (London: Longman, Green, 1874–1875).

34. Henry David Thoreau, "October 11, 1840," in *The Journal of Henry David Thoreau: 1837–1861*, ed. Damion Searls (New York: New York Review of Books, 2009).

35. Thoreau, *Journal of Henry David Thoreau*, 422.

36. Thoreau, *Journal of Henry David Thoreau*, 423.

37. Thoreau, *Journal of Henry David Thoreau*, 68.

38. Henry David Thoreau, *Thoreau: Walden and Other Writings*, ed. Joseph Wood Krutch (New York: Bantam Books, 1962), 241.

39. Thoreau, *Journal of Henry David Thoreau*, 187.

40. Thoreau, *Thoreau: Walden*, 316.

41. Thoreau, *Thoreau: Walden*, 316.

42. Thoreau, *Thoreau: Walden*, 316.

43. Thoreau, *Thoreau: Walden*, 319.

44. Thoreau, *Thoreau: Walden*, 319.

45. John Tyndall, "Address Delivered before the British Association Assembled at Belfast, with Additions" (London: Longmans, Green, 1874), 43, http://www.victorian web.org/science/science_texts/belfast.html.

46. Tyndall, "Address Delivered before the British Association," 44.

47. Tyndall, "Address Delivered before the British Association," 44.

48. Tyndall, "Address Delivered before the British Association," 44.

49. Charles Darwin, "Prefatory Notice," in Hermann Müller, *The Fertilisation of Flower*, trans. and ed. D'Arcy W. Thompson, with a preface by Charles Darwin (London: Macmillan, 1883), vii-x.

Chapter 5

1. Darwin, *Autobiography*, 126.

2. Einstein, "Motives for Research."

3. Max Weber, "Science as a Vocation," in *From Max Weber: Essays in Sociology*, ed. and trans. H. H. Gerth and C. Wright Mills (New York: Oxford University Press, 1946), 135.

4. Weber, "Science as a Vocation," 134–135.

5. Einstein, "Motives for Research."

6. Einstein, "Motives for Research."

7. Darwin, *Diary of the Voyage of the Beagle*.

8. Charles Darwin, *Charles Darwin's Beagle Diary*, ed. Richard Darwin Keynes (Cambridge: Cambridge University Press, 1988), 44.

9. Darwin, *Autobiography*, 78.

10. Adrian Desmond and James Moore, *Darwin: The Life of a Tormented Evolutionist* (London: W. W. Norton, 1991), 352.

11. Charles Darwin, "Geology," in *A Manual of Scientific Enquiry Prepared for the Use of Her Majesty's Navy and Adapted for Travellers in General*, ed. John F. W. Herschel (London: John Murray, 1849), 195.

12. Darwin, "Geology," 156.

13. Aristotle, *Nichomachean Ethics*, 1110a.

14. Aristotle, *Nichomachean Ethics*, 1003b30.

15. John Dewey, *Human Nature and Conduct: An Introduction to Social Psychology* (New York: Henry Holt, 1922), 25.

16. Dewey, *Human Nature and Conduct*, 27.

17. Dewey, *Human Nature and Conduct*, 28.

18. Dewey, *Human Nature and Conduct*, 31.

19. Eleanor Rosch, "Natural Categories," *Cognitive Psychology* 4 (1973): 328–350; Paul Kay et al., *World Color Survey* (Stanford, CA: Center for the Study of Language and Information Publications, 2009).

20. Dewey, *Human Nature and Conduct*, 32.

21. Darwin, *Autobiography*, 144.

22. Darwin, *Autobiography*, 137–138.

23. Simon Baron-Cohen et al., "The Autism Spectrum Quotient (AQ): Evidence from Asperger Syndrome/High-Functioning Autism, Males and Females, Scientists and Mathematicians," *Journal of Autism and Developmental Disorders* 31, no. 1 (February 2001): 5–17.

24. Pennock, "Research Funding and the Virtue of Scientific Objectivity."

25. Lord Kelvin, "Lecture to the Institution of Civil Engineers," May 3, 1883.

26. Robert K. Merton, David L. Sills, and Stephen M. Stigler, "The Kelvin Dictum and Social Science: An Excursion into the History of an Idea," *Journal of the History of the Behavioral Sciences* 20 (October 1984): 319–331.

27. Darwin, *Origin of Species*, 629.

28. Darwin, *Autobiography*, 145.

29. Darwin, *Autobiography*, 141.

30. Charles Darwin, "Letter to Dr. Abbott," November 16, 1871, quoted in *The Life and Letters of Charles Darwin*, ed. Francis Darwin (New York: D. Appleton and Sons, 1919), https://www.gutenberg.org/files/2087/2087-h/2087-h.htm.

31. Browne, *Charles Darwin*, 400.

32. Preferred terminology changes over time; my colleague Richard Bellon has a wonderful book that examines how Victorian science and religion understood the virtue of patience, and subtle differences and interconnections between it, perseverance, and humility. Richard Bellon, *A Sincere and Teachable Heart: Self-Denying Virtue in British Intellectual Life, 1736–1859* (Leiden, The Netherlands: Koninklijke Brill, 2015).

33. Cynthia Dwork, quoted in Kevin Harnett, "How Humans Can Force the Machines to Play Fair," *Wired*, November 26, 2016, https://www.wired.com/2016/11/humans -can-force-machines-play-fair/.

34. John Dalton, quoted in Arnold Thackray, *John Dalton: Critical Assessments of His Life and Science* (Cambridge, MA: Harvard University Press, 1833), 175.

35. William Whewell, "On the Influence of the History of Science upon Intellectual Education," in *Lectures on Education Delivered at the Royal Institution of Great Britain* (London: John W. Parker and Son, 1854), 6.

36. Whewell, "On the Influence of the History of Science," 20.

37. Carlo Rubbia, "Asking Nature," in *Passionate Minds*, ed. Lewis Wolpert and Alison Richards (Oxford: Oxford University Press, 1977), 196.

38. Ernst Straus, "Memoir," in *Einstein: A Centenary Volume*, ed. A. P. French (Boston: Harvard University Press, 1979), 31–32.

39. Greg Ross, "First Things First," *Futility Closet*, April 8, 2014, http://www.futility closet.com.

Chapter 6

1. Imre Lakatos, *Proofs and Refutations: The Logic of Mathematical Discovery*, ed. John Worrall and Elie Zahar (Cambridge: Cambridge University Press, 1976), 30.

2. Steven Weinberg, *Dreams of a Final Theory* (New York: Pantheon Books, 1992), 13.

3. Darwin, *Autobiography*, 141.

4. Darwin, *Autobiography*, 145.

5. Charles Sanders Peirce, "The Scientific Attitude and Fallibilism," in *Philosophical Writings of Peirce*, ed. Justus Buchler (New York: Dover, 1955), 59.

6. Darwin, *Autobiography*, 141.

7. Richard Dawkins, *A Devil's Chaplin: Reflections on Hope, Lies, Science, and Love* (Boston: Houghton Mifflin, 2009), 59–60; Emma Townsend, "Richard Dawkins: Strident, Do They Mean Me?," *Independent*, October 4, 2009, www.independent.co .uk/arts-entertainment/books/features/richard-dawkins-strident-do-they-mean-me -1796244.html.

8. Lord Kelvin, quoted in D. H. Saxon, "Preface: In Appreciation of Lord Kelvin," in *Kelvin, Thermodynamics and the Natural World*, ed. M. W. Collins et al. (Southampton: WIT, 2016), xxii.

9. Einstein, "Foreword," xv.

10. Galileo Galilei, quoted in François Arago, *Biographies of Distinguished Scientific Men*, trans. Baden Powell, Robert Grant, and William Fairbairn (Boston: Ticknor & Fields, 1859), 365, https://quod.lib.umich.edu/m/moa/AGG6094.0001.001/377?rgn=full+text;view=image.

11. Isaac Newton, *Principia*, trans. Andrew Motte (Berkeley: University of California Press, 1934), xviii.

12. Richard Feynman, *The Character of Physical Laws* (Boston: MIT Press, 1967), 58.

13. Robert Merton, *On the Shoulders of Giants: A Shandean Postscript* (Chicago: University of Chicago Press, 1965).

14. Lakatos, *Proofs and Refutations*, 30.

15. Philosophers of science will find most of the next few sections on confirmation theory to be familiar, but I have tried to explain the positions and arguments in sufficient detail that readers who are not familiar with philosophy of science can get a sense of the issues. However, complete newcomers may find this compact overview to be too dense and should feel free to jump to the conclusions if the details are not of interest.

16. Karl Popper, *The Logic of Scientific Discovery* (New York: Harper & Row, 1959).

17. W. V. Quine, "Main Trends in Recent Philosophy: Two Dogmas of Empiricism," *Philosophical Review* 60, no. 1 (1951): 20–43; Pierre Maurice Marie Duhem, *The Aim and Structure of Physical Theory* (Princeton, NJ: Princeton University Press, 1954).

18. Robert T. Pennock, "Bayesianism, Ravens and Evidential Relevance," *Annals of the Japan Association for Philosophy of Science* 13, no. 1 (2004): 1–26.

19. Robert T. Pennock, "Evidential Relevance and the Grue Paradox," *Philosophy of Science—Japan* 31, no. 1 (1998): 101–119.

20. Wesley Salmon, *Scientific Explanation and the Causal Structure of the World* (Princeton, NJ: Princeton University Press, 1984).

21. Robert T. Pennock, "Epistemic and Ontic Theories of Explanation and Confirmation," *Philosophy of Science—Japan* 28 (1995): 31–45.

22. Thomas Kuhn, *The Structure of Scientific Revolutions* (Chicago: University of Chicago Press, 1962), 5.

23. Thomas Kuhn, *The Essential Tension: Selected Studies in Scientific Tradition and Change* (Chicago: University of Chicago Press, 1977).

24. Kuhn, *Essential Tension*, 356.

25. Kuhn, *Essential Tension*, 357.

26. Kuhn, "Objectivity, Value Judgment, and Theory Choice," in *Essential Tension*, 320–339.

27. Kuhn, "Objectivity, Value Judgment, and Theory Choice," 358; Thomas Kuhn, "Afterwords," in *World Changes: Thomas Kuhn and the Nature of Science*, ed. Paul Horwich (Cambridge, MA: MIT Press, 1993), 338.

28. Kuhn, "Objectivity, Value Judgment, and Theory Choice," 359.

29. Kuhn, "Objectivity, Value Judgment, and Theory Choice," 362.

30. Kuhn, "Objectivity, Value Judgment, and Theory Choice," 362–363.

31. Kuhn, "Objectivity, Value Judgment, and Theory Choice," 363.

32. Kuhn, "Objectivity, Value Judgment, and Theory Choice," 367.

33. Kuhn, "Objectivity, Value Judgment, and Theory Choice," 367.

34. Kuhn, "Objectivity, Value Judgment, and Theory Choice," 367.

35. Kuhn, "Objectivity, Value Judgment, and Theory Choice," 367.

36. Kuhn, "Objectivity, Value Judgment, and Theory Choice," 365.

37. Imre Lakatos, "Falsification and the Methodology of Scientific Research Programs," in *Criticism and the Growth of Knowledge*, ed. Imre Lakatos and Alan Musgrave (Cambridge: Cambridge University Press, 1970), 91–195.

38. Kuhn, "Objectivity, Value Judgment, and Theory Choice," 363.

39. Richard Dawkins, "Science, Delusion and the Appetite for Wonder," *Reports of the National Center for Science Education* 17, no. 1 (January/February 1997): 8–14.

40. Dawkins, "Science, Delusion and the Appetite for Wonder," 8–14.

41. Chuck Salter, "Jeffrey Wigand: The Whistle-Blower," *Fast Company*, April 30, 2001, https://www.fastcompany.com/65027/jeffrey-wigand-whistle-blower2002.

42. James Stephens, *The Crock of Gold* (Charleston: BiblioBazaar, 2006 [1912]), 13.

43. Victor Hugo, *Ninety-Three*, trans. Aline Delano (Boston: Little, Brown, 1889).

44. Hugo, *Ninety-Three*.

45. Marie Curie, quoted in *Today in Science History*, https://todayinsci.com/C/Curie_Marie/CurieMarie-Quotations.htm.

46. Feynman, *Pleasure of Finding Things Out*.

47. Feynman, *Pleasure of Finding Things Out*.

48. Plato, *Apology*, trans. Benjamin Jowett, The Project Gutenberg Edition, http://www.gutenberg.org/files/1656/1656-h/1656-h.htm.

49. Thomas Henry Huxley, "Letter to Charles Kingsley," September 23, 1860, in *Life and Letters of Thomas Henry Huxley* (New York: Appleton, 1900), 235.

Chapter 7

1. Joseph Wood Krutch, *The Modern Temper: A Study and a Confession* (New York: Harcourt, Brace, 1956 [1929]).

2. George Prothero, quoted in "The Late Mr. Darwin," *The Illustrated London News*, April 29, 1882, 418.

3. Robert T. Pennock, "Following Humbly Where Nature Leads: How Scientists Embody Humility," *Slate*, What Do We Know? Essays and Opinions, August 10, 2015.

4. Genesis 1:1 (King James Version).

5. Thomas Henry Huxley, "Letter to F. Dyster," January 30, 1859, in *The Life and Letters of Thomas Henry Huxley* (New York: Appleton, 1900).

6. Andrew Dickson White, *A History of the Warfare of Science with Theology in Christendom* (New York: George Braziller, 1896).

7. White, *History of the Warfare*, 42.

8. John William Draper, *History of the Conflict between Religion and Science* (New York: D. Appleton, 1874).

9. Andrew Dickson White, "Letter to Gerrit Smith," September 1, 1862, http://rmc.library.cornell.edu/cornell150/exhibition/white/AndrewDWhitelettergerritSmith.pdf.

10. White, *History of the Warfare*, 42.

11. White, *History of the Warfare*, vi.

12. White, *History of the Warfare*, xii.

13. Martin Luther, quoted in Ralph M. Blake, Curt J. Ducasse, and Edward H. Madden, *Theories of Scientific Method: The Renaissance through the Nineteenth Century* (Seattle: University of Washington Press, 1966), 29–30.

14. Robert Bellarmine, quoted in Blake, Ducasse, and Madden, *Theories of Scientific Method*, 29–30.

15. Bellarmine, quoted in Blake, Ducasse, and Madden, *Theories of Scientific Method*, 29–30.

16. Galileo, *Considerazioni circa l'opiniorne copernicana*, in Le opere, V, 351 ff, quoted in *Theories of Scientific Method*, 45.

17. Galileo Galilei, "Letter to the Grand Duchess Christina of Tuscany," 1615, in *Internet History Sourcebooks Project*, ed. Paul Halsall, https://sourcebooks.fordham .edu/mod/galileo-tuscany.asp.

18. Robert T. Pennock, "The Pre-modern Sins of Intelligent Design," in *Oxford Handbook of Science and Religion*, ed. Phillip Clayton (Oxford: Oxford University Press, 2006), 732–748.

19. Robert Boyle, "Notion of Nature," quoted in J. E. McGuire, "Boyle's Conception of Nature," *Journal of the History of Ideas* 33, no. 4 (Oct.–Dec. 1972): 533.

20. Boyle wrote: "It more sets off the wisdom of God in the fabric of the universe, that he can make so vast a machine perform all those many things, which he designed it should, by the mere contrivance of brute matter managed by certain laws of local motion and upheld by his ordinary and general concourse, than if he employed from time to time an intelligent overseer, such as nature is fancied to be, to regulate, assist, and control the motions of the parts." Robert Boyle, "Free Inquiry," quoted in Timothy Shanahan, "God and Nature in the Thought of Robert Boyle," *Journal of the History of Philosophy* 26 (1988): 558.

21. Charles Kingsley, quoted in Darwin, *On the Origin of Species*, 2nd ed., chap. 14.

22. Hume, *Enquiries*, 270.

23. 2 Corinthians 5:7 (King James Version).

24. Hebrews 11:1 (King James Version).

25. For this perspective, see Judith Andre, *Worldly Virtue: Moral Ideals and Contemporary Life* (Lanham, MD: Lexington Books, 2015).

26. Albert Einstein, "The World as I See It," *Forum and Century* 84 (1931): 193–194, in Albert Einstein, *Ideas and Opinions* (New York: Bonanza Books, 1954), 8–11.

27. Tertullian, *De praescriptione baereticorum*, VII, *Tertulliani opera*, ed. Nic. Rigaltius (Paris, 1664), 204–205, quoted in David C. Lindberg and Ronald L. Numbers, *When Science & Christianity Meet* (Chicago: University of Chicago Press, 2003), 11.

28. Augustine, *Confessions*, in *The Basic Writings of Saint Augustine*, trans. J. G. Pilkington, ed. Whitney J. Oakes (New York: Random House, 1948), X.35, vol. 1, 174 (with one change of wording), and IV.16, 56.

29. Martin Luther, *A Commentary on St. Paul's Letter to the Galatians* (London: James Cundee, 1807), 149.

30. 1 Corinthians 1:25 (King James Version).

31. Galileo Galilei, "Letter to the Grand Duchess Christina of Tuscany," 1615, https://sourcebooks.fordham.edu/halsall/mod/galileo-tuscany.asp.

32. Alan Cowell, "After 350 Years, Vatican Says Galileo Was Right: It Moves," *New York Times*, October 31, 1992, https://www.nytimes.com/ ... /after-350-years-vatican-says-galileo-was-right-it-moves.html.

33. Art Swift, "In US, Belief in Creationist View of Humans Hits a New Low," Gallup, May 22, 2017.

34. Robert K. Merton, *Science, Technology and Society in Seventeenth-Century England* (New York: Harper & Row, 1970).

35. Steven Shapin, "Understanding the Merton Thesis," *Isis* 79, no. 4 (1988): 598.

36. Albert Einstein, "Science and Religion," in *Ideas and Opinions* (New York: Bonanza, 1954), 41–49, 46.

37. Martin Luther King, Jr., "A Tough Mind and a Tender Heart," in *Strength to Love* (Minneapolis: Fortress, 1977 [1959]).

38. Ian Barbour, *Religion in an Age of Science*, vol. 1 (San Francisco: Harper-Collins, 1990).

39. Dalai Lama, *The Universe in a Single Atom: The Convergence of Science and Spirituality* (New York: Morgan Road, 2005), 76.

40. Richard Rohr, *Religion & Ethics Newsweekly*, PBS Interview, WPSU, 2011, http://www.pbs.org/wnet/religionandethics/2011/11/11/november-11-2011-richard-rohr/9902/.

41. William James, *The Varieties of Religious Experience* (Cambridge, MA: Harvard University Press, 1985 [1902]), 7.

42. D. Elton Trueblood, "A Radical Experiment," 1947, http://www.quaker.org/pamphlets/wpl1947a.html.

43. Geoffrey Cantor, *Quakers, Jews, and Science: Religious Responses to Modernity and the Sciences in Britain, 1650–1900* (Oxford: Oxford University Press, 2005).

44. Frans de Waal, *The Bonobo and the Atheist: In Search of Humanism among the Primates* (New York: W. W. Norton, 2013), chap. 4.

45. David Sloan Wilson, *Evolution for Everyone: How Darwin's Theory Can Change the Way We Think about Our Lives* (New York: Delta, 2007), 320–321.

46. Nancy Pearcey, *Total Truth: Liberating Christianity from Its Cultural Captivity* (Wheaton, IL: Crossway, 2004).

47. Francis Schaeffer, "Address at the University of Notre Dame," April 1981.

48. Bertrand Russell, *Religion and Science* (Oxford: Oxford University Press, 1935), 14.

49. Maria Mitchell, quoted in Helen Wright, *Sweeper in the Sky: The Life of Maria Mitchell, First Woman Astronomer in America* (New York: Macmillan, 1949), 168.

50. Alexander Pope, *An Essay on Man*, ed. Henry Morley (London: Cassell, 1891).

51. Pope, *Essay on Man*.

52. Plato, *Phaedrus*, 229c–230a.

53. Pope, *Essay on Man*.

54. Erasmus Darwin, quoted in "A Fool Defined," *Medical and Surgical Reporter* 55 (August 14, 1886): 224.

55. Charles Darwin, in E. Ray Lankester, "Charles Robert Darwin," in *Library of the World's Best Literature, Ancient and Modern*, vol. 2, ed. C. D. Warner (New York: R. S. Peale & J. A. Hill, 1896), 4385–4393, http://darwin-online.org.uk/content/frameset? itemID=F2113&viewtype=text&pageseq=1.

56. Jean-François Lyotard, *The Postmodern Condition: A Report on Knowledge* (Manchester: Manchester University Press, 1979).

57. Bruno Latour and Steve Woolgar, *Laboratory Life: The Social Construction of Scientific Facts* (Princeton, NJ: Princeton University Press, 1986): 235–236.

58. Bruno Latour, "For David Bloor…and Beyond: A Reply to David Bloor's 'Anti-Latour,'" *Studies in the History and Philosophy of Science* 30, no. 1 (1999): 113–129.

59. Paul R. Gross and Norman Levitt, *Higher Superstition: The Academic Left and Its Quarrels with Science* (Baltimore: Johns Hopkins University Press, 1994).

60. For some representative examples see the articles in Noretta Koertge, *A House Built on Sand: Exposing Postmodernist Myths about Science* (New York: Oxford University Press, 1998).

61. Alan D. Sokal, "Transgressing the Boundaries: Towards a Transformative Hermeneutics of Quantum Gravity," *Social Text* 46/47, Science Wars (1996): 217.

62. Keith Parsons, *The Science Wars: Debating Scientific Knowledge and Technology* (Amherst, NY: Prometheus, 2003); Robert T. Pennock, *Tower of Babel: The Evidence against the New Creationism* (Cambridge, MA: MIT Press, 1999).

63. Philip E. Johnson, "The Creationist and the Sociobiologist: Two Stories about Illiberal Education," *California Law Review* 80, no. 4 (July 1992): 1071–1090.

64. Philip E. Johnson, "Paul Feyerabend's Choice for Freedom," review of *Killing Time, the Autobiography of Paul Feyerabend, Books and Culture*, March/April 1996.

65. I trace the postmodern roots of IDC in Robert T. Pennock, "The Postmodern Sin of Intelligent Design Creationism," *Science & Education* 19, nos. 6–8 (2010): 757–778.

66. John 1:1 (King James Version).

67. Kitzmiller v. Dover Area School Dist., 400 F. Supp. 2d 707 (M.D. Pa. 2005).

68. Jacob Bronowski, quoted in J. Wakeman, *World Authors 1950–1970* (New York: H. W. Wilson, 1975), 221–223.

69. Sune K. Bergström, "Banquet Speech Accepting Nobel Prize in Physiology or Medicine," December 10, 1982, in *Les Prix Nobel. The Nobel Prizes*, ed. Wilhelm Odelberg (Stockholm: Nobel Foundation, 1969).

70. G. H. Hardy, *A Mathematician's Apology* (Edmonton: University of Alberta Mathematical Sciences Society, 1940), 12.

71. Mark Littmann, "Courses in Science Writing as Literature," *Public Understanding of Science* 14 (2005): 103–112.

72. Keller, *Feeling for the Organism*, xii.

73. Dean Simonton, *Creativity in Science: Chance, Logic, Genius, and Zeitgeist* (Cambridge: Cambridge University Press, 2004).

74. We will see an example of this error from the scientific side in the next chapter when we discuss gluttonous reductionism.

75. Thomas Henry Huxley, "Reflection #296," in *Aphorisms and Reflections from the Works of T. H. Huxley*, ed. Thomas Henry Huxley and Henrietta A. Huxley (London: Macmillan, 1907), 143.

76. Rachel Carson, "National Book Award Address" (Hotel Commodore, January 27, 1952), quoted in Linda Lear, *Rachel Carson: Witness for Nature* (Boston: Houghton Mifflin Harcourt, 1997), 219.

77. Jack Selzer, ed., *Understanding Scientific Prose* (Madison: University of Wisconsin Press, 1993).

78. Philip Steadman, *Vermeer's Camera: Uncovering the Truth behind the Masterpieces* (Oxford: Oxford University Press, 2002).

79. Steadman, *Vermeer's Camera*.

80. David Hockney, *Secret Knowledge: Rediscovering the Lost Techniques of Old Masters* (New York: Avery, 2006).

81. Kurt Anderson, "Reverse-Engineering a Genius (Has a Vermeer Mystery Been Solved?)," *Vanity Fair*, November 29, 2013.

82. Teller, quoted in Anderson, "Reverse-Engineering a Genius (Has a Vermeer Mystery Been Solved?)," *Vanity Fair*, November 29, 2013.

83. Put another way, postmodernism does not remove privilege, but bases its epistemology upon it, leading to an Orwellian endless war of all against all.

84. Albert Einstein, "A Mathematician's Mind," in *Ideas and Opinions* (New York: Bonanza, 1954), 26.

85. Max Delbrück, "Nobel Lecture," The Nobel Foundation, 1969, https://www .nobelprize.org/nobel_prizes/medicine/laureates/1969/delbruck-lecture.html.

86. Joseph Wood Krutch, *The Modern Temper: A Study and a Confession* (New York: Harcourt, Brace, 1956 [1929]).

87. Carl Sagan, *Contact* (New York: Simon & Schuster, 1985).

88. Sinclair Lewis, *Arrowsmith* (New York: Grosset & Dunlap, 1925).

89. American Friends Service Committee, *Speak Truth to Power: A Quaker Search for an Alternative to Violence* (Philadelphia: American Friends Service Committee, 1955).

90. Max Horkheimer, *Critical Theory: Selected Essays* (New York: Continuum, 1982), 244.

91. American Friends Service Committee, *Speak Truth to Power*, 35.

92. Henry David Thoreau, "On the Duty of Civil Disobedience," 1849, www.guten berg.org/files/71/71-h/71-h.htm.

Chapter 8

1. Edmund Husserl, *Ideas: General Introduction to Pure Phenomenology*, trans. W. R. Boyce Gibson (London: George Allen & Unwin, 1931), 86.

2. Genesis 3:5 (King James Version).

3. Murray Gell-Mann, *The Quark and the Jaguar: Adventures in the Simple and the Complex* (New York: W. H. Freeman, 1994), 80.

4. Gell-Mann, *Quark and the Jaguar*, 80.

5. Fanelli, "How Many Scientists Fabricate and Falsify Research?"

6. Kenneth E. Goodpaster, *Conscience and Corporate Culture* (Malden, MA: Blackwell, 2007).

7. J. E. Wikström, "Opening Remarks," in *Ethics for Science Policy: Proceedings of a Nobel Symposium Held at Södergarn, Sweden, August 20–25 1979*, ed. Torgny Segerstedt (Oxford: Pergamon, 1979), xiii.

8. Mark Fiege, "The Atomic Scientists, the Sense of Wonder, and the Bomb," *Environmental History* 12, no. 3 (2007): 581, https://doi.org/10.2307/25473133.

9. Robert Oppenheimer, "Testimony before United States Atomic Energy Commission," April 13, 1954, 95/266, http://www.osti.gov/includes/opennet/includes/Oppen heimer%20hearings/Vol%20II%20Oppenheimer.pdf.

10. Irving Langmuir, "Banquet Speech," The Nobel Foundation, 1932, https://www.nobelprize.org/nobel_prizes/chemistry/laureates/1932/langmuir-speech.html.

11. Kurt Vonnegut, *Cat's Cradle* (New York: Delacourt, 1963), 20.

12. Robert K. Musil, "There Must Be More to Love Than Death: A Conversation with Kurt Vonnegut," *The Nation*, August 2–9, 1980.

13. Musil, "There Must Be More."

14. Alexander Pope, *The Works of Alexander Pope*, vol. 5, ed. Dr. Johnson (London: Strahan and Preston, 1812), 231.

15. Oliver Goldsmith, *The Works of Oliver Goldsmith*, vol. 2, ed. Peter Cunningham (New York: Harper and Brothers, 1881), 117.

16. Thomas Hobbes, *Leviathan, or The Matter, Form and Power of a Commonwealth Ecclesiastical and Civil*, ed. Michael Oakeshott (New York: Collier, 1977), 51.

17. Kevin Mitnick, "Kevin Mitnick Loved the Cyberspace Break-in Game," interview by Peter Finnocchiaro, *Salon*, August 20, 2011, https://www.salon.com/2011/08/20/kevin_mitnick_interview/.

18. Aristotle, *Nichomachean Ethics*, 1106b34–35.

19. Aristotle, *Nichomachean Ethics*, 1106b31.

20. Anastasious Ladikos, "Revisiting the Virtue of Courage in Aristotle," *Phronimon* 5, no. 2 (2004): 77–92.

21. Aristotle, *Nicomachean Ethics*, 1106b18–23.

22. Musil, "There Must Be More,"129–130.

23. Edward O. Wilson, *Naturalist* (Washington, DC: Island, 1994), 219.

24. Wilson, *Naturalist*, 219.

25. Darwin, *Autobiography*, 141.

26. Darwin, *Autobiography*, 142.

27. Albert Einstein, "Letter to Jost Winteler," July 8, 1901, in Isaacson, *Einstein*, 22.

28. Albert Einstein, "Letter to Mileva Maric," December 12, 1901," in Isaacson, *Einstein*, 7.

29. C. A. Timiazeff, *The Life of Plants*, trans. Anna Sheremeteva (London: Longmans, Green, 1912), 6.

30. Enrico Fermi, quoted in Robert Jungk, *Brighter Than a Thousand Suns: A Personal History of the Atomic Scientists* (Boston: Houghton Mifflin Harcourt, 1958), 202.

31. Charles Babbage, quoted in Philip and Emily Morrison, *Charles Babbage and His Calculating Engines: Selected Writings by Charles Babbage and Others* (New York: Dover, 1961), xxiii.

32. Robert T. Pennock, "Naturalism, Evidence and Creationism: The Case of Phillip Johnson," *Biology and Philosophy* 11, no. 4 (1996): 543–559.

33. Robert T. Pennock, "God of the Gaps: The Argument from Ignorance and the Limits of Methodological Naturalism," in *Scientists Confront Creationism and Intelligent Design*, ed. Andrew Petto and Laurie Godfrey (New York: W. W. Norton, 2007), 309–338.

34. Francis Crick, *The Astonishing Hypothesis: The Scientific Search for the Soul* (New York: Scribner, 1994), 3.

35. Pennock, "Research Funding and the Virtue of Scientific Objectivity."

36. In this section and the next, I repeat a diagnosis and recommendation I gave in Robert T. Pennock, "Fostering a Culture of Scientific Integrity: Legalistic vs. Scientific Virtue-Based Approaches," *Professional Ethics Report* 28, no. 2 (2015): 1–3.

37. E.g., C. K. Gunsalus, "Institutional Structure to Ensure Research Integrity," *Academic Medicine* 68, no. 9 (1993): S33–S38; Ronald Atlas, "Responsible Conduct by Life Scientists in an Age of Terrorism," *Science and Engineering Ethics* 15, no. 3 (2009): 292–301.

38. Sandra L. Titus, James A. Wells, and Lawrence J. Rhoades, "Repairing Research Integrity," *Nature* 453, no. 7198 (2008): 981–982.

39. Titus, Wells, and Rhoades, "Repairing Research Integrity," 980.

40. Titus, Wells, and Rhoades, "Repairing Research Integrity," 980–982.

41. Nicholas H. Steneck, "Research Universities and Scientific Misconduct: History, Policies, and the Future," *Journal of Higher Education* 65, no. 3 (1994): 310–330; Nicholas H. Steneck and Ruth Ellen Bulger, "The History, Purpose, and Future of Instruction in the Responsible Conduct of Research," *Academic Medicine* 82, no. 9 (2007): 829–834.

42. Robert T. Pennock, "Scientific Integrity and Science Museums," *Museums and Social Issues* 1, no. 1 (2006): 7–18.

43. I will take up the relationship of scientific flourishing to human flourishing in chapter 10.

44. One advantage of the scientific virtue approach is that it provides a way to systematize some of these disparate aspects of the subject matter. A scientific virtue approach can be helpful in analyzing traditional issues in RCR such as just authorship attribution (e.g., Robert T. Pennock, "Inappropriate Authorship in Collaborative

Scientific Research," *Public Affairs Quarterly* 10, no. 4 [October 1996]: 379–393), socially controversial subjects such as human cloning (e.g., Robert T. Pennock, "The Virtuous Scientist Meets the Human Clone," in *New Ethical Challenges in Science and Technology,* Sigma Xi Forum 2000 Proceedings [2001]: 117–124), responsible research funding and conflict of interest (e.g., Pennock, "Research Funding and the Virtue of Scientific Objectivity"), and general issues such as the responsibility to defend the integrity of scientific methods (e.g., Pennock, "Scientific Integrity and Science Museums"). It also helps highlight other professional responsibilities that deserve greater attention, including peer review, dissemination, professional development, mentoring, and education. It helps make sense of interests and conflicts of interest. It can even help put issues of scientists' social responsibility (which also goes beyond the traditional legalistic framework) in a new light, as such issues involve relationships between scientific and broader human values. These and other aspects of the scientific virtue approach deserve further attention, but here my purpose was just to highlight its general utility for developing a culture of integrity.

45. Kenneth D. Pimple, "Six Domains of Research Ethics: A Heuristic Framework for the Responsible Conduct of Research," *Science and Engineering Ethics* 8, no. 2 (2002): 198.

46. This is not the place to discuss ways that virtue-based RCR training can be implemented. For individuals with a philosophical bent, one can begin from first principles and consider what follows from the virtue-theoretical concepts, but it can work equally well to organize the material around the lives of exemplary scientists—individuals with practical wisdom who may serve as role models in the manner that Aristotle suggested. See Robert T. Pennock and Michael O'Rourke, "Developing a Scientific Virtue-Based Approach to Science Ethics Training," *Science & Engineering Ethics* 23, no. 1 (2017): 243–262, https://doi.org/10.1007/s11948-016-9757-2.

47. Boeing, "MB-1/AIR-2 Genie Missile," http://www.boeing.com/history/products /mb-1-air-2-genie-missile.page.

48. Dwight D. Eisenhower, "Atoms for Peace—Address before the General Assembly of the United Nations on Peaceful Uses of Atomic Energy," December 8, 1953, http://www.presidency.ucsb.edu/ws/?pid=9774.

Chapter 9

1. Albert Einstein, "Letter to Otto Juliusburger," September 29, 1942. Albert Einstein Archives #38–238, quoted in Isaacson, *Einstein*, 15.

2. Anatole France, *The Crime of Sylvestre Bonnard*, 1860, http://www.gutenberg.org /files/2123/2123-h/2123-h.htm.

3. Richard Feynman, "What Is and What Should Be the Role of Scientific Culture in Modern Society," in *The Pleasure of Finding Things Out: The Best Short Works of Richard Feynman*, ed. Jeffrey Robbins (Cambridge, MA: Helix/Perseus, 1999), 103.

4. National Research Council, *Successful K-12 STEM Education: Identifying Effective Approaches in Science, Technology, Engineering, and Mathematics* (Washington, DC: National Academies Press, 2011), https://doi.org/10.17226/13158.

5. Jon Miller, "The Conceptualization and Measurement of Civic Scientific Literacy for the Twenty-First Century," in *Science and the Educated American: A Core Component of Liberal Education*, ed. Jerrold Meinwald and John G. Hildebrand (Cambridge, MA: American Academy of Arts & Sciences, 2010), 241.

6. Miller, "Conceptualization and Measurement," 243.

7. Robert T. Pennock, "On Teaching Evolution and the Nature of Science," in *Evolutionary Science and Society: Educating a New Generation*, ed. Joel Cracraft and Roger Bybee, (Colorado Springs, CO: *Biological Sciences Curriculum Study*, 2005), 1–12.

8. E.g., Carol A. Brewer and Diane Smith, eds., *Vision & Change in Undergraduate Biology Education: A Call to Action* (Washington, DC: American Association for the Advancement of Science, 2009), http://visionandchange.org/files/2013/11/aaas-VISchange-web1113.pdf; College Board, *AP Biology Curriculum Framework 2012–2013* (New York: College Board, 2011); Committee on a Conceptual Framework for New K-12 Science Education Standards, *A Framework For K-12 Science Education Practices, Crosscutting Concepts, and Core Ideas* (Washington, DC: National Academies Press, National Research Council, 2012); Next Generation Science Standards, 2013, http://www.nextgenscience.org/next-generation-science-standards.

9. Robert T. Pennock, "Models, Simulations, Instantiations and Evidence: The Case of Digital Evolution," *Journal of Experimental and Theoretical Artificial Intelligence* 19, no. 1 (2007): 29–42.

10. Elena B. Speth et al., "Using Avida-ED for Teaching and Learning about Evolution in Undergraduate Introductory Biology Courses," *Evolution Education & Outreach* 2, no. 3 (2009): 415–428; Amy Lark, Gail Richmond, and Robert T. Pennock, "Modeling Evolution in the Classroom: The Case of the Fukushima Butterflies," *American Biology Teacher* 76, no. 7 (2014): 450–454; Robert T. Pennock, "Experimental Evolution in the Classroom: The Active LENS Project," in *Envisioning the Future of Undergraduate STEM Education: Research and Practice Symposium*, April 2016, http://www.enfusestem.org/projects/experimental-evolution-in-the-classroom-the-active-lens-project/; James J. Smith et al., "An Avida-ED Digital Evolution Curriculum for Undergraduate Biology," *Evolution: Education and Outreach* 9, no. 9 (2016): 1–11; Amy M. Lark et al., "Exploring the Relationship between Experiences with Digital Evolution and Students' Scientific Understanding and Acceptance of Evolution," *American Biology Teacher* 80, no. 2 (February 2018): 74–86, https://doi.org/10.1525/abt.2018.80.2.74.

11. Plato, "Meno," in *Plato: The Collected Dialogues*, ed. Edith Hamilton and Huntington Cairns (Princeton, NJ: Princeton University Press, 1961), 354.

12. Allan Gotthelf, "Darwin on Aristotle," *Journal of the History of Biology* 32, no. 1 (Spring 1999): 3–30.

13. Charles Darwin, "Letter to William Ogle," February 22, 1882. Darwin Correspondence Project, "Letter no. 13,697," http://www.darwinproject.ac.uk/DCP-LETT -13697.

14. Robert T. Pennock, "Can Darwinian Mechanisms Make Novel Discoveries? Learning from Discoveries Made by Evolving Neural Networks," *Foundations of Science* 5, no. 2 (2000): 225–238.

15. MacIntyre, *After Virtue*, 191.

16. Edmund Burke, *A Philosophical Inquiry into the Origin of Our Ideas of the Sublime and Beautiful, with an Introductory Discourse concerning Taste and Several Other Additions*, §1, Pt. 1 (London: John Nimmo, 1887 [1756]), 101.

17. Samuel Coleridge, quoted in F. P. Lock, *Burke's Reflections on the Revolution in France* (Boston: Allen & Unwin, 1985), 174.

18. Leo Szilard, quoted in Fiege, *Atomic Scientists*, 584.

19. Robert N. Emde et al., "The Moral Self of Infancy: Affective Core and Procedural Knowledge," *Developmental Review* 11 (1991): 251–270.

20. Robert N. Emde, "From a Baby Smiling: Reflections on Virtues in Development," in *Developing the Virtues: Integrating Perspectives*, ed. Julia Annas, Darcia Narvaez, and Nancy E. Snow (Oxford: Oxford University Press, 2016), 69–94.

21. Emde et al., "The Moral Self of Infancy," 254.

22. Alison Gopnik, Andrew N. Meltzoff, and Patricia K. Kuhl, *The Scientist in the Crib: What Early Learning Tells Us about the Mind* (New York: William Morrow, 2000).

23. Albert Einstein, quoted in Isaacson, *Einstein*, 9.

24. Keller, *Feeling for the Organism*, 70.

25. Feynman, "Making of a Scientist," 17.

26. Darwin, *Autobiography*, 72.

27. Darwin, *Autobiography*, 78.

28. Darwin, *Autobiography*, 66.

29. Aristotle's sense of "scientific" knowledge is, as we have seen, somewhat different from ours. Aristotle, *Nichomachean Ethics*, VI, 1141b.

30. Julia Annas, "Virtue as a Skill," *International Journal of Philosophical Studies* 3, no. 2 (1995): 227–243.

31. Benjamin Franklin, *The Autobiography of Benjamin Franklin*, ed. Frank Wood-worth Pine (New York: Henry Holt, 1916).

32. Franklin, *Autobiography*.

33. John Dewey, "Science as Subject-Matter and as Method," *Science*, n.s., 31, no. 787 (January 28, 1910): 121–127.

34. Isidor Rabi, quoted in Donald Sheff, "Izzy, Did You Ask a Good Question Today?," opinion, *New York Times*, January 19, 1988.

35. Feynman, *Pleasure of Finding Things Out*.

36. Seymore Papert, quoted in "Seymour Papert, Artificial Intelligence Guru—Obituary," *The Telegraph*, August 11, 2016, http://www.telegraph.co.uk/obituaries/2016/08/11/seymour-papert-artificial-intelligence-guru--obituary/.

37. Avida-ED was primarily designed for undergraduates, but it can and has been used for graduate courses and also in high school. See Robert T. Pennock, "Learning Evolution and the Nature of Science Using Evolutionary Computing and Artificial Life," *McGill Journal of Education* 42, no. 2 (2007): 211–224.

38. Maria Mitchell, "The Need of Women in Science," *The Victoria Magazine* 28 (November–April, 1877): 187–192, 191–192.

39. Maria Mitchell, quoted in Margaret Moore Booker, *Among the Stars: The Life of Maria Mitchell; Astronomer, Educator, Women's Rights Activist* (Nantucket, MA: Mill Hill, 2007), 16.

40. Mitchell, quoted in Booker, *Among the Stars*, 274.

41. Booker, *Among the Stars*, 276.

42. Booker, *Among the Stars*, 285.

43. Mitchell, quoted in Booker, *Among the Stars*, 316.

44. Maria Mitchell, *Maria Mitchell: Life, Letters and Journals*, ed. Phebe Mitchell Kendall (1896).

45. Mitchell, *Maria Mitchell*.

46. Mitchell, "Need of Women in Science," 189.

47. Booker, *Among the Stars*, 310.

48. Booker, *Among the Stars*, 311.

49. Keller, *Feeling for the Organism*, 76–77.

50. Booker, *Among the Stars*, 309.

51. Mitchell, "Need of Women in Science," 189.

52. Booker, *Among the Stars*, 320.

Chapter 10

1. Freeman Dyson, *Disturbing the Universe* (New York: Basic Books, 1979), 6.

2. Anton Chekhov, *Note-Book of Anton Chekhov* (New York: B. W. Huebsch, 1921), 18.

3. Valery N. Soyfer, "The Consequences of Political Dictatorship for Russian Science," *Nature Reviews Genetics* 2, no. 9 (September 2001): 723–729.

4. Hume, *Treatise*, 252.

5. Hume, *Treatise*, 260.

6. Snow, *Two Cultures*, 14.

7. Frances Wright, *Course of Popular Lectures* (New York: Office of the Free Enquirer, 1829), 119, https://archive.org/details/courseofpopularl00wrigrich.

8. Celia Morris Eckhardt, *Fanny Wright: Rebel in America* (Cambridge, MA: Harvard University Press, 1984).

9. Orestes Augustus Brownson, *An Oration on Liberal Studies*, delivered before the Philomathian Society of Mount Saint Mary's College, June 29, 1853 (Baltimore: Hedian & O'Brien, 1853), 20–21.

10. Benjamin Jowett, quoted in Logan Pearsall Smith, *Unforgotten Years* (Boston: Little, Brown, 1938), https://www.gutenberg.ca/ebooks/smithlp-unforgottenyears/smithlp-unforgottenyears-00-e.html.

11. Jowett, quoted in Smith, *Unforgotten Years*.

12. Le Corbusier, quoted in Philippe Boudon, *Lived-in Architecture: Pessac Revisited*, trans. Gerald Onn (Boston: MIT Press, 1969).

13. Alistair Cooke, quoted in David Wilkinson, *Science, Religion, and the Search for Extraterrestrial Intelligence* (Oxford: Oxford University Press, 2013), 3.

14. William Samuel Johnson, "No. 150. Adversity Useful to the Acquisition of Knowledge," *The Rambler*, August 24, 1751, http://www.johnsonessays.com/the-rambler/no-150-adversity-useful-to-the-acquisition-of-knowledge/.

15. William Bruce Cameron, *Informal Sociology: A Casual Introduction to Sociological Thinking* (New York: Random House, 1963), 13.

16. Robert K. Merton, *The Sociology of Science: Theoretical and Empirical Investigations*, ed. Norman W. Storer (Chicago: University of Chicago Press, 1973), 332.

17. Merton, *Sociology of Science*, 332.

18. Merton, *Sociology of Science*, 333.

19. Aristotle, "On Virtues and Vices," in Jonathan Barnes, *The Complete Works of Aristotle*, vol. 2, trans. J. Solomon (Princeton, NJ: Princeton University Press, 1984), 8:26.

20. Charles Darwin, "Letter to Edward Bibbens Aveling," October 13, 1880. Darwin Correspondence Project, Letter no. 12757.

21. Paul Feyerabend, "How to Defend Society against Science," *Radical Philosophy* 11, no. 1 (January 1975): 4.

22. Feyerabend, "How to Defend Society against Science," 4.

23. Feyerabend, "How to Defend Society against Science," 4–5.

24. Feyerabend, "How to Defend Society against Science," 3–8.

25. Feyerabend, "How to Defend Society against Science," 4.

26. Feyerabend, "How to Defend Society against Science," 6.

27. Merton, "Science and the Social Order," 330–331.

28. Merton, *Sociology of Science*, 328.

29. Feyerabend, "How to Defend Society against Science," 7.

30. Feyerabend, "How to Defend Society against Science," 7.

31. Feyerabend, "How to Defend Society against Science," 7.

32. Feyerabend, "How to Defend Society against Science," 4.

33. Feyerabend, "How to Defend Society against Science," 8.

34. Feyerabend, "How to Defend Society against Science," 8.

35. John Horgan, "Was Philosopher Paul Feyerabend Really 'Science's Worst Enemy'?," Scientific American (October 24, 2016). https://blogs.scientificamerican.com/cross-check/was-philosopher-paul-feyerabend-really-science-s-worst-enemy/.

36. Ron Suskind, "Faith, Certainty and the Presidency of George W. Bush," *New York Times Magazine*, October 17, 2004, http://www.nytimes.com/2004/10/17/magazine/faith-certainty-and-the-presidency-of-george-w-bush.html.

37. Mark Danner, "Words in a Time of War: Taking the Measure of the First Rhetoric-Major President" (commencement address given to graduates of the Department of Rhetoric, University of California, Berkeley, 2007), http://web.archive.org/web/20150206043442/https://www.mtholyoke.edu/acad/intrel/speech/danner.htm.

38. Karl Popper, *The Open Society and Its Enemies*, vols. 1–2 (London: George Routledge and Sons, 1945).

39. Philip Kitcher, *Science, Truth, and Democracy* (Oxford: Oxford University Press, 2003); Philip Kitcher, *Science in a Democratic Society* (Amherst, NY: Prometheus, 2011).

40. Kitcher, *Science, Truth, and Democracy*, 149.

41. Sune K. Bergström, "Banquet Speech."

42. Robert Wilson, "R. R. Wilson's Congressional Testimony," April 1969, http://history.fnal.gov/testimony.html.

43. Wilson, "R. R. Wilson's Congressional Testimony."

44. George Washington, "First Annual Message to Congress on the State of the Union," January 8, 1790.

45. Dyson, *Disturbing the Universe*, 6.

46. Jawaharlal Nehru, *The Discovery of India* (Oxford: Oxford University Press, 1985 [1946]), 510.

47. Nehru, *Discovery of India*, 59.

48. Nehru, *Discovery of India*, 511.

49. Nehru, *Discovery of India*, 31.

50. Nehru, *Discovery of India*, 512.

51. India's Constitution of 1949 with Amendments through 2012, part IVA, §51Ah, http://extwprlegs1.fao.org/docs/pdf/ind132810.pdf.

52. Nehru, *Discovery of India*, 512.

Bibliography

American Friends Service Committee. *Speak Truth to Power: A Quaker Search for an Alternative to Violence.* Philadelphia: American Friends Service Committee, 1955.

Anderson, Kurt. "Reverse-Engineering a Genius (Has a Vermeer Mystery Been Solved?)." *Vanity Fair,* November 29, 2013.

Andre, Judith. *Worldly Virtue: Moral Ideals and Contemporary Life.* Lanham, MD: Lexington Books, 2015.

Annas, Julia. "Virtue as a Skill." *International Journal of Philosophical Studies* 3, no. 2 (1995): 227–243.

Arago, François. *Biographies of Distinguished Scientific Men.* Translated by Baden Powell, Robert Grant, and William Fairbairn. Boston: Ticknor & Fields, 1859.

Aristotle. *Nicomachean Ethics.* In *The Complete Works of Aristotle,* translated by Ross Urmson, edited by Jonathan Barnes. Princeton, NJ: Princeton University Press, 1985.

Aristotle. "On Virtues and Vices." In *The Complete Works of Aristotle,* Vol. 2, translated by J. Solomon, edited by Jonathan Barnes. Princeton, NJ: Princeton University Press, 1984.

Armitage, David, and Jo Guldi. "Bonfire of the Humanities." *Aeon Magazine,* October 2, 2014.

Atlas, Ronald. "Responsible Conduct by Life Scientists in an Age of Terrorism." *Science and Engineering Ethics* 15, no. 3 (2009): 292–301.

Augustine. *Confessions.* In *The Basic Writings of Saint Augustine,* translated by J. G. Pilkington, edited by Whitney J. Oakes. New York: Random House, 1948.

Bacon, Francis. *The Advancement of Learning.* London: MacMillan, 1898.

Bacon, Francis. *The New Organon, or True Directions Concerning the Interpretation of Nature.* In *The Philosophical Works of Francis Bacon,* edited by John M. Robertson, 256–387. London: George Routledge and Sons, 1905.

Baehr, Jason. *The Inquiring Mind: On Intellectual Virtues & Virtue Epistemology*. Oxford: Oxford University Press, 2011.

Baier, Kurt. *The Moral Point of View*. Ithaca, NY: Cornell University Press, 1958.

Barbour, Ian. *Religion in an Age of Science*, Vol. 1. San Francisco: Harper-Collins, 1990.

Baron-Cohen, Simon, Sally Wheelwright, Richard Skinner, Joanne Martin, and Emma Clubley. "The Autism Spectrum Quotient (AQ): Evidence from Asperger Syndrome/ High-Functioning Autism, Males and Females, Scientists and Mathematicians." *Journal of Autism and Developmental Disorders* 31, no. 1 (February 2001): 5–17.

Bartley, W. W., III. "What Was Wrong with Darwin?" *New York Review of Books*, September 15, 1977. http://www.nybooks.com/articles/1977/09/15/what-was-wrong-with -darwin/.

Beasley, Malcolm R., Supriyo Datta, Herwig Kogelnik, Herbert Kroemer, and Don Monroe. "Report of the Investigation Committee on the Possibility of Scientific Misconduct in the Work of Hendrik Schon and Coauthors." Lucent Technologies, September 2002.

Bellon, Richard. *A Sincere and Teachable Heart: Self-Denying Virtue in British Intellectual Life, 1736–1859*. Leiden, The Netherlands: Koninklijke Brill, 2015.

Bergström, Sune K. "Banquet Speech Accepting Nobel Prize in Physiology or Medicine," December 10, 1982. In *The Nobel Prizes*, edited by Wilhelm Odelberg. Stockholm: Nobel Foundation, 1983.

Blake, Ralph M., Curt J. Ducasse, and Edward H. Madden. *Theories of Scientific Method: The Renaissance through the Nineteenth Century*. Seattle: University of Washington Press, 1966.

Blount, Zachary D., Christina Z. Borland, and Richard E. Lenski. "Historical Contingency and the Evolution of a Key Innovation in an Experimental Population of *Escherichia coli*." *Proceedings of the National Academy of Sciences* 105, no. 23 (June 2008): 7899–7906. https://doi.org/10.1073/pnas.0803151105.

Booker, Margaret Moore. *Among the Stars: The Life of Maria Mitchell; Astronomer, Educator, Women's Rights Activist*. Nantucket, MA: Mill Hill Press, 2007.

Boudon, Philippe. *Lived-In Architecture: Pessac Revisited*. Translated by Gerald Onn. Boston: MIT Press, 1969.

Boulding, Kenneth. "The Two Cultures." In *Technology in Western Civilization*, Vol. 2, edited by Melvin Kranzberg and Carroll W. Pursell, 686–695. London: Oxford University Press, 1967.

Brewer, Carol A., and Diane Smith, eds. *Vision & Change in Undergraduate Biology Education: A Call to Action*. Washington, DC: American Association for the Advancement of Science, 2009. http://visionandchange.org/files/2013/11/aaas-VISchange-web1113.pdf.

Browne, Janet. *Charles Darwin: The Power of Place*. Princeton, NJ: Princeton University Press, 2002.

Brownson, Orestes Augustus. *An Oration on Liberal Studies*. Delivered before the Philomathian Society of Mount Saint Mary's College, June 29, 1853. Baltimore: Hedian & O'Brien, 1853.

Burke, Edmund. *A Philosophical Inquiry into the Origin of Our Ideas of the Sublime and Beautiful, with an Introductory Discourse concerning Taste and Several Other Additions*. London: John Nimmo, 1887 [1756].

Bush, Vannevar. *Science: The Endless Frontier*. Washington, DC: U.S. Government Printing Office, 1945.

Calaprice, Alice. *The Ultimate Quotable Einstein*. Princeton, NJ: Princeton University Press, 2011.

Cameron, William Bruce. *Informal Sociology: A Casual Introduction to Sociological Thinking*. New York: Random House, 1963.

Cantor, Geoffrey. *Quakers, Jews, and Science: Religious Responses to Modernity and the Sciences in Britain, 1650–1900*. Oxford: Oxford University Press, 2005.

Carnap, Rudolph. *Der Logische Aufbau der Welt*. Hamburg: Felix Meiner Verlag, 1928.

Carnap, Rudolph. "The Two Concepts of Probability." *Philosophy and Phenomenological Research* 5, no. 4 (June 1945): 513–532.

Chekhov, Anton. *Note-Book of Anton Chekhov*. New York: B. W. Huebsch, 1921.

Cline, Barbara Lovett. *Men Who Made a New Physics: Physicists and the Quantum Theory*. Chicago: University of Chicago Press, 1987.

Cohen, H. Floris. *The Scientific Revolution: A Historiographic Analysis*. Chicago: University of Chicago Press, 1995.

Cohen, William A. *Drucker on Marketing: Lessons from the World's Most Influential Business Thinker*. New York: McGraw Hill, 2013.

College Board. *AP Biology Curriculum Framework 2012–2013*. The College Board. 2011. http://media.collegeboard.com/digitalServices/pdf/ap/10b_2727_AP_Biology_CF_WEB_110128.pdf.

Committee on a Conceptual Framework for New K-12 Science Education Standards. *A Framework for K-12 Science Education Practices, Crosscutting Concepts, and Core Ideas*. Washington, DC: National Academies Press, National Research Council, 2012.

Cornforth, John W. "Scientists as Citizens." *Australian Journal of Chemistry* 46 (1993): 265–275.

Cowell, Alan. "After 350 Years, Vatican Says Galileo Was Right: It Moves." *New York Times*, October 31, 1992. https://www.nytimes.com/…/after-350-years-vatican-says -galileo-was-right-it-moves.html.

Cressey, Daniel. "Paper That Says Human Hand Was 'Designed by Creator' Sparks Concern." *Nature* 531 (March 2016). https://www.nature.com/news/paper-that-says -human-hand-was-designed-by-creator-sparks-concern-1.19499\.

Crick, Francis. *The Astonishing Hypothesis: The Scientific Search for the Soul*. New York: Scribner, 1994.

Cringley, Robert X. "The Television Program Transcripts: Part 1." *Triumph of the Nerds*, PBS, 1996. http://www.pbs.org/nerds/part1.html.

Dalai Lama. *The Universe in a Single Atom: The Convergence of Science and Spirituality*. New York: Morgan Road, 2005.

Dalton, John. In *John Dalton: Critical Assessments of His Life and Science*, by Arnold Thackray. Cambridge, MA: Harvard University Press, 1833.

Danner, Mark. "Words in a Time of War: Taking the Measure of the First Rhetoric-Major President." Commencement address given to graduates of the Department of Rhetoric, University of California, Berkeley, 2007. http://web.archive.org/web/201 50206043442/https://www.mtholyoke.edu/acad/intrel/speech/danner.htm.

Darwin, Charles. *The Autobiography of Charles Darwin*. Edited by Nora Barlow. New York: W. W. Norton, 1958.

Darwin, Charles. *Charles Darwin's Beagle Diary*. Edited by Richard Darwin Keynes. Cambridge: Cambridge University Press, 1988.

Darwin, Charles. *The Descent of Man, and Selection in Relation to Sex (1871)*. In *From So Simple a Beginning: The Four Great Books of Charles Darwin*, edited by Edward O. Wilson, 767–1248. New York: W. W. Norton, 1958.

Darwin, Charles. "Geology." In *A Manual of Scientific Enquiry Prepared for the Use of Her Majesty's Navy and Adapted for Travellers in General*, edited by John F. W. Herschel, 156–195. London: John Murray, 1849.

Darwin, Charles. "Letter to Edward Bibbens Aveling," October 13, 1880. Darwin Correspondence Project, "Letter no. 12757." http://www.darwinproject.ac.uk/DCP -LETT-12757.

Darwin, Charles. "Letter to H. E. Strickland," February 4, 1849. Darwin Correspondence Project, "Letter no. 1221." http://www.darwinproject.ac.uk/DCP-LETT-1221.

Darwin, Charles. "Letter to J. D. Hooker," May 10, 1848. Darwin Correspondence Project, "Letter no. 1174." http://www.darwinproject.ac.uk/DCP-LETT-1174.

Darwin, Charles. "Letter to John Stephens Henslow," April 1, 1848. Darwin Correspondence Project, "Letter no. 1167." http://www.darwinproject.ac.uk/DCP-LETT-1167.

Darwin, Charles. "Letter to T. H. Huxley," July 9, 1857. Darwin Correspondence Project, "Letter no. 2122." https://www.darwinproject.ac.uk/letter/DCP-LETT-2122.xml.

Darwin, Charles. "Letter to W. D. Fox," October 24, 1852. Darwin Correspondence Project, "Letter no. 1489." http://www.darwinproject.ac.uk/DCP-LETT-1489.

Darwin, Charles. "Letter to William Ogle," February 22, 1882. Darwin Correspondence Project, "Letter no. 13697." http://www.darwinproject.ac.uk/DCP-LETT-13697.

Darwin, Charles. *The Life and Letters of Charles Darwin*. Edited by Francis Darwin. New York: D. Appleton and Sons, 1919. https://www.gutenberg.org/files/2087/2087-h/2087-h.htm.

Darwin, Charles. *On the Origin of Species (1859)*. In *From So Simple a Beginning: The Four Great Books of Charles Darwin*, edited by Edward O. Wilson, 441–760. New York: W. W. Norton, 1958.

Darwin, Charles. "Prefatory Notice." In Hermann Müller, *The Fertilisation of Flowers*, translated and edited by D'Arcy W. Thompson. London: Macmillan, 1883.

Daston, Lorraine, and Peter Galison. *Objectivity*. New York: Zone Books, 2007.

Dawkins, Richard. "Science, Delusion and the Appetite for Wonder." *Reports of the National Center for Science Education* 17, no. 1 (January/February 1997): 8–14.

Delbrück, Max. "Nobel Lecture." The Nobel Foundation, 1969. https://www.nobelprize.org/nobel_prizes/medicine/laureates/1969/delbruck-lecture.html.

Desmond, Adrian, and James Moore. *Darwin: The Life of a Tormented Evolutionist*. London: W. W. Norton, 1991.

de Waal, Frans. *The Bonobo and the Atheist: In Search of Humanism among the Primates*. New York: W. W. Norton, 2013.

Dewey, John. *Human Nature and Conduct: An Introduction to Social Psychology*. New York: Henry Holt, 1922. https://www.gutenberg.org/files/41386/41386-h/41386-h.htm.

Dewey, John. "Science as Subject-Matter and as Method." *Science*, n.s., 31, no. 787 (January 28, 1910): 121–127.

Douglas, Heather. "The Irreducible Complexity of Objectivity." *Synthese* 138, no. 3 (February 2004): 453–473.

Draper, John William. *History of the Conflict between Religion and Science*. New York: D. Appleton, 1874.

Duhem, Pierre Maurice Marie. *The Aim and Structure of Physical Theory*. Princeton, NJ: Princeton University Press, 1954.

Dyson, Freeman. *Disturbing the Universe.* New York: Basic Books, 1979.

Eckhardt, Celia Morris. *Fanny Wright: Rebel in America.* Cambridge, MA: Harvard University Press, 1984.

Einstein, Albert. "Foreword." In Galileo Galilei, *Dialogue concerning the Two Chief World Systems*, translated by Stillman Drake. Berkeley: University of California Press, 2001.

Einstein, Albert. *Ideas and Opinions.* New York: Bonanza Books, 1954.

Einstein, Albert. "Letter to Jost Winteler," July 8, 1901. In *The Collected Papers of Albert Einstein.* Vol. 1, *The Early Years, 1879–1902*, edited by John Stachel, David C. Cassidy, and Robert Schulmann, 309–310. Princeton, NJ: Princeton University Press, 1987.

Einstein, Albert. "A Mathematician's Mind." In *Ideas and Opinions*, 25–26. New York: Bonanza Books, 1954.

Einstein, Albert. "Motives for Research." In *Zu Max Plancks sechzigstem Geburtstag. Ansprachen, gehalten am 26. April 1918 in der Deutschen Physikalischen Gesellschaft von E. Warburg, M. v. Laue, A. Sommerfeld und A. Einstein*, 29–32. Karlsruhe: C. F. Müllersche Hofbuchhandlung, 1918. http://alberteinstein.info/vufind1/images/einstein/ear01 /view/1/4009_000000616.pdf.

Einstein, Albert. "Science and Religion." In *Ideas and Opinions*, 41–49. New York: Bonanza, 1954.

Eisenhower, Dwight D. "Atoms for Peace—Address Before the General Assembly of the United Nations on Peaceful Uses of Atomic Energy," December 8, 1953. http://www .presidency.ucsb.edu/ws/?pid=9774.

Emde, Robert N. "From a Baby Smiling: Reflections on Virtues in Development." In *Developing the Virtues: Integrating Perspectives*, edited by Julia Annas, Darcia Narvaez, and Nancy E. Snow, 69–94. Oxford: Oxford University Press, 2016.

Emde, Robert N., Z. Biringen, R. B. Clyman, and D. Oppenheim. "The Moral Self of Infancy: Affective Core and Procedural Knowledge." *Developmental Review* 11 (1991): 251–270.

Fanelli, Daniele. "How Many Scientists Fabricate and Falsify Research? A Systematic Review and Meta-analysis of Survey Data." *PLOS ONE* 4, no. 5 (July 25, 2009): 311–313. https://doi.org/10.1371/journal.pone.0005738.

Feyerabend, Paul. "How to Defend Society against Science." *Radical Philosophy* 11, no. 1 (January 1975): 3–8.

Feynman, Richard. "Cargo Cult Science." Caltech commencement address, 1974. In *Classic Feynman: All the Adventures of a Curious Character*, edited by Ralph Leighton, 490–497. New York: W. W. Norton, 2006.

Feynman, Richard. *The Character of Physical Laws.* Boston: MIT Press, 1967.

Feynman, Richard P. "The Making of a Scientist." In *Classic Feynman: All the Adventures of a Curious Character,* edited by Ralph Leighton, 13–19. New York: W. W. Norton, 2006.

Feynman, Richard. Nobel Prize Banquet Speech. The Nobel Foundation, 1965. https://www.nobelprize.org/nobel_prizes/physics/laureates/1965/feynman-speech .html.

Feynman, Richard P. "The Pleasure of Finding Things Out." Interview. *NOVA*, WGBH Educational Foundation, 1983.Feynman, Richard. *Surely You're Joking, Mr. Feynman? (Adventures of a Curious Character).* Edited by Edward Hutchings. New York: W. W. Norton, 1985.

Feynman, Richard. "What Is and What Should Be the Role of Scientific Culture in Modern Society." In *The Pleasure of Finding Things Out: The Best Short Works of Richard Feynman,* edited by Jeffrey Robbins, 98–116. New York: Helix Books/Perseus Books, 1999.

Fiege, Mark. "The Atomic Scientists, the Sense of Wonder, and the Bomb." *Environmental History* 12, no. 3 (2007): 578–613. https://doi.org/10.2307/25473133.

France, Anatole. *The Crime of Sylvestre Bonnard.* 1860. http://www.gutenberg.org/files /2123/2123-h/2123-h.htm.

Franklin, Benjamin. *The Autobiography of Benjamin Franklin.* Edited by Frank Woodworth Pine. New York: Henry Holt, 1916.

Friedman, Michael. "Rational Reconstruction, Explication, and the Rejection of Metaphysics." In *Carnap's Ideal of Explication and Naturalism,* edited by P. Wagner and Michael Beaney, 190–204. London: Palgrave Macmillan, 2012.

Gaither, Carl C., and Alma E. Cavazos-Gaither, eds. *Gaither's Dictionary of Scientific Quotations.* New York: Springer, 2012.

Galileo Galilei. "Letter to the Grand Duchess Christina of Tuscany, 1615." In Internet History Sourcebooks Project, edited by Paul Halsall. https://sourcebooks.fordham .edu/mod/galileo-tuscany.asp.

Gell-Mann, Murray. *The Quark and the Jaguar: Adventures in the Simple and the Complex.* New York: W. H. Freeman, 1994.

Goldsmith, Oliver. *The Works of Oliver Goldsmith,* Vol. 2. Edited by Peter Cunningham. New York: Harper and Brothers, 1881.

Goodpaster, Kenneth E. *Conscience and Corporate Culture.* Malden, MA: Blackwell, 2007.

Gopnik, Alison, Andrew N. Meltzoff, and Patricia K. Kuhl. *The Scientist in the Crib: What Early Learning Tells Us about the Mind.* New York: William Morrow, 2000.

Gotthelf, Allan. "Darwin on Aristotle." *Journal of the History of Biology* 32, no.1 (Spring 1999): 3–30.

Gross, Paul R., and Norman Levitt. *Higher Superstition: The Academic Left and Its Quarrels with Science*. Baltimore: Johns Hopkins University Press, 1994.

Gunsalus, C. K. "Institutional Structure to Ensure Research Integrity." *Academic Medicine* 68, no. 9 (1993): S33–S38.

Hardy, G. H. *A Mathematician's Apology*. University of Alberta Mathematical Sciences Society, 1940. https://www.math.ualberta.ca/mss/misc/A%20Mathematician%27s%20Apology.pdf.

Harman, Oren, and Michael Dietrich. *Rebels, Mavericks and Heretics in Biology*. New Haven, CT: Yale University Press, 2008.

Harnett, Kevin. "How Humans Can Force the Machines to Play Fair." *Wired*, November 26, 2016. https://www.wired.com/2016/11/humans-can-force-machines-play-fair/.

Highfield, Roger. "Al-Qa'eda's Atom Plans Were Spoof Science." *The Telegraph*, November 20, 2001.

Hobbes, Thomas. *Leviathan, or The Matter, Form, and Power of a Commonwealth Ecclesiastical and Civil*. Edited by Michael Oakeshott. New York: Collier Books, 1977.

Hockney, David. *Secret Knowledge: Rediscovering the Lost Techniques of Old Masters*. New York: Avery, 2006.

Hoffmann, Roald. "McPherson Lecture." Michigan State University, September 25, 2002.

Horgan, John. "Was Philosopher Paul Feyerabend Really 'Science's Worst Enemy'?" Scientific American, October 24, 2016. https://blogs.scientificamerican.com/cross-check/was-philosopher-paul-feyerabend-really-science-s-worst-enemy/.

Horkheimer, Max. *Critical Theory: Selected Essays*. New York: Continuum, 1982.

Hugo, Victor. *Ninety-Three*. Translated by Aline Delano. Boston: Little, Brown, 1889. http://www.gutenberg.org/files/49372/49372-h/49372-h.htm.

Hume, David. *Enquiries concerning Human Understanding and the Principles of Morals*. Edited by L. A. Selby-Bigge. Oxford: Oxford University Press, 1975 [1777].

Hume, David. "Of the Standard of Taste." In *The Philosophical Works of David Hume*, Vol. 3, edited by T. H. Green and T. H. Grose. London: Longman, Green, 1874–1875.

Hume, David. *Treatise of Human Nature*. Edited by L. A. Selby-Bigge. Oxford: Oxford University Press, 1978 [1888].

Hursthouse, Rosalind, and Glen Pettigrove. "Virtue Ethics." In *The Stanford Encyclopedia of Philosophy*, edited by Edward N. Zalta, Winter 2016 edition. https://plato.stanford.edu/archives/win2016/entries/ethics-virtue/.

Husserl, Edmund. *Ideas: General Introduction to Pure Phenomenology*. Translated by W. R. Boyce Gibson. London: George Allen & Unwin, 1931.

Huxley, Thomas Henry. "Darwinian Hypothesis." In *Darwiniana: Essays*, edited by T. H. Huxley, 1–21. New York: D. Appleton, 1896.

Huxley, Thomas Henry. *Lay Sermons, Addresses and Reviews*. London: MacMillan, 1880.

Huxley, Thomas Henry. "Letter to Charles Kingsley," September 23, 1860. In *The Life and Letters of Thomas Henry Huxley*. New York: Appleton, 1900.

Huxley, Thomas Henry. "Letter to F. Dyster," January 30, 1859. In *The Life and Letters of Thomas Henry Huxley*. New York: Appleton, 1900.

Huxley, Thomas Henry. "Reflection #296." In *Aphorisms and Reflections from the Works of T. H. Huxley*, edited by Thomas Henry Huxley and Henrietta A. Huxley. London: Macmillan, 1907.

India's Constitution of 1949 with Amendments through 2012. Part IVA, §51Ah. http://extwprlegs1.fao.org/docs/pdf/ind132810.pdf.

Isaacson, Walter. *Einstein: His Life and Universe*. New York: Simon & Schuster, 2008.

Jacobson, Louis. "Yes, Donald Trump Did Call Climate Change a Chinese Hoax." *Politifact*, June 3, 2016. http://www.politifact.com/truth-o-meter/statements/2016/jun/03/hillary-clinton/yes-donald-trump-did-call-climate-change-chinese-h/.

James, William. *The Varieties of Religious Experience*. Cambridge, MA: Harvard University Press, 1985 [1902].

Johnson, Philip E. "The Creationist and the Sociobiologist: Two Stories about Illiberal Education." *California Law Review* 80, no. 4, (July 1992).

Johnson, Philip E. "Paul Feyerabend's Choice for Freedom." Review of *Killing Time, the Autobiography of Paul Feyerabend*. *Books and Culture*, March/April 1996.

Johnson, William Samuel. "No. 150. Adversity Useful to the Acquisition of Knowledge." *Samuel Johnson's Essays*, 1751. http://www.johnsonessays.com/the-rambler/no-150-adversity-useful-to-the-acquisition-of-knowledge/.

Jungk, Robert. *Brighter Than a Thousand Suns: A Personal History of the Atomic Scientists*. Boston: Houghton Mifflin Harcourt, 1958.

Kay, Paul, Brent Berlin, Luisa Maffi, William R. Merrifield, and Richard Cook. *World Color Survey*. Stanford, CA: Center for the Study of Language and Information Publications, 2009.

Keller, Evelyn Fox. *A Feeling for the Organism: The Life and Work of Barbara McClintock*. New York: Henry Holt, 1983.

Kelvin, Lord. "Lecture to the Institution of Civil Engineers," May 3, 1883.

Kemble, Edwin C. *Physical Science, Its Structure and Development*, Vol. 1. Cambridge, MA: MIT Press, 1966.

King, Martin Luther, Jr. "A Tough Mind and a Tender Heart." In *Strength to Love*. Minneapolis: Fortress, 1977 [1959].

Kitcher, Philip. *Science, Truth, and Democracy*. Oxford: Oxford University Press, 2003.

Kitcher, Philip. *Science in a Democratic Society*. Amherst, NY: Prometheus, 2011.

Kitzmiller v. Dover Area School Dist., 400 F. Supp. 2d 707 (M.D. Pa. 2005).

Koertge, Noretta, ed. *A House Built on Sand: Exposing Postmodernist Myths about Science*. New York: Oxford University Press, 1998.

Koertge, Noretta. *Scientific Values and Civic Virtues*. Oxford: Oxford University Press, 2005.

Koshland, Daniel E., Jr. "Crazy, but Correct: How a Non-conformist Theory Beat Scepticism and Got Into the Textbooks." *Nature* 432 (November 25, 2004). https://doi.org/10.1038/432447a.

Krutch, Joseph Wood. *The Modern Temper: A Study and a Confession*. New York: Harcourt, Brace, 1956 [1929].

Kuhn, Thomas. "Afterwords." In *World Changes: Thomas Kuhn and the Nature of Science*, edited by Paul Horwich, 311–341. Cambridge, MA: MIT Press, 1993.

Kuhn, Thomas. *The Essential Tension: Selected Studies in Scientific Tradition and Change*. Chicago: University of Chicago Press, 1977.

Kuhn, Thomas. "Objectivity, Value Judgment, and Theory Choice." In *The Essential Tension: Selected Studies in Scientific Tradition and Change*, 320–339. Chicago: University of Chicago Press, 1977.

Kuhn, Thomas. *The Structure of Scientific Revolutions*. Chicago: University of Chicago Press, 1962.

Kvanvig, Jonathan L. *The Intellectual Virtues and the Life of the Mind: On the Place of the Virtues in Contemporary Epistemology*. Savage, MD: Rowman & Littlefield, 1992.

Ladikos, Anastasious. "Revisiting the Virtue of Courage in Aristotle." *Phronimon* 5, no. 2 (2004): 77–92.

Lakatos, Imre. "Falsification and the Methodology of Scientific Research Programs." In *Criticism and the Growth of Knowledge*, edited by Imre Lakatos and Alan Musgrave, 91–195. Cambridge: Cambridge University Press, 1970.

Lakatos, Imre. "History of Science and Its Rational Reconstructions." In *PSA: Proceedings of the Biennial Meeting of the Philosophy of Science Association 1970*, 91–136.

Chicago: University of Chicago Press on behalf of the Philosophy of Science Association, 1970.

Lakatos, Imre. *Proofs and Refutations: The Logic of Mathematical Discovery.* Edited by John Worrall and Elie Zahar. Cambridge: Cambridge University Press, 1976.

Langmuir, Irving. "Banquet Speech." The Nobel Foundation, 1932. https://www.nobelprize.org/nobel_prizes/chemistry/laureates/1932/langmuir-speech.html.

Lankester, E. Ray. "Charles Robert Darwin." In *Library of the World's Best Literature, Ancient and Modern,* Vol. 2, edited by C. D. Warner, 4385–4393. New York: R. S. Peale & J. A. Hill, 1896. http://darwin-online.org.uk/content/frameset?itemID=F2113&viewtype=text&pageseq=1.

Lark, Amy M., Gail Richmond, Louise S. Mead, James J. Smith, and Robert T. Pennock. "Exploring the Relationship between Experiences with Digital Evolution and Students' Scientific Understanding and Acceptance of Evolution." *American Biology Teacher* 80, no. 2, (February 2018): 74–86. https://doi.org/10.1525/abt.2018.80.2.74.

Lark, Amy M., Gail Richmond, and Robert T. Pennock. "Modeling Evolution in the Classroom: The Case of the Fukushima Butterflies." *American Biology Teacher* 76, no. 7 (2014): 450–454.

Latour, Bruno. "For David Bloor … and Beyond: A Reply to David Bloor's 'Anti-Latour.'" *Studies in the History and Philosophy of Science* 30, no. 1 (1999): 113–129.Latour, Bruno, and Steve Woolgar. *Laboratory Life: The Social Construction of Scientific Facts.* Princeton, NJ: Princeton University Press, 1986.

Lear, Linda. *Rachel Carson: Witness for Nature.* Boston: Houghton Mifflin Harcourt, 1997.

Lederman, Leon. *The God Particle: If the Universe Is the Answer, What Is the Question?* Boston: Houghton Mifflin, 1993.

Lewandowsky, Stephan, Werner Stritzke, Klaus Oberauer, and M. Morales. "Misinformation and the 'War on Terror': When Memory Turns Fiction into Fact." In *Terrorism and Torture: An Interdisciplinary Perspective,* edited by W. G. K. Stritzke, S. Lewandowsky, D. Denemark, J. Clare, and F. Morgan, 179–203. Cambridge: Cambridge University Press, 2009.

Lewis, Sinclair. *Arrowsmith.* New York: Grosset & Dunlap, 1925.

Lindberg, David C., and Ronald L. Numbers. *When Science & Christianity Meet.* Chicago: University of Chicago Press, 2003.

Littmann, Mark. "Courses in Science Writing as Literature." *Public Understanding of Science* 14 (2005): 103–112.

Lloyd, Elisabeth A. "Objectivity and the Double Standard for Feminist Epistemologies." *Synthese* 104, no. 3, Feminism and Science (September 1995): 351–381.

Lock, F. P. *Burke's Reflections on the Revolution in France.* Boston: Allen & Unwin, 1985.

Longino, Helen. *Science as Social Knowledge: Values and Objectivity in Scientific Inquiry.* Princeton, NJ: Princeton University Press, 1990.

Lorenz, Konrad. *On Aggression.* Translated by Marjorie Kerr Wilson. London: Routledge, 1963.

Luther, Martin. *A Commentary on St. Paul's Letter to the Galatians.* London: James Cundee, 1807.

Lyne, Andrew. In "Hunt for Alien Worlds." *NOVA,* February 18, 1997. http://www.pbs.org/wgbh/nova/transcripts/2407alien.html.

Lyotard, Jean-François. *The Postmodern Condition: A Report on Knowledge.* Manchester, UK: Manchester University Press, 1979.

MacIntyre, Alasdair. *After Virtue: A Study in Moral Theory.* Notre Dame, IN: University of Notre Dame Press, 1997.

Mailes, Matthew, Andrew G. Lyne, and S. L. Shemar. "A Planet Orbiting the Neutron Star PSR1829–10." *Nature* 352 (July 25, 1991): 311–313. https://doi.org/10.1038/352311a0.

Marchant, James. *Alfred Russel Wallace: Letters and Reminiscences,* Vol. 1. London: Cassell, 1916. http://www.gutenberg.org/files/15997/15997-h/15997-h.htm.

Marcy, Geoff. In "Hunt for Alien Worlds." *NOVA,* February 18, 1997. http://www.pbs.org/wgbh/nova/transcripts/2407alien.html.

McCright, Aaron M., and Riley E. Dunlap. "Defeating Kyoto: The Conservative Movement's Impact on U.S. Climate Change Policy." *Social Problems* 5, no. 3 (2003): 348–373.

McGuire, J. E. "Boyle's Conception of Nature." *Journal of the History of Ideas* 33, no. 4 (Oct.–Dec. 1972): 523–542.

Merton, Robert K. "The Normative Structure of Science." In *The Sociology of Science: Theoretical and Empirical Investigations,* 267–278. Chicago: University of Chicago Press, 1973.

Merton, Robert K. *On the Shoulders of Giants: A Shandean Postscript.* Chicago: University of Chicago Press, 1965.

Merton, Robert K. *Science, Technology and Society in Seventeenth-Century England.* New York: Harper & Row, 1970.

Merton, Robert K. "Science and the Social Order." *Philosophy of Science* 5 (1938): 321–337.

Merton, Robert K. *The Sociology of Science: Theoretical and Empirical Investigations.* Edited and with an introduction by Norman W. Storer. Chicago: University of Chicago Press, 1973.

Merton, Robert K., David L. Sills, and Stephen M. Stigler. "The Kelvin Dictum and Social Science: An Excursion into the History of an Idea." *Journal of the History of the Behavioral Sciences* 20 (October 1984): 319–331.

Miller, Jon. "The Conceptualization and Measurement of Civic Scientific Literacy for the Twenty-First Century." In *Science and the Educated American: A Core Component of Liberal Education*, edited by Jerrold Meinwald and John G. Hildebrand, 241–255. Cambridge, MA: American Academy of Arts & Sciences, 2010.

Mitchell, Maria. *Maria Mitchell: Life, Letters and Journals.* Edited by Phebe Mitchell Kendall. 1896. http://www.gutenberg.org/cache/epub/10202/pg10202-images.html.

Mitchell, Maria. "The Need of Women in Science." *The Victoria Magazine* 28, November–April 1877, 187–192.

Mitnick, Kevin. "Kevin Mitnick Loved the Cyberspace Break-in Game." Interview by Peter Finnocchiaro. *Salon*, August 20, 2011. https://www.salon.com/2011/08/20/kevin_mitnick_interview/.

Molière. *The Imaginary Invalid.* 1673.

Morrison, Philip, and Emily Morrison. *Charles Babbage and His Calculating Engines: Selected Writings by Charles Babbage and Others.* New York: Dover, 1961.

Musil, Robert K. "There Must Be More to Love Than Death: A Conversation with Kurt Vonnegut." *The Nation*, August 2–9, 1980, 128–132.

National Institutes of Health. "Notice Announcing the Removal of the Funding Pause for Gain-of-Function Research Projects." December 19, 2017. https://grants.nih.gov/grants/guide/notice-files/NOT-OD-17-071.html.

National Research Council. *Successful K-12 STEM Education: Identifying Effective Approaches in Science, Technology, Engineering, and Mathematics.* Washington, DC: National Academies Press, 2011. https://doi.org/10.17226/13158.

Nehru, Jawaharlal. *The Discovery of India.* Oxford: Oxford University Press, 1985 [1946].

Newton, Isaac. *Principia.* Translated by Andrew Motte. Berkeley: University of California Press, 1934.

Newton, Isaac. "Quaestiones quaedam philosophicae" [Certain philosophical questions]. *The Newton Project.* 2003. http://www.newtonproject.ox.ac.uk/view/texts/normalized/THEM00092.

Next Generation Science Standards. 2013. http://www.nextgenscience.org/next-gen eration-science-standards.

Niebuhr, Reinhold. "Ideology and the Scientific Method." In *The Essential Reinhold Niebuhr: Selected Essays and Addresses,* edited by Robert Mcafee Brown, 205–217. New Haven, CT: Yale University Press, 1986.

NOVA. *The Best Mind since Einstein.* December 21, 1993.

Odelberg, Wilhelm, ed. *Les Prix Nobel. The Nobel Prizes.* Stockholm: Nobel Foundation, 1969.

Oppenheimer, Robert. Testimony before United States Atomic Energy Commission, April 13, 1954. 95/266. http://www.osti.gov/includes/opennet/includes/Oppenheimer %20hearings/Vol%20II%20Oppenheimer.pdf.

Oxford Dictionaries. "Word of the Year 2016." https://en.oxforddictionaries.com/word -of-the-year/word-of-the-year-2016.

Pais, Abraham. *Subtle Is the Lord: The Science and Life of Albert Einstein.* Oxford: Oxford University Press, 2005.

Papert, Seymore. "Seymour Papert, Artificial Intelligence Guru—Obituary." *The Telegraph,* August 11, 2016. http://www.telegraph.co.uk/obituaries/2016/08/11/seymour -papert-artificial-intelligence-guru--obituary/.

Parrington, John. *The Deeper Genome: Why There Is More to the Human Genome Than Meets the Eye.* Oxford: Oxford University Press, 2015.

Parsons, Keith. *The Science Wars: Debating Scientific Knowledge and Technology.* Amherst, NY: Prometheus, 2003.

Pearcey, Nancy. *Total Truth: Liberating Christianity from Its Cultural Captivity.* Wheaton, IL: Crossway, 2004.

Peirce, Charles Sanders. "The Scientific Attitude and Fallibilism." In *Philosophical Writings of Peirce,* edited by Justus Buchler, 42–59. New York: Dover, 1955.

Pennock, Robert T. "Bayesianism, Ravens and Evidential Relevance." *Annals of the Japan Association for Philosophy of Science* 13, no. 1 (2004): 1–26.

Pennock, Robert T. "Can Darwinian Mechanisms Make Novel Discoveries? Learning from Discoveries Made by Evolving Neural Networks." *Foundations of Science* 5, no. 2 (2000): 225–238.

Pennock, Robert T. "Epistemic and Ontic Theories of Explanation and Confirmation." *Philosophy of Science—Japan* 28 (1995): 31–45.

Pennock, Robert T. "Evidential Relevance and the Grue Paradox." *Philosophy of Science—Japan* 31, no. 1 (1998): 101–119.

Pennock, Robert T. "Experimental Evolution in the Classroom: The Active LENS Project." In *Envisioning the Future of Undergraduate STEM Education: Research and Practice Symposium*. April 2016. http://www.enfusestem.org/projects/experimental-evolution-in-the-classroom-the-active-lens-project/.

Pennock, Robert T. "Following Humbly Where Nature Leads: How Scientists Embody Humility." *Slate*, What Do We Know? Essays and Opinions, August 10, 2015.

Pennock, Robert T. "Fostering a Culture of Scientific Integrity: Legalistic vs. Scientific Virtue-Based Approaches." *Professional Ethics Report* 28, no. 2 (2015): 1–3.

Pennock, Robert T. "God of the Gaps: The Argument from Ignorance and the Limits of Methodological Naturalism." In *Scientists Confront Creationism and Intelligent Design*, edited by Andrew Petto and Laurie Godfrey, 309–338. New York: W. W. Norton, 2007.

Pennock, Robert T. "Inappropriate Authorship in Collaborative Scientific Research." *Public Affairs Quarterly* 10, no. 4 (October 1996): 379–393.

Pennock, Robert T. "Learning Evolution and the Nature of Science Using Evolutionary Computing and Artificial Life." *McGill Journal of Education* 42, no. 2 (2007): 211–224.

Pennock, Robert T. "Models, Simulations, Instantiations and Evidence: The Case of Digital Evolution." *Journal of Experimental and Theoretical Artificial Intelligence* 19, no. 1 (2007): 29–42.

Pennock, Robert T. "Moral Darwinism: Ethical Evidence for the Descent of Man." *Biology & Philosophy* 10 (1995): 287–307.

Pennock, Robert T. "Naturalism, Evidence and Creationism: The Case of Phillip Johnson." *Biology and Philosophy* 11, no. 4 (1996): 543–559.

Pennock, Robert T. "On Teaching Evolution and the Nature of Science." In *Evolutionary Science and Society: Educating a New Generation*, edited by Joel Cracraft and Roger Bybee, 1–12. Colorado Springs, CO: Biological Sciences Curriculum Study, 2005.

Pennock, Robert T. "The Postmodern Sin of Intelligent Design Creationism." *Science & Education* 19, nos. 6–8 (2010): 757–778.

Pennock, Robert T. "The Pre-modern Sins of Intelligent Design." In *Oxford Handbook of Science and Religion*, edited by Phillip Clayton, 732–748. Oxford: Oxford University Press, 2006.

Pennock, Robert T. "Research Funding and the Virtue of Scientific Objectivity." *Academic Integrity* 5, no. 2 (Spring 2002): 3–6.

Pennock, Robert T. "Scientific Integrity and Science Museums." *Museums and Social Issues* 1, no. 1 (2006): 7–18.

Pennock, Robert T. *Tower of Babel: The Evidence against the New Creationism*. Cambridge, MA: MIT Press, 1999.

Pennock, Robert T. "The Virtuous Scientist Meets the Human Clone." In *New Ethical Challenges in Science and Technology. Sigma Xi Forum 2000 Proceedings*, 2001, 117–124.

Pennock, Robert T., and Jon D. Miller. "The Values of Exemplary and Promising Younger Scientists: A Preliminary Analysis." American Association for the Advancement of Science Symposium paper, February 13, 2016.

Pennock, Robert T., and Michael O'Rourke. "Developing a Scientific Virtue-Based Approach to Science Ethics Training." *Science & Engineering Ethics* 23, no. 1 (2017): 243–262. https://doi.org/10.1007/s11948-016-9757-2.

Pimple, Kenneth D. "Six Domains of Research Ethics: A Heuristic Framework for the Responsible Conduct of Research." *Science and Engineering Ethics* 8, no. 2 (2002): 191–205.

Plato. "Apology." Translated by Benjamin Jowett. The Project Gutenberg Edition. http://www.gutenberg.org/files/1656/1656-h/1656-h.htm.

Plato. "Meno," translated by W. K. C. Guthrie. In *Plato: The Collected Dialogues*, edited by Edith Hamilton and Huntington Cairns. Princeton, NJ: Princeton University Press, 1961.

Plato. "Theaetetus," translated by F. M. Cornford. In *Plato: The Collected Dialogues*, edited by Edith Hamilton and Huntington Cairns. Princeton, NJ: Princeton University Press, 1961.

Pliny the Elder. *Naturalis Historia*. Edited by John Bostock and H. T. Riley. London: Taylor and Francis, 1855. http://www.perseus.tufts.edu/hopper/text?doc=Perseus:text:1999.02.0137:book=23:chapter=77&highlight=proof+against+all+poisons%2C.

Pope, Alexander. *An Essay on Man*. Edited by Henry Morley. London: Cassell, 1891.

Pope, Alexander. *The Works of Alexander Pope*, Vol. 5. Edited by Dr. Johnson. London: Strahan and Preston, 1812.

Popper, Karl. *The Logic of Scientific Discovery*. New York: Harper & Row, 1959.

Popper, Karl. *The Open Society and Its Enemies*, Vols. 1 and 2. London: George Routledge and Sons, 1945.

Prothero, George. Quoted in "The Late Mr. Darwin." *The Illustrated London News*, April 29, 1882, 416–418.

Quine, W. V. "Main Trends in Recent Philosophy: Two Dogmas of Empiricism." *Philosophical Review* 60, no. 1 (1951): 20–43.

Rawls, John. *A Theory of Justice*. Cambridge, MA: Harvard University Press, 1971.

Reich, Eugenie Samuel. *Plastic Fantastic: How the Biggest Fraud in Physics Shook the Scientific World*. New York: Palgrave Macmillan, 2009.

Rohr, Richard. PBS Interview. *Religion & Ethics Newsweekly*, WPSU, 2011. http://www.pbs.org/wnet/religionandethics/2011/11/11/november-11-2011-richard-rohr/9902/.

Rosch, Eleanor. "Natural Categories." *Cognitive Psychology* 4 (1973): 328–350.

Ross, Greg. "First Things First." *Futility Closet*, April 8, 2014. http://www.futility closet.com.

Rubbia, Carlo. "Asking Nature." In *Passionate Minds*, edited by Lewis Wolpert and Alison Richards. Oxford: Oxford University Press, 1977.

Russell, Bertrand. *Religion and Science*. Oxford: Oxford University Press, 1935.

Sagan, Carl. *Contact*. New York: Simon & Schuster, 1985.

Salmon, Wesley. *Scientific Explanation and the Causal Structure of the World*. Princeton, NJ: Princeton University Press, 1984.

Salter, Chuck. "Jeffrey Wigand: The Whistle-Blower." *Fast Company*, April 30, 2001. https://www.fastcompany.com/65027/jeffrey-wigand-whistle-blower2002.

Salzberg, Steven L. "Creationism Slips Into a Peer-Reviewed Journal." *Reports of the National Center for Science Education* 28, no. 3 (May–June 2008): 12–19. https://ncse.com/library-resource/creationism-slips-into-peer-reviewed-journal.

Saxon, D. H. "Preface: In Appreciation of Lord Kelvin." In *Kelvin, Thermodynamics and the Natural World*, edited by M. W. Collins, R. C. Dougal, C. S. Konig, and I. S. Ruddock, ix–xxiv. Southampton: WIT, 2016.

Schaeffer, Francis. "Address at the University of Notre Dame," April 1981.

Scheffler, Israel. *Science and Subjectivity*. Indianapolis: Bobbs-Merrill, 1967.

Selzer, Jack, ed. *Understanding Scientific Prose*. Madison: University of Wisconsin Press, 1993.

Shanahan, Timothy. "God and Nature in the Thought of Robert Boyle." *Journal of the History of Philosophy* 26 (1988): 547–569.

Shapin, Steven. "Understanding the Merton Thesis." *Isis* 79, no. 4 (1988): 594–605.

Sheff, Donald. "Izzy, Did You Ask a Good Question Today?" Opinion, *New York Times*, January 19, 1988.

Sigma Xi. "Constitution 1886." https://www.sigmaxi.org/about/history.

Simonton, Dean. *Creativity in Science: Chance, Logic, Genius, and Zeitgeist*. Cambridge: Cambridge University Press, 2004.

Smith, James J., Wendy R. Johnson, Amy M. Lark, Louise S. Mead, Michael J. Wiser, and Robert T. Pennock. "An Avida-ED Digital Evolution Curriculum for Undergraduate Biology." *Evolution: Education and Outreach* 9, no. 9 (2016): 1–11.

Smith, Logan Pearsall. *Unforgotten Years.* Boston: Little, Brown, 1938. https://www.gutenberg.ca/ebooks/smithlp-unforgottenyears/smithlp-unforgottenyears-00-e.html.

Snow, C. P. *The Two Cultures and the Scientific Revolution.* New York: Cambridge University Press, 1959.

Snow, Nancy E. *Virtue as Social Intelligence: An Empirically Grounded Theory.* New York: Routledge, 2010.

Sokal, Alan D. "Transgressing the Boundaries: Towards a Transformative Hermeneutics of Quantum Gravity." *Social Text* 46/47, Science Wars (1996): 217–252.

Sosa, Ernest. *Knowledge in Perspective: Selected Essays in Epistemology.* Cambridge: Cambridge University Press, 1991.

Soyfer, Valery N. "The Consequences of Political Dictatorship for Russian Science." *Nature Reviews Genetics* 2, no. 9 (September 2001): 723–729.

Speth, Elena B., Tammy Long, Robert T. Pennock, and Diane Ebert-May. "Using Avida-ED for Teaching and Learning about Evolution in Undergraduate Introductory Biology Courses." *Evolution: Education and Outreach* 2, no. 3 (2009): 415–428.

Stanford University. "Openness in Research." October 10, 2017. https://doresearch.stanford.edu/policies/research-policy-handbook/conduct-research/openness-research.

Steadman, Philip. *Vermeer's Camera: Uncovering the Truth behind the Masterpieces.* Oxford: Oxford University Press, 2002.

Steneck, Nicholas H. "Research Universities and Scientific Misconduct: History, Policies, and the Future." *Journal of Higher Education* 65, no. 3 (1994): 310–330.

Steneck, Nicholas H., and Ruth Ellen Bulger. "The History, Purpose, and Future of Instruction in the Responsible Conduct of Research." *Academic Medicine* 82, no. 9 (2007): 829–834.

Stephens, James. *The Crock of Gold.* Charleston: BiblioBazaar, 2006 [1912].

Straus, Ernst. "Memoir." In *Einstein: A Centenary Volume,* edited by A. P. French, 31–32. Cambridge, MA: Harvard University Press, 1979.

Suppe, Frederick. "The Search for Philosophic Understanding of Scientific Theories." In *The Structure of Scientific Theories,* 2nd ed., edited by Frederick Suppe. Urbana: University of Illinois Press, 1977.

Suskind, Ron. "Faith, Certainty and the Presidency of George W. Bush." *New York Times Magazine*, October 17, 2004. http://www.nytimes.com/2004/10/17/magazine /faith-certainty-and-the-presidency-of-george-w-bush.html.

Swift, Art. "In US, Belief in Creationist View of Humans Hits a New Low." *Gallup News*, May 22, 2017. http://news.gallup.com/poll/210956/belief-creationist-view -humans-new-low.aspx.

Tertullian. *De praescriptione baereticorum*, VII. In *Tertulliani opera*, edited by Nic. Rigaltius, 204–205. Paris, 1664.

Thomas, Lewis. "Natural Science." In *The Lives of a Cell: Notes of a Biology Watcher*. New York: The Viking Press, 1974.

Thoreau, Henry David. *The Journal of Henry David Thoreau: 1837–1861*. Edited by Damion Searls. New York: New York Review of Books, 2009.

Thoreau, Henry David. "On the Duty of Civil Disobedience." 1849. www.gutenberg .org/files/71/71-h/71-h.htm.

Thoreau, Henry David. *Thoreau: Walden and Other Writings*. Edited by Joseph Wood Krutch. New York: Bantam Books, 1962.

Timiazeff, C. A. *The Life of Plants*. Translated by Anna Sheremeteva. London: Longmans, Green, 1912.

Timmer, John. "Science Education Group Decides It's Time to Tackle Climate Change." *Ars Technica*, January 16, 2012. https://arstechnica.com/science/2012/01 /science-education-group-decides-its-time-to-tackle-climate-change/?comments=1

Titus, Sandra L., James A. Wells, and Lawrence J. Rhoades. "Repairing Research Integrity." *Nature* 453, no. 7198 (2008): 980–982.

Trout, Jack. "Peter Drucker on Marketing." *Forbes*, July 3, 2006. https://www.forbes .com/2006/06/30/jack-trout-on-marketing-cx_jt_0703drucker.html#184a57eb555c.

Trueblood, D. Elton. "A Radical Experiment." 1947. http://www.quaker.org/pam phlets/wpl1947a.html.

Truzzi, Marcello. "Editorial." *The Zetetic* 1, no. 1 (Fall/Winter 1976): 3–6.

Tyndall, John. "Address Delivered before the British Association Assembled at Belfast, with Additions." London: Longmans, Green, 1874. http://www.victorianweb.org/sci ence/science_texts/belfast.html.

U.S. Department of Health and Human Services. "United States Government Policy for Oversight of Life Sciences Dual Use Research of Concern." http://www.phe.gov/s3 /dualuse/Documents/us-policy-durc-032812.pdf.

Van Fraassen, Bas. *The Scientific Image*. Oxford: Clarendon Press, 1980.

Vonnegut, Kurt. *Cat's Cradle.* New York: Delacourt Press, 1963.

Wakeman, J. *World Authors 1950–1970.* New York: H. W. Wilson, 1975.

Washington, George. "First Annual Message to Congress on the State of the Union." January 8, 1790.

Weber, Max. "Science as a Vocation." In *From Max Weber: Essays in Sociology,* translated and edited by H. H. Gerth and C. Wright Mills, 129–156. New York: Oxford University Press, 1946.

Weinberg, Steven. *Dreams of a Final Theory.* New York: Pantheon Books, 1992.

Whewell, William. "On the Influence of the History of Science upon Intellectual Education." In *Lectures on Education Delivered at the Royal Institution of Great Britain.* London: John W. Parker, 1854.

White, Andrew Dickson. *A History of the Warfare of Science with Theology in Christendom.* New York: George Braziller, 1955 [1896].

White, Andrew Dickson. "Letter to Gerrit Smith," September 1, 1862. http://rmc .library.cornell.edu/cornell150/exhibition/white/AndrewDWhiteletterGerritSmith .pdf.

White House Office of Science and Technology Policy and Department of Health and Human Services. "Doing Diligence to Assess the Risks and Benefits of Life Sciences Gain-of-Function Research." October 17, 2014. https://obamawhitehouse .archives.gov/blog/2014/10/17/doing-diligence-assess-risks-and-benefits-life-sci ences-gain-function-research.

Wikström, J. E. "Opening Remarks." In *Ethics for Science Policy: Proceedings of a Nobel Symposium Held at Södergarn, Sweden, 20–25 August 1979,* edited by Torgny Segerstedt, xiii–xiv. Oxford: Pergamon Press, 1979.

Wilford, John Nobel. "Mary Leakey, 83, Dies; Traced Human Dawn." *New York Times,* December 10, 1966. http://www.nytimes.com/1996/12/10/world/mary-leakey-83-dies -traced-human-dawn.html.

Wilkinson, David. *Science, Religion, and the Search for Extraterrestrial Intelligence.* Oxford: Oxford University Press, 2013.

Wilson, David Sloan. *Evolution for Everyone: How Darwin's Theory Can Change the Way We Think about Our Lives.* New York: Delta, 2007.

Wilson, Edward O. *Consilience: The Unity of Knowledge.* New York: Vintage Books, 1998.

Wilson, Edward O. *Naturalist.* Washington, DC: Island, 1994.

Wilson, Robert. "R. R. Wilson's Congressional Testimony." April 1969. http://history .fnal.gov/testimony.html.

Wright, Frances. *Course of Popular Lectures*. New York: Office of the Free Enquirer, 1829.

Wright, Helen. *Sweeper in the Sky: The Life of Maria Mitchell, First Woman Astronomer in America*. New York: Macmillan, 1949.

Zagzebski, Linda Trinkaus. *Virtues of the Mind: An Inquiry into the Nature of Virtue and the Ethical Foundations of Knowledge*. New York: Cambridge University Press, 1996.

Ziman, John M. *Public Knowledge: An Essay concerning the Social Dimensions of Science*. New York: Cambridge University Press, 1968.

Ziman, John M. *Real Science: What It Is and What It Means*. New York: Cambridge University Press, 2000.

Index